COMMUNITY ECOLOGY

COMMUNITY ECOLOGY

Peter J. Morin

Department of Ecology, Evolution & Natural Resources
Rutgers University
New Brunswick, New Jersey

b

Blackwell
Science

©1999 by Blackwell Science, Inc.

Editorial Offices:
Commerce Place, 350 Main Street, Malden,
    Massachusetts 02148, USA
Osney Mead, Oxford OX2 0EL, England
25 John Street, London WC1N 2BL, England
23 Ainslie Place, Edinburgh EH3 6AJ, Scotland
54 University Street, Carlton, Victoria 3053,
    Australia

Other Editorial Offices:
Blackwell Wissenschafts-Verlag GmbH,
    Kurfürstendamm 57, 10707 Berlin, Germany
Blackwell Science KK, MG Kodenmacho
    Building, 7-10 Kodenmacho Nihombashi,
    Chuo-ku, Tokyo 104, Japan

DISTRIBUTORS:
USA
    Blackwell Science, Inc.
    Commerce Place
    350 Main Street
    Malden, Massachusetts 02148
    (Telephone orders: 800-215-1000 or 781-
        388-8250; fax orders: 781-388-8270)

Canada
    Login Brothers Book Company
    324 Saulteaux Crescent
    Winnipeg, Manitoba, R3J 3T2
    (Telephone orders: 204-224-4068)

Australia
    Blackwell Science Pty, Ltd.
    54 University Street
    Carlton, Victoria 3053
    (Telephone orders: 03-9347-0300;
        fax orders: 03-9349-3016)

Outside North America and Australia
    Blackwell Science, Ltd.
    c/o Marston Book Services, Ltd.
    P.O. Box 269
    Abingdon
    Oxon OX14 4YN
    England
    (Telephone orders: 44-01235-465500;
        fax orders: 44-01235-465555)

Acquisitions: Nancy Hill-Whilton
Production: Irene Herlihy
Manufacturing: Lisa Flanagan
Cover photo by Timon McPhearson
Cover design by Lynn McPhearson
Typeset by Best-set Typesetter Ltd., Hong Kong
Printed and bound by Braum-Brumfield, Inc.
Printed in the United States of America
99 00 01 02 5 4 3 2 1

The Blackwell Science logo is a trade mark of
Blackwell Science Ltd., registered at the United
Kingdom Trade Marks Registry

Library of Congress Cataloging-in-Publication
Data
Morin, Peter Jay. 1953–
    Community ecology/Peter J. Morin.
        p. cm.
    Includes bibliographical references.
    ISBN 0–86542–350–4
    1. Biotic communities.   I. Title.
QH541.M574   1999
577.8′2—dc21   98-38821
                                         CIP

CONTENTS

Preface

This book is based on the lectures that I have given in a community ecology course offered at Rutgers University over the last 15 years. The audience is typically first-year graduate students who come to the course with a diversity of backgrounds in biology, ecology, and mathematics. I have tried to produce a book that will be useful both to upper-level undergraduates and to graduate students. The course is structured around lectures on the topics covered here, supplemented with readings and discussions of original research papers; some readings are classic studies, and others are more recent. Throughout the course, the guiding theme is that progress in community ecology comes from the interplay between theory and experiments.

I find that the examples and case studies highlighted here are particularly useful for making important points about key issues and concepts in community ecology. I have tried to maintain a balance between describing the classic studies that every student should know and emphasizing recent work that has the potential to change the way that we think about communities. Limits imposed by space, time, and economy mean that the coverage of important studies could not even begin to be encyclopedic. I apologize to the many excellent hard-working ecologists whose work I was unable to include. I also encourage readers to suggest their favorite examples or topics that would make this book more useful.

Early drafts of most of these chapters were written while I was a visiting scientist at the Centre for Population Biology, Imperial College at Silwood Park, Ascot, England. Professor John Lawton was an ideal host during those stays, and he deserves special thanks for making those visits possible. The Centre for Population Biology is a stimulating place to work and write while free from the distractions of one's home university.

During the prolonged period during which this book took form, several of my graduate students, past and present, took the time to read most of the chapters and make careful comments on them. For that I thank Sharon Lawler, Jill McGrady-Steed, Mark Laska, Christina Kaunzinger, Jeremy Fox, Yoko Kato, Marlene Cole, and Timon McPhearson. Other colleagues at other universities, including Norma Fowler, Mark McPeek, Tom Miller, and Jim Clark, commented on various drafts of different chapters. Any errors or omissions remain my responsibility.

Simon Rallison of Blackwell originally encouraged me to begin writing this book. Along the way the process was facilitated by the able editorial efforts of Jane Humphreys, Nancy Hill-Whilton, and Irene Herlihy. Jennifer Rosenblum and Jill Connor provided frequent editorial feedback and the necessary prodding to keep the project going. They have been patient beyond all reason.

Finally, Marsha Morin deserves special praise for putting up with my many moods while this project slowly took form. I could not have completed it without her support and understanding.

P. J. M.

PART I

COMMUNITIES: BASIC PATTERNS AND ELEMENTARY PROCESSES

CHAPTER 1

Communities

Ecology is the science of communities. A study of the relations of a single species to the environment conceived without reference to communities and, in the end, unrelated to the natural phenomena of its habitat and community associations is not properly included in the field of ecology.

—Victor Shelford (1929)

OVERVIEW

This chapter briefly describes how ecological communities are defined and classified and introduces some of the properties and interactions that community ecologists study. The major interspecific interactions, or elementary processes, between pairs of species include competition, predation, and mutualism. Complex indirect interactions can arise among chains of three or more interacting species. Important community properties include the number of species present, measures of diversity, which reflect both the number and relative abundances of species, and statistical distributions that describe how the species differ in abundance.

Observations of natural patterns and explorations of mathematical models have inspired generalizations about the underlying causes of community organization. One pattern important in the historical development of community ecology concerns an apparent limit to the similarity of coexisting species. The case of limiting similarity provides a cautionary example of the way in which community patterns are initially recognized, explained in terms of causal mechanisms, and eventually evaluated. Community patterns are the consequence of a hierarchy of processes that interact in complex ways to mold the diversity of life on earth.

COMMUNITIES

Our best estimates suggest that somewhere between 1.5 million and 30 million different species of organisms live on earth today (Erwin 1982; May 1990). The small fraction of this enormous global collection of species that can be found at any particular place is an **ecological community**. One important goal of community ecology is to understand the origin, maintenance, and consequences of biological diversity within local communities. Different processes, operating on very different timescales, can influence the number and identity of species in communities. Long-term evolutionary processes operating over timescales spanning millions of years can produce different numbers of species in different locations. Short-term ecological interactions can either exclude or facilitate species over shorter timescales ranging from a few hours to many years. This book provides an overview of community patterns and the processes that create them.

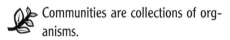 Communities are collections of organisms.

Like many fields of modern biology, community ecology began as a descriptive science. Early community ecology was preoccupied with identifying and listing the species found in particular localities (Clements 1916; Elton 1966). These surveys revealed some of the basic community patterns that continue to fascinate ecologists. In many communities, a few dominant species are much more common than others. Dominant species often play an important role in schemes used to identify and categorize communities. But why should some species be so much more common than others? Communities also change over time, often in ways that are quite repeatable. But what processes drive temporal patterns of community change, and why are those patterns so regular within a given area? Different communities can also contain very different numbers of species. A hectare of temperate forest in New Jersey might contain up to 30 tree species (Robichaud and Buell 1973), while a similar-sized plot of rain forest in Panama might contain over 200 tree species

 Some basic questions.

(Hubbell and Foster 1983). More than 10 hypotheses have been proposed to explain the striking latitudinal gradient in biodiversity that contributes to the differences between temperate and tropical communities (Pianka 1988). Although there are many reasonable competing explanations for the commonness and rarity of species and for latitudinal differences in biodiversity, the exact causes of these very basic patterns remain speculative. Related questions address the consequences of biodiversity for community processes. Do communities with many species function differently from those with fewer species? How do similar species manage to coexist in diverse communities?

The central questions in community ecology are disarmingly simple. Our ability to answer these questions says something important about our understanding of the sources of biological diversity and the processes that maintain biodiversity in increasingly stressed and fragmented natural ecosystems. Answering these questions allows us to wisely manage artificial communities, which include the major agricultural systems that we depend on for food and for biologically produced materials, and to restore the natural communities that we have damaged either through habitat destruction or overexploitation.

Ecologists use a variety of approaches to explore the sources of community patterns. Modern community ecology has progressed beyond basic description of patterns, and often experiments can identify which processes create particular patterns (Hairston 1989). However, some patterns and their underlying processes are experimentally intractable because the organisms driving those processes are so large, long lived, or wide ranging that experimental manipulations are impossible. Consequently, community ecologists must rely on information from many sources, including mathematical models, statistical comparisons, and experiments, to understand what maintains patterns in the diversity of life. The interplay among description, experiments, and mathematical models is a hallmark of modern community ecology.

Before describing how ecologists identify and classify communities, it is important to recognize that the term *community* means different things to different ecologists. Most definitions of ecological communities include the idea of a collection of species found 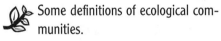 Some definitions of ecological communities. in a particular place. The definitions part company over whether those species must interact in some significant way to be considered community members. For instance, Robert Whittaker's (1975) definition of a community as

> an assemblage of populations of plants, animals, bacteria and fungi that live in an environment and interact with one another, forming together a distinctive living system with its own composition, structure, environmental relations, development, and function (pp. 1–2)

clearly emphasizes both physical proximity of community members and their various interactions. In contrast, Robert Ricklefs' (1990) definition does not stress interactions but does emphasize that communities are often identified by prominent features of the biota (dominant species) or physical habitat:

> [T]he term has often been tacked on to associations of plants and animals that are spatially delimited and that are dominated by one or more prominent species or by a physical characteristic. (p. 656)

Other succinct definitions include those by Peter Price (1984), "the organisms that interact in a given area[,]" and by John Emlen (1977), "A biological community is a collection of organisms in their environment[,] (p. 341)" which emphasize the somewhat arbitrary nature of communities as sets of organisms found in a particular place.

Charles Elton's (1927) definition, while focused on animals, differs from the previous ones in drawing an analogy between the roles that various individuals play in human communities and the functional roles of organisms in ecological communities:

> One of the first things with which an ecologist has to deal is the fact that each different kind of habitat contains a characteristic set of animals. We call these animal associations, or better, animal communities, for we shall see later on that they are not mere assemblages of species living together, but form closely-knit communities or societies comparable to our own. (p. 5)

Elton's emphasis on the functional roles of species remains crucial to our understanding of how functions and processes within communities change in response to natural or anthropogenic changes in community composition.

For our purposes, **community ecology** will include the study of patterns and processes involving at least two species at a particular location. This broad definition embraces topics such as predator-prey interactions and interspecific competition that are traditionally considered part of **population ecology**. Population ecology focuses primarily on patterns and processes involving single-species groups of individuals. Of course, any separation of the ecology of populations and communities must be highly artificial since natural populations always occur in association with other species in communities of varying complexity and often interact with many other species as competitors, consumers, prey, or mutually beneficial associates.

Community ecology vs. population ecology.

Most communities are extraordinarily complex. That complexity makes it difficult even to assemble a complete species list for a particular locale (e.g., Elton 1966; Martinez 1991). The problem is compounded by the fact that the taxonomy of smaller organisms, especially bacteria, protists, and many invertebrates, remains poorly known (Wilson 1992). Consequently, community ecologists often focus their attention on conspicuous, readily identified sets of species that are ecologically or taxonomically similar. One important subset of the community is the **guild**, a collection of species that use similar resources in similar ways (Root 1967). There are no taxonomic restrictions on guild membership, which depends only on the similarity of resource use. For example, the granivore guild in deserts of the southwestern United States consists of a taxonomically disparate group of birds, rodents, and insects that all consume seeds as their primary source of food (Brown and Davidson 1977). Another term, **taxocene** (Hutchinson 1978), refers to a set of taxonomically related species within a community. Ecologists often refer to lizard, bird, fish, and plant communities, but these assemblages are really various sorts of taxocenes. Unlike the guild, membership in a taxocene is restricted to taxonomically similar organisms. Although ecologists often study taxocenes rather than guilds, the use of the term taxocene to describe such associations has been slow to catch on.

 Some useful subsets of communities.

Other useful abstractions refer to subsets of the community with similar feeding habits. **Trophic levels** provide a way to recognize subsets of species within communities that acquire energy in similar ways. Abstract examples of trophic levels include primary producers, herbivores, primary carnivores (which feed on herbivores), and decomposers (which consume dead organisms from all trophic levels). With the exception of most primary producers, many species acquire energy and matter from more than one adjacent trophic level, making it difficult to unambiguously assign many species to a particular trophic level. While trophic levels are a useful abstraction and have played a prominent role in the development of ecological theory (Lindeman 1942; Hairston et al. 1960), the problem of assigning real species to a particular trophic level can limit the concept's operational utility (Polis 1991).

Other descriptive devices help to summarize the feeding relations among organisms within communities. **Food chains** and **food webs** describe patterns of material and energy flow in communities, usually by diagramming the feeding links between consumers and the species that they consume. In practice, published examples of food webs usually describe feeding relations among a very small subset of the species in the complete community (Paine 1988). More complete descriptions of feeding connections in natural communities can be dauntingly complex and difficult to interpret (Winemiller 1990). Patterns in the organization of food webs are a topic considered in Chapter 6.

Ecosystems consist of one or more communities, together with their abiotic surroundings. Ecosystem ecologists often come closer than community ecologists to studying the workings of entire communities, although they often do so by lumping many species into large functional groups such as producers and decomposers. Ecosystem ecologists manage to study whole communities only by ignoring many of the details of population dynamics, focusing instead on fluxes and cycles of important substances like carbon, nitrogen, phosphorus, and water. There is an increasing awareness that distinctions between community and ecosystem ecology are just as artificial as distinctions between population and community ecology (Vitousek 1990). The processes of energy and material flow that interest ecosystem ecologists are certainly affected in no small way by interactions among species. Conversely, feedbacks between species and pools of abiotic nutrients may play an important role in affecting the dynamics of species in food chains (DeAngelis et al. 1989). Certain species that physically alter the environment through their presence or behavior effectively function as ecosystem engineers (Jones et al. 1994). Examples include modifications of stream courses by beavers, and changes in light, humidity, and physical structure created by dominant forest trees.

 Communities vs. ecosystems.

COMMUNITIES AND THEIR MEMBERS

Community ecologists recognize and classify communities in a variety of ways. Most of these approaches have something to do with the number and identity of species found in the community. Regardless of the criteria used, some communities are easier to delineate than others. Ecologists use several different approaches to delineate communities: 1) physically, by discrete habitat boundaries, 2) taxonomically, by the identity of a dominant indicator species, 3) interactively, by the existence of strong interactions among species, 4) or statistically, by patterns of association among species.

 Different ways of identifying communities.

Physically defined communities include assemblages of species found in a particular place or habitat. To the extent that the boundaries of the habitat are easily recognized, so are the boundaries of the community. Some spatially discrete habitats, such as lakes, ponds, rotting fruits, and decaying carcasses, contain equally discrete communities of resident organisms. Less discrete communities may grade gradually into other communities, defying a simple spatial delimitation. For example, forests grade relatively imperceptibly into savannas and then into grasslands, without any clear boundaries. Whittaker and

 Some communities are defined by habitat.

Niering's (1965) study of plant communities along an elevational gradient in southeastern Arizona illustrates the gradual transition between different kinds of terrestrial communities (Figure 1.1). The Sonoran desert scrub and subalpine forest communities at the base and summit of the Santa Catalina Mountains are quite distinct, with giant cactus present in the desert scrub and evergreen fir trees abundant at the summit; however, the transitions between these endpoints are gradual and contain intervening communities.

Biomes are basic categories of communities that differ in their physical environments and in the lifestyles of their dominant organisms. A list of the major biomes of the world recognized by Whittaker (1975) is shown in Table 1-1. The composition of the list betrays Whittaker's keen interest in terrestrial plants, since most of the biomes describe differences among assemblages of terrestrial plants and their associated biota. Had the list been drawn up by a limnologist or a marine ecologist, more kinds of aquatic biomes certainly would have been recognized. Biomes are a useful shorthand for describing certain kinds of communities and, as such, help to facilitate communication among ecologists. The global distribution of terrestrial biomes is strongly influenced by annual precipitation and average temperature (Holdridge 1947), as summarized in Figure 1.2.

TABLE 1-1. Major biomes of the world.

1. Tropical rain forests	19. Arctic-alpine semideserts
2. Tropical seasonal forests	20. True deserts
3. Temperate rain forests	21. Arctic-alpine deserts
4. Temperate deciduous forests	22. Cool temperate bogs
5. Temperate evergreen forests	23. Tropical freshwater swamp forests
6. Taiga forests	24. Temperate freshwater swamp forests
7. Elfinwoods	25. Mangrove swamps
8. Tropical broadleaf woodlands	26. Salt marshes
9. Thornwoods	27. Freshwater lentic communities (lakes and ponds)
10. Temperate woodlands	28. Freshwater lotic communities (rivers and streams)
11. Temperate shrublands	29. Marine rocky shores
12. Savannas	30. Marine sandy beaches
13. Temperate grasslands	31. Marine mudflats
14. Alpine shrublands	32. Coral reefs
15. Alpine grasslands	33. Marine surface pelagic
16. Tundras	34. Marine deep pelagic
17. Warm semidesert scrubs	35. Continental shelf benthos
18. Cool semideserts	36. Deep-ocean benthos

Source: Whittaker (1975).

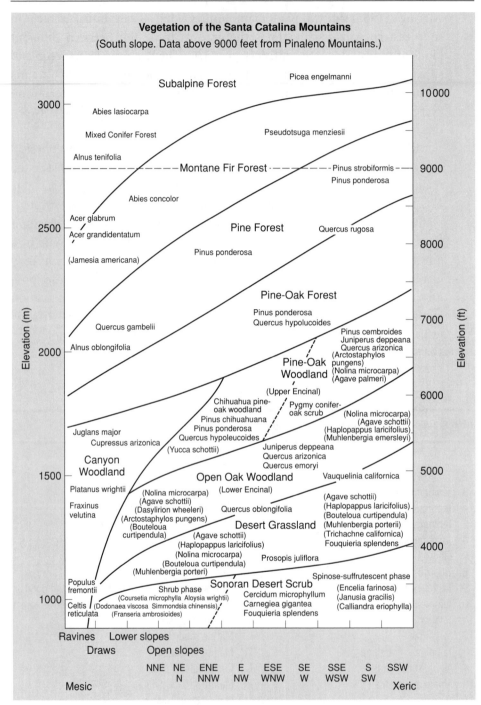

FIGURE 1.1. *Changes in plant species composition along an elevational gradient in the Santa Catalina Mountains of southeastern Arizona. Changes in elevation result in changes in both temperature and rainfall that lead to differences in the identity of predominant plant species. (Adapted from Whittaker and Niering, 1965, with permission of the Ecological Society of America.)*

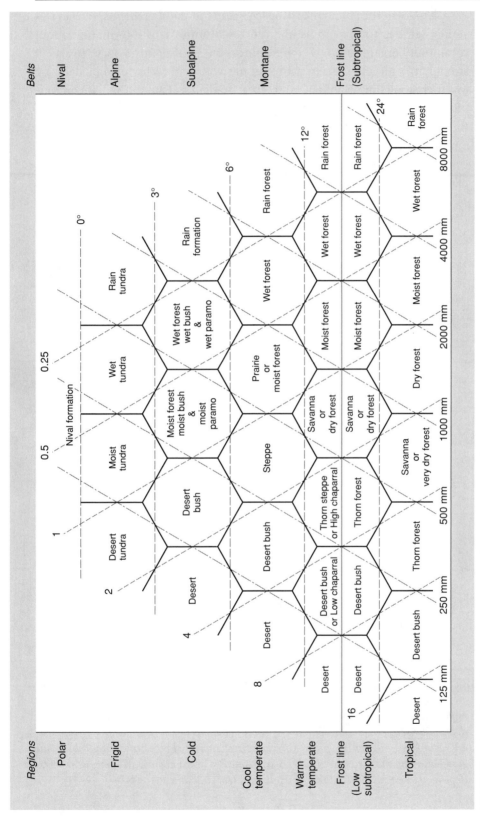

FIGURE 1.2. Relation between average annual temperature, rainfall, and the presence of particular terrestrial biomes characterized by different kinds of vegetation. Annual rainfall (in mm) is indicated along the base of the chart. Increasing elevation or latitude is indicated by increasing height along both sides of the graph. (Adapted with permission from Holdridge, 1947. © 1947 American Association for the Advancement of Science).

Changes in the abundance of species along physical gradients, such as elevation, temperature, or moisture, can reveal important information about community organization. If communities consist of tightly associated sets of strongly interacting species, those species will tend to increase or decrease together along important environmental gradients (Figure 1.3A). If communities are loosely associated sets of weakly interacting species, abundances of those species will tend to vary independently, or individually, along important gradients (Figure 1.3B). Most of the information gathered to address community patterns along gradients describes single trophic levels, usually plants, and seems consistent with a loose model of community organization

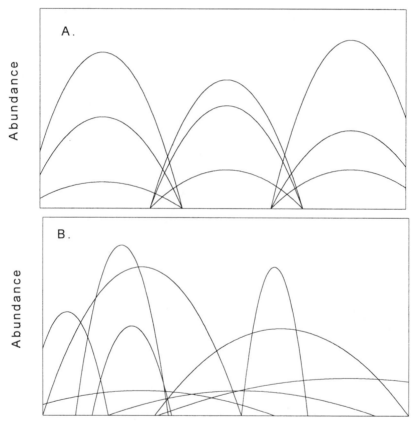

Environmental Gradient (Moisture, Temperature, Altitude)

FIGURE 1.3. *Two hypothetical patterns of abundance for sets of species along an environmental gradient. (A) Groups of tightly integrated and strongly competing species that respond as an entire community to environmental variation. Strong competition creates sharp breaks in species composition. (B) Species responding individually to environmental variation, with no integrated correlated response of the entire community to the gradient. (Modified from Whittaker,* Communities and Ecosystems, *2nd ed. © 1975. Reprinted by permission of Prentice-Hall, Inc., Upper Saddle River, NJ.)*

(Whittaker 1967). However, the kinds of tight associations between species that would yield the pattern seen in Figure 1.3A are far more likely to occur between trophic levels, such as for species-specific predator-prey, parasite-host, or mutualistic relations. Descriptions of associations between plants and their specialized herbivores (see Futuyma and Gould 1979) or between herbivores and their specialized predators or parasites might yield a pattern more like that seen in Figure 1.3A. Strangely, such studies are rare, perhaps because the taxonomic biases of ecologists restrict their attention to particular groups of organisms that tend to fall within single trophic levels.

Taxonomically defined communities usually are recognized by the presence of one or more conspicuous species that either dominate the community through sheer biomass or otherwise contribute importantly to the physical attributes of the community. Examples would include the beech (*Fagus*) and maple (*Acer*) forests of the northeastern United States and the longleaf pine (*Pinus palustris*) and wiregrass (*Aristida*) savannas of the southeastern United States. In both cases, the predominance of one or two plant species defines the community. In some cases, the dominant, or most abundant, species that identifies a particular community type also plays an important role in defining the physical structure of the community (Jones et al. 1994).

 Other communities are recognized by dominant species.

Statistically defined communities consist of sets of species whose abundances are significantly correlated, positively or negatively, over space or time. The approach makes use of overall patterns in the identity and abundance of species to quantify similarities and differences among communities. One way to describe the species composition of a community is to simply list the identity and abundance of each species. But how do you compare these lists? For long lists containing many species, such comparisons are difficult to make by just reading down the list and making species-by-species comparisons. Imagine instead a geometric space defined by S independent axes, each of which represents the abundance of a different species (Figure 1.4). The species composition of a particular community is represented by a point whose coordinates correspond to the abundance of each species, $(n_1, n_2, \ldots, n_s)$, where n_i is some measure of the abundance of species i. While it is difficult to visualize species composition in more than three dimensions (more than three species), in principle the mathematical and geometric interpretations of this approach generalize for any number of species, S. Species composition then has a geometric interpretation as a directional vector, or arrow, in S-dimensional space, as shown in Figure 1.4.

Statistical associations can also identify communities.

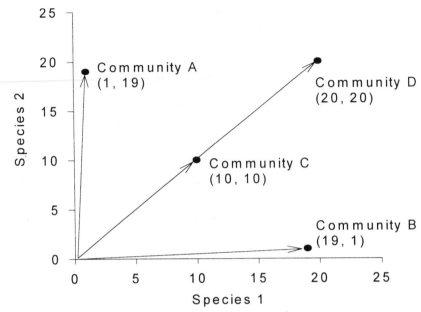

FIGURE 1.4. *A geometric representation of species composition as a vector in a space defined by axes that describe the abundances of different species measured in a comparable sample area. This simple example focuses only on communities of two hypothetical species. Note that both communities A and B have identical values of species richness, S = 2, and species diversity, H′ = 0.199, but they clearly differ in species composition, as shown by the different directions of the arrows. Communities C and D have identical relative abundances of the two species, but one community contains twice the number of individuals as the other. This approach generalizes to patterns for any value of species richness, although it is difficult to visualize for S > 3.*

One advantage of the geometric approach is that it clearly distinguishes among communities with similar numbers of species that differ in the identity of common and rare species. In such cases, community composition vectors point in different directions in the space defined by the abundances of different species in the communities being compared. Comparisons involving more than three species rely on various statistical techniques, mostly involving ways of classifying or ordering communities based on the identity and abundance of species. The development of effective statistical techniques for the description of species composition has been a major goal of mathematical ecology. Many of the techniques employ multivariate statistics to derive concise descriptors of community composition that can be interpreted in terms of differences among communities in the abundance of particular sets of species. The computational details of these techniques, which are collectively termed **ordination,** fall outside the scope of this book, but Gauch (1982) and Pielou (1984) provide excellent summaries geared toward the interests of ecologists.

Two examples of ordinated sets of communities are shown in Figure 1.5. In each case, overall species composition is represented by an index, or score, for a community along a set of coordinate axes. The score for a community along one axis is a linear function of the species composition in each community, with the general form $a_{11}n_{11} + a_{12}n_{12} + \ldots + a_{ij}n_{ij} + \ldots a_{1S}n_{1S}$, where the a_{ij}'s are constants selected to maximize the variation among communities represented in this new space, and the n_{ij}'s represent the abundance of the jth species in the ith community. For different axes the coefficients a_{ij} will also differ, so that the axes, and the patterns of species occurrence that they describe, are statistically independent. Often two or three ordination axes, with different sets of coefficients, are sufficient to describe the majority of the variation in species composition among communities. Figure 1.5A shows patterns of similarity in a large number of sampled stands of vegetation, based on abundances of 101 plant species. Stands of similar composition fall near each other in this two-dimensional space, whereas increasingly different stands are separated by larger distances. Figure 1.5B shows the results of a similar approach applied to the zooplankton species found in a large number of Canadian lakes. Lakes with similar species composition have similar locations in the set of coordinates used to describe species composition. In both cases, the position of a community with respect to the coordinate axes says something about the abundance of a few key species that vary in abundance among communities, that is, the species that make these communities recognizably different. The advantage of these approaches is that information about a large number of species can be distilled into measures of position along one to several coordinate axes. The resulting classification usually does not identify the proximal factors leading to the predominance of one species versus another in a particular community. Such information usually comes from direct experimental studies of interspecific interactions.

Interactively defined communities consist of those subsets of species in a particular place or habitat whose interactions significantly influence their abundances. Only some, and perhaps none, of the species in a physically defined community may constitute an interactively defined community. Hairston (1981) used this approach to point out that only a small subset of the species of salamanders found in the mountains of North Carolina could be shown to interact and affect each other's abundance. Of the seven common species of plethodontid salamanders in his study plots, only the two most common species, *Plethodon jordani* and *Plethodon glutinosus*, significantly affected each other's abundance. The remaining five species, while taxonomically and ecologically similar to the others, remained unaffected by the

 Interactions can also be used to define communities.

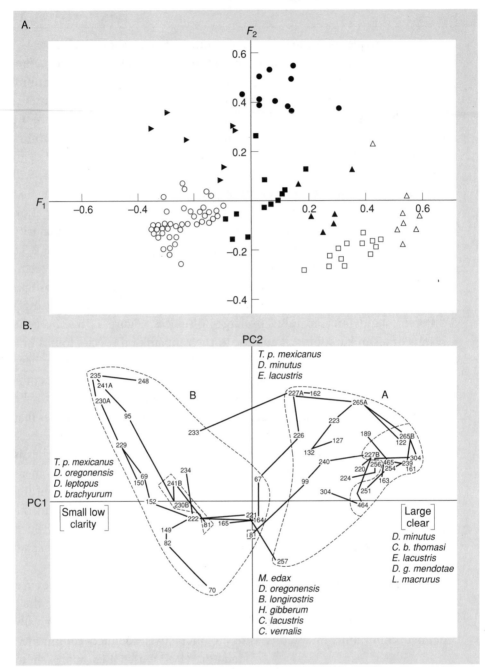

FIGURE 1.5. *Examples of statistically classified or ordinated communities. (A) Plant assemblages growing on sand dunes. Different symbols correspond to different habitat types. Positions of each community represent the frequency (abundance) of 101 plant species. (Reprinted from Orloci, 1966, with permission of Blackwell Science Ltd.) (B) Zooplankton assemblages from a large number of Canadian lakes. Each number corresponds to a particular lake. Similarity in species composition is represented by proximity in a complex space defined by weighted functions of the original abundances of various species in field samples. The axes can be interpreted as indicating a predominance of some species as opposed to others, or as gradients in physical factors that are correlated with the abundance of particular species. (Adapted from Sprules, 1977, with permission of the NRC Research Press.)*

abundance of the two most common species. The key point is that the assignment of membership in a guild or taxocene based on similarity of resource use or taxonomy is no guarantee that species will really interact.

COMMUNITY PROPERTIES

Given that you can identify communities using some repeatable criteria, what is the best way to compare complex systems composed of many species that can interact in many ways? The potentially bewildering complexity of communities encourages ecologists to use various descriptors to condense and summarize information about the number, identity, and relative abundance of species. No single number, index, or graph can provide a complete description of a community, but some of these measures provide a useful way of comparing different communities.

Species Richness

Robert May (1975) observed that "one single number that goes a long way toward characterizing a biological community is simply the total number of species present, S_T" (p. 82). This number, often called **species richness**, is synonymous with our most basic notions of biodiversity. It is, in practice, a difficult number to obtain, partly because we simply do not have complete taxonomic information about many of the groups of organisms found in even the most-studied communities. Even

Species richness is an important community attribute.

if we did have the ability to unambiguously identify all the species found in a particular place, there would still be the practical problem of deciding whether we had searched long and hard enough to say that all the species in that place had been found. Therefore, in practice, species richness is evaluated for groups that are taxonomically well known and readily sampled according to some repeatable unit of effort. One way to decide whether enough sampling effort has been made is to plot the cumulative number of species found against the amount of sampling effort. Beyond a certain amount of effort, the species versus effort curve should reach an asymptote. That asymptote provides a reasonable estimate of the number of species present. Comparisons among communities that have been sampled with different amounts of effort can be made by using rarefaction curves (Sanders 1968; Hurlbert 1971). These are essentially catch per unit effort curves that permit comparisons among communities scaled to the same amount of effort.

Species richness is more than a convenient descriptive device. There is increasing evidence that it is related to important functional attributes of communities. Recent experimental work indicates that primary production, resistance to natural disturbances, and resistance to invasion all increase as species richness increases (Tilman and Downing 1994; Naeem et al. 1994; Tilman et al. 1996; Tilman 1997; McGrady-Steed et al. 1997).

Diversity

Although species richness provides an important basis for comparisons among communities, it says nothing about the relative commonness and rarity of species. Various diversity indices have been proposed to account for variation in both the number of species in a community and the way that individuals within the community are distributed among species (Magurran 1988). One measure is the Shannon-Weaver index of diversity,

$$H' = \sum_{i=1}^{S} -p_i \cdot \ln(p_i)$$

where S is the total number of species present in a sample and p_i is the fraction of the total number of individuals in the sample that belong to species i. For instance, imagine that two communities have the same species richness, but individuals are evenly distributed among species in the first community and unevenly distributed among species in the second. A satisfying measure of species diversity would give the first community a higher measure of diversity.

> Species diversity measures species richness and relative abundance.

The comparisons get complicated when comparing communities that vary in both species richness and the evenness of distribution of individuals among species. For this reason, it is often preferable to break species diversity down into its two components, species richness and evenness. Evenness is usually defined as

$$J = H'/H_{\max}$$

where H' is the observed value of species diversity and $H_{\max}$ is the value that would be obtained if individuals were evenly distributed among the number of species found in the community (if the values of p_i were identical for each species). Species diversity indices are seductive in that they offer a simple way to describe the complexity present in a community. Their main drawback is that they gloss over potentially important information about the actual identities of the species.

The amount of diversity found within a single type of habitat is sometimes called **alpha diversity** (Whittaker 1975). Within a region, the turnover in species composition among different habitats will contribute additional diversity to a region. This interhabitat component of diversity is called **beta diversity**.

Species Abundance Relations

Graphical ways of summarizing the relative abundances of species in a sample have a long tradition of use in community ecology. Many communities display well-defined patterns, which may or may not have important ecological significance. Examples of some of the more historically important species-abundance distributions are shown in Figure 1.6. Each distribution has an underlying statistical distribution, which can be derived by making some assumptions about the way in which species interact in communities. In each case, the importance value of each species, usually a measure of the fraction of total number of individuals or biomass in the sample accounted for by each species, is plotted against the importance rank of each species, where a rank of 1 corresponds to the most important species and a rank of s corresponds to the least important (least abundant) species in a sample of s species.

 Species abundance relations graphically summarize community patterns.

Three of the more important species-abundance relations that have attracted the attention of ecologists are the broken stick distribution, the geometric series, and the lognormal distribution (Whittaker 1975; May 1975). Each distribution can be derived by making particular assumptions about the way that species divide up resources within a community. For example, the geometric series can be obtained by assuming that a dominant species accounts for some fraction, k, of the total number of individuals in a sample, and each successively less abundant species accounts for a fraction k of the remaining number of individuals. This leads to the following formula for the abundance of the ith species:

$$n_i = Nk(1-k)^{i-1},$$

where N is the total number of individuals in the sample and i runs from 1 for the most abundant species to s for the least. The fraction k is usually approximated by n_1/N.

The problem with using these statistical distributions to infer the existence of underlying processes is that even if collections of species are found to fit a particular distribution, there is no guarantee that the species in fact interact in the fashion assumed by the underlying model (Cohen 1968). Largely for this reason, the study of species abundance patterns no longer figures prominently in community ecology, although there are occasional efforts to revive interest in particular patterns (e.g., Sugihara 1980). These distributions are described here primarily because they played an important role in the historical development of community ecology and because they continue to provide a useful alternate way of describing patterns of abundance within communities.

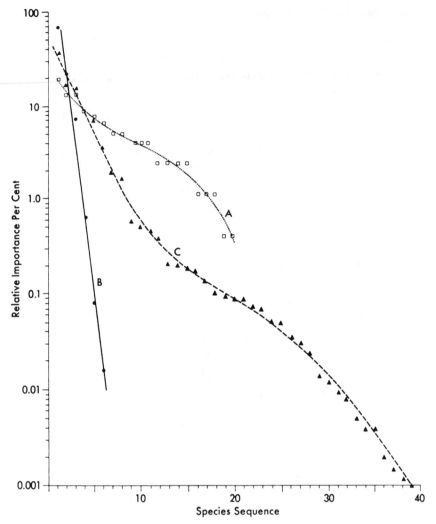

FIGURE 1.6. *Examples of three common species-abundance relations that fit different collections of species. (A) Nesting birds in a West Virginia forest, following a broken stick distribution. (B) Vascular plants in a subalpine fir forest in Tennessee, following a geometric series. (C) Vascular plants in a deciduous cove forest in Tennessee, following the lognormal distribution. (Reprinted from Whittaker,* Communities and Ecosystems, *2nd ed. © 1975. Reprinted by permission of Prentice-Hall, Inc., Upper Saddle River, NJ.)*

Species Composition

We have already seen how the species composition of a particular community can be represented by a point whose coordinates correspond to the abundance of each species (see Figures 1.4 and 1.5). This geometric representation conveys more information than either species richness or species diversity measures, but that information comes with a somewhat greater difficulty of interpretation. It differs from measures of richness or diversity in that both the identity and abundance of particular species are considered to be important attributes.

INTERSPECIFIC INTERACTIONS

Rather than attempting to infer the influence of interspecific interactions on community patterns from indirect means, such as the species abundance relations described above, community ecologists often directly study how various interactions affect patterns of abundance. Interspecific interactions are the basic elementary processes that can influence species abundances and the community composition. Figure 1.7 shows how interactions between a pair of species can be categorized by assigning positive or negative signs to the net effect that a population of each species has on the population size of the other (Burkholder 1952; Price 1984). More complex interactions involving chains of three or more species can be represented similarly (Holt 1977). Abrams (1987) has

Species interact in a limited number of ways.

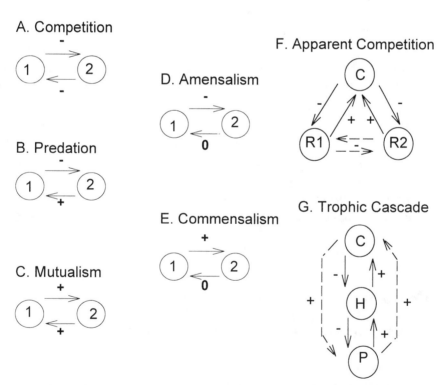

FIGURE 1.7. *Examples of direct and indirect interactions among species in communities. Direct effects are indicated by solid lines, with signs corresponding to the signs of interactions between the species. Net indirect effects are indicated by broken lines. C, consumer; R1, resource 1; R2, resource 2; P, primary producer, H, herbivore.*

criticized the approach of classifying interspecific interactions by the signs of net effects because the sign of the interactions can depend on the responses used to classify interactions, such as population growth rates, population size, or relative fitness. However, as long as the criteria used to describe how one species affects another are explicit, the approach often has heuristic value.

Predation, **parasitism**, and **herbivory** all involve a −/+ interaction between a pair of species, in which the net effect of an individual consumer on an individual prey is negative, and the effect of the consumed prey on the predator is positive. These interactions share the common features of consumer-resource interactions, where all or part of the resource species is consumed by the other. Predation and other −/+ interactions drive processes of energy and material flow through food webs. **Competition** involves a mutually negative (−/−) interaction between a pair of species. **Amensalism** is a one-sided competitive interaction (0/−), in which one species has a negative effect on another, but the other has no detectable effect on the first. **Mutualism** involves a mutually positive (+/+) interaction between a pair of species, in which each has a positive effect on the other. **Commensalism** is a one-sided mutualistic (0/+) interaction, in which one species has a positive effect on another species, but the second species has no net effect on the first.

Of course, communities are more complex entities than simple pairs of interacting species. Interactions among pairs of species can be transmitted indirectly through chains of species to others. Such indirect effects have their own terminology; some of the simpler scenarios are outlined in Figure 1.7. For example, consider two prey species, A and B, that are consumed by a third predator species. Assume that neither prey species competes with the other, but that more predators will persist when both prey species are present than when only one prey species is present. The net result will be that predation is more intense on both prey when they co-occur. This scenario, termed **apparent competition** by Holt (1977), results when each prey has an indirect negative effect on the other by virtue of its direct positive effect on the abundance of a shared predator. There are many intriguing variations on this theme, which are explored in greater detail in Chapter 8.

COMMUNITY PATTERNS AS THE INSPIRATION FOR THEORY: ALTERNATE HYPOTHESES AND THEIR CRITICAL EVALUATION

The major organizing themes in community ecology have been inspired by the discovery of particular patterns, and different ideas about the causes of those patterns play an important role in the development of theories of community organization. Progress toward the development of predictive theories of community ecology has sometimes been sidetracked by focusing on

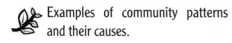
Examples of community patterns and their causes.

patterns that were not clearly related to particular processes. Also, some patterns may arise from multiple processes, and important processes may be difficult to identify by observation alone. In some cases, what initially appeared to be an important community pattern eventually proved to be indistinguishable from a random pattern!

One community-level pattern that has yielded important insights into the roles of interspecific interactions in community organization is the striking vertical zonation of marine organisms in the rocky intertidal zone. One particularly well-studied example of this zonation concerns two species of barnacles found on the rocky coast of Scotland. The smaller of the two species, *Chthamalus stellatus*, is consistently found higher in the intertidal zone than the larger species, *Balanus balanoides*. Such differences in zonation were historically attributed entirely to physiological differences between the barnacles, presumably reflecting differences in the ability of the two species to withstand desiccation at low tide and immersion at high tide. However, observations and a careful series of experimental transplants and removals showed that several factors, including interspecific competition, predation, and physiological constraints, produce the pattern (Connell 1961). Both species initially settle within a broadly overlapping area of the intertidal zone, but overgrowth by the larger barnacle, *Balanus*, smothers and crushes the smaller *Chthamalus*, excluding it from the lower reaches of the intertidal zone. Other experiments show that predation by the snail *Thais* influences the lower limit of the *Balanus* distribution, whereas different tolerances to desiccation during low tide set the upper limits of both barnacle distributions. Consequently, a rather simple pattern of vertical zonation ultimately proves to depend on a complex interaction among competition, predation, and physiological tolerances. This example illustrates the important role of natural community patterns as a source for ideas about the processes that organize communities. It also emphasizes that inductive reasoning alone may not provide an accurate explanation for a given pattern, especially when there are several competing hypotheses that could account for that pattern.

Not all community patterns are as readily recognized and understood as the intertidal zonation of barnacles. Some of the patterns that preoccupied ecologists for decades have eventually been recognized as artifacts that offer little insight into community-level processes. Differences in the body sizes of ecologically similar coexisting species provide a telling case in point. The story begins with observations about the body sizes of aquatic insects in the family Corixidae, called water boatmen (Figure 1.8). Hutchinson (1959) noted that three European species, *Corixa affinis*, *Corixa macrocephala*, and *Corixa punctata*, have segregated distributions, such that the largest species, *C. punctata*,

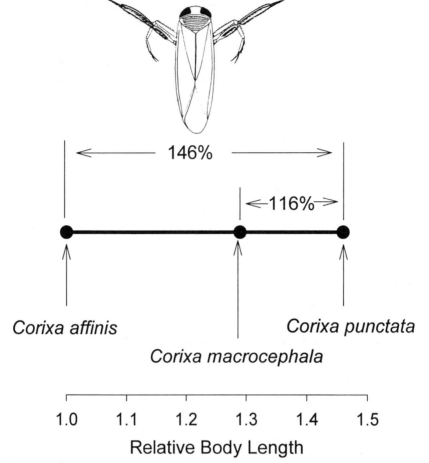

FIGURE 1.8. *Corixids, a kind of common aquatic hemipteran insect, inspired Hutchinson's (1959) concept of limiting morphological similarity of coexisting species. Relative sizes of the three species considered by Hutchinson are indicated by their positions along a scale that corresponds to relative body size.*

occurs with either *C. affinis* or *C. macrocephala*, whereas the two smaller species do not coexist in the same pond. *C. punctata* is larger than either of the species with which it coexists by about 116% to 146%. Hutchinson suggested that species that differ sufficiently in size or other life history features may also differ sufficiently in resource use to avoid competitive exclusion. Examination of other taxa indicated that coexisting species tended to differ in some aspect of size by a factor of about 1.3, or 130%. Hutchinson did not mention that the two species that fail to coexist also differ in size by a factor of 1.46/1.16, or 1.259, which is clearly within the range observed for the two pairs of species that do coexist! Also, many sets of inanimate objects, including cooking utensils and musical instruments (Horn and May 1977), also fit the 1.3 rule to a good approximation, which casts considerable doubt on the pattern holding deep ecological significance.

Competitive exclusion of species that are too similar in size, and therefore too similar in resource use, is one possible explanation for the differences in body size that Hutchinson observed, but alternative explanations exist. One possibility is that differences in the sizes of coexisting species might be no greater than expected for any randomly selected sets of species (Strong et al. 1979), that is, no greater than expected by chance. Clearly, some differences in the sizes of any set of species would be expected to occur regardless of the intensity of their interactions, since by definition species must differ in some way for taxonomists to recognize them as separate entities. The crucial question is whether those differences are any greater than would be expected to occur by chance (Simberloff and Boecklin 1981). Determinations of the randomness or nonrandomness of the sizes of coexisting species are by no means straightforward (Colwell and Winkler 1984), but some studies suggest that observed size differences among coexisting species may be no greater than those expected in randomly selected sets of noninteracting species.

Another way to assess the ecological significance of size differences among coexisting species would be to experimentally measure whether species that differ greatly in body size compete less intensely than species of similar size. Experimental studies of competition among corixids in other aquatic systems suggest that substantial morphological differences among species do not prevent competition. Both Istock (1973) and Pajunen (1982) have shown that even when coexisting corixid species differ substantially in size, they still compete strongly. Pajunen (1982) suggested that his corixid species only manage to coexist by virtue of their ability to disperse among pools as adults and to rapidly recolonize pools after competitive extinctions. Co-occurrence of similarly sized species may be fleeting and illusory, rather than a persistent consequence of differences in resource use. Strangely, no one has directly tested whether the intensity of competition among corixid species depends on similarity in size or some other aspect of morphology.

Studies of another group of aquatic insects also offer little support for the idea that morphological similarity is a good predictor of competition's intensity. Juliano and Lawton (1990a,b) examined patterns of co-occurrence for several species of larval dytiscid beetles, which prey on other aquatic organisms. Size differences among coexisting species were no greater than expected by chance. Experimental manipulations of these species failed to identify a clear relation between body size and competition. In fact, competition among these species was generally quite weak, despite their similar requirements as small aquatic predators.

Hutchinson's corixids and the concept of limiting morphological similarity provide a cautionary tale about the kinds of patterns that intrigue community ecologists and the need to critically evaluate the explanations proposed for those patterns. The search for general mechanisms that might explain such patterns is one of the main goals of community ecology. Examples of other kinds of patterns in multispecies assemblages include geographic patterns of diversity and species richness, repeatable patterns in the structure of guilds, and recurring patterns observed in the architecture of food webs. Discovery of these patterns depends on careful observational studies of natural systems, but it is important to remember that each pattern may result from multiple processes that can only be disentangled by experiments.

Community Patterns Are a Consequence of a Hierarchy of Interacting Processes

Community ecologists recognize that many factors affect the species composition of a given community, with no single factor providing a complete explanation for observed patterns (Schoener 1986). The factors can interact in a complex hierarchical fashion, as sketched in Figure 1.9. For example, the composition of a regional species pool of potential community members sets an upper limit on the species composition of a new community developing in a given place, as might happen after the creation of a new lake or after removal of an established natural community by a catastrophic disturbance. Membership in the regional species pool is constrained by physiological tolerances, historical factors, and the evolutionary processes responsible for the generation of different numbers of species in different taxonomic groups or habitats. Species generally do not occur in areas that tax their physiological limits. Successful introductions of species into areas far from their normal ranges show that accidents of biogeography can exclude whole groups of species from some geographic regions (Elton 1958). For example, salamanders are absent from Australia and sub-Saharan Africa, although many species possess physiological adaptations that allow them to inhabit climatically similar regions on other continents.

 Community patterns arise from a hierarchy of processes.

Dispersal and habitat selection sift and filter species from the regional species pool to set the identity of those species available to colonize a given community. The idea of community assembly as a filtering process has been developed for plant assemblages by Paul Keddy (1992), and it applies equally well to other kinds of organisms. These factors act to make communities nonrandom subsets of the regional species pool. Habitat selection can be influenced by the species already present in the community. Finally, interspecific interactions, or the lack thereof, influence the subsequent success or failure of species that actually arrive at a community. The following chapters will consider how various patterns arise in communities by first considering

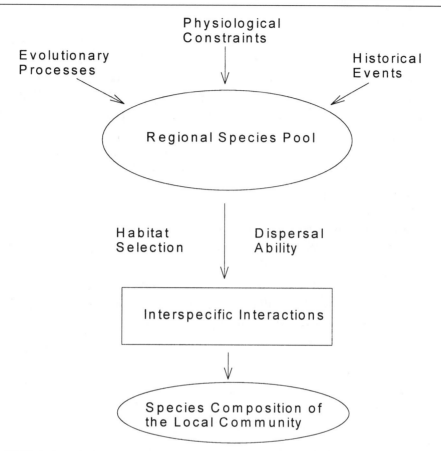

FIGURE 1.9. *The species composition of a local community at any time is a consequence of many factors interacting in a hierarchical fashion. The composition of the species pool of potential community members depends on past evolutionary and historical events, as well as physiological constraints. Dispersal ability and habitat selection influence which members of the species pool arrive in a particular location. Interspecific interactions among those species that manage to arrive in a particular place further inhibit or facilitate the inclusion of species in the community.*

how interspecific interactions affect the success or failure of species as community members. Subsequent chapters explore some of the processes that influence which species interact and how those interactions vary over space and time.

CONCLUSIONS

The many definitions of ecological communities all identify collections of species found in particular locations. Useful commonly studied subsets of communities include guilds, taxocenes, and trophic levels. Species richness and species diversity are two important community attributes. Species-abundance relations, sometimes called dominance-diversity curves, provide a graphical way of describing species richness and the relative abundance of species in communities. The concept of species composition includes these ideas, as well as coupling the identity of particular species to patterns of relative abundance. Communities can be identified by physical habitats, by dominant organisms, by statistical associations among species, or by the identification of sets of interacting species. Fundamental interspecific interactions, such as competition, predation, and mutualism, contribute to important community patterns. Some patterns, such as vertical zonation of species in intertidal communities, can be shown to result from interactions among species and physiological constraints. Other patterns, such as the suggested regularity of morphological differences among closely related coexisting species, may not be easily linked to interspecific interactions. Community patterns can have multiple alternate explanations, which may not be completely understood by simple inspection and inductive reasoning. It does seem likely, though, that community patterns result from a complex hierarchy of interacting processes.

CHAPTER 2

Competition: Mechanisms, Models, and Niches

OVERVIEW

Interspecific competition is any mutually negative interaction between two or more species that does not involve mutual predation. This chapter begins by describing different mechanisms of interspecific competition. Competition can occur via one or more of six distinct mechanisms. Simple descriptive models of competition for animals, plants, and microbes are summarized to emphasize how models can be used to predict conditions favoring the coexistence of competitors. Mechanistic models of competition are also briefly introduced as a way to link patterns of resource utilization to competitive ability. The chapter concludes by linking the process of competition to ideas about how species differ in their use of resources. Differences in resource use are often described in terms of the ecological niches of species. Attempts to test simple models of competition experimentally, and empirical explorations of links between observed competition and patterns predicted by niche theory, provide the motivation for the overview of experimental studies of competition in Chapter 3.

INTERSPECIFIC COMPETITION

One way to define **interspecific competition** is as a mutually negative (−/−) interaction between two or more species within the same guild or trophic

Competition defined.

level. Cases of mutual predation are usually not classified as competitive interactions, although they also share the −/− sign structure. Negative competitive interactions manifest themselves as reduced abundance, decreased fitness, or as a decrease in some fitness component, such as body size, growth rate, fecundity, or survivorship. The assumption is that decreases in fitness components will eventually cause a reduced abundance of affected species, although this assumption is seldom tested.

For much of its early history, community ecology was virtually synonymous with the study of interspecific competition. As community ecology matured, explanations for community patterns became more pluralistic and seldom relied on single processes to account for patterns. Competition's perceived role in community organization remains important, but less dominant than in the past.

Studies of the impact of interspecific competition on community structure take many forms. Most ecologists make important distinctions between observational approaches, which search for patterns produced by interspecific competition in natural communities without manipulating the abundances of competitors, and experimental approaches, which observe how species respond to direct manipulations of potential competitors. The decision to use one or the other of these approaches may simply reflect the investigator's style and training, but it can also depend on constraints imposed by the natural history of the study organism. Some ecologists feel that experimental approaches are more direct and provide stronger inferences than other approaches. Other ecologists feel that observational approaches play an essential role in understanding how competition affects experimentally intractable organisms.

 Observations vs. experiment in studies of competition.

The relative merits of different approaches have been discussed and debated extensively. For example, the observed distributions of bird species among islands of the Bismarck Archipelago have been variously interpreted either as evidence for complementary distributions resulting from competition (Diamond 1975; Diamond and Gilpin 1982; Gilpin and Diamond 1982) or as patterns attributable solely to chance events (Connor and Simberloff 1978). Since the birds are virtually impossible to manipulate experimentally, experimental approaches are not likely to resolve the dispute. However, in other systems, simple experimental manipulations can provide compelling evidence of ongoing competition among species (Connell 1961). The essence of these discussions can be appreciated by reading and comparing the writings of Strong et al. (1984), Diamond (1986), and Hairston (1989), among many others.

One common observational approach to the study of interspecific competition involves searching for negative correlations between the abundances of ecologically similar species. Such complementary distributions are then attributed to the present or past effects of interspecific competition, as long as other mechanisms that might produce the same pattern can be ruled out. The extreme case of such distributions is often likened to a checkerboard pattern, where units of habitat contain either one species or another. Another observational approach uses interspecific differences in morphology or resource use to infer possible competitive interactions. Particularly regular or nonrandom patterns of morphology or resource use are then interpreted as evidence that species must differ by some fixed amount in order to avoid competitive exclusion. This approach is central to arguments about the competitive significance of **character displacement**, where differences in the morphology of ecologically similar species are greater in sympatry than in allopatry (Lack 1947; Brown and Wilson 1956). Observational approaches can be used with a great variety of organisms, including species that are experimentally intractable because of long generation times (e.g., trees, whales) or high motility that complicates experimental manipulations of competitors (e.g., birds). The chief disadvantage of purely observational studies is that complementary distributions of species or differences in morphology or resource use need not be caused solely by competition.

Experiments that directly assess responses to manipulations of competitors have the advantage of providing strong inferences about whether competition is responsible for a pattern. If a pattern (e.g., abundance, resource use) changes in response to the addition or removal of competitors, the interpretation of ongoing competition is clear. One disadvantage of experimental studies is that it may be difficult or unethical to manipulate species that are either long-lived or rare. Also, response times of long-lived species to competitor removals may be very slow relative to the time scale over which most studies are conducted. Field experiments seldom continue for more than a few years. Experimental studies are best suited for small or sedentary organisms that can be readily manipulated and that will respond to competitors over short time frames.

When experiments are impossible, observational studies can sometimes be made more compelling by determining whether patterns attributed to competition differ from those expected by chance. Such null model approaches attempt to test whether observed patterns, such as complementary distributions, size ratios, or differences in resource use, are statistically different from patterns that would arise among organisms that do not compete. Null model approaches have many of the same advantages as purely observational studies. The main drawback is that ecologists seldom agree on exactly how to best formulate a null model that will unambiguously predict the patterns expected to be produced by chance events (see Colwell and Winkler 1984). A few examples of this approach are outlined in Chapter 3.

MECHANISMS OF INTERSPECIFIC COMPETITION

Competition includes a variety of interactions between species that can proceed by several different mechanisms. Historically, ecologists distinguished between **exploitative** competition, which operates indirectly by the depletion of some shared resource, and **interference** competition, which involves direct interactions between species, such as territorial interactions or chemical interference. A similar distinction was made between **scramble** competition, usually involving resource utilization, and **contest** competition, which, as the name implies, involves some sort of behavioral interaction. The problem with all these dichotomous categories was that some competitive interactions did not fit unambiguously into one category or the other.

 Competitive mechanisms.

As an alternative to dichotomous classifications of competitive interactions, Thomas Schoener (1983) suggested that six different mechanisms are sufficient to account for most instances of interspecific competition. The six mechanisms of competition that Schoener proposed are as follows:

1. Consumption
2. Preemption
3. Overgrowth
4. Chemical interactions (allelopathy)
5. Territoriality
6. Encounter competition

Consumptive competition happens when one species inhibits another by consuming a shared resource. Competition between granivorous rodents and ants for seeds is an example of this kind of interaction (Brown and Davidson 1977).

Preemptive competition occurs primarily among sessile organisms and results when a physical resource, such as open space required for settlement, is occupied by one organism and made unavailable to others. Many examples of competition among sessile rocky intertidal organisms for space fall into this category (see, for example, Connell 1961).

Overgrowth competition occurs, quite literally, when one organism grows directly over another, with or without physically contacting the other organism. Overgrowth competition does not require direct contact between the organisms. For example, forest trees overgrowing smaller plants and intercepting light results in the exclusion of shade-intolerant species (see, for example, Chapman 1945). In other cases, particularly among encrusting marine organisms like bryozoans and corals, competition results from direct contact and overgrowth, which also inhibits access to some important resource, such as light, food, or oxygen (Buss 1986; Connell 1979).

Chemical competition amounts to chemical warfare between competitors, in which chemical growth inhibitors or toxins produced by some species inhibit or kill other species. Some of the best examples of chemical competition come from studies of **allelopathy** in plants, in which chemicals produced by some plants inhibit the growth or seed germination of other plants (Keever 1950; Muller et al. 1964). Other kinds of organisms, including the aquatic tadpoles of frogs, can interact via growth inhibitors associated with gut symbionts (Griffiths et al. 1993).

Territorial competition results from the aggressive behavioral exclusion of organisms from specific units of space that are defended as territories. The strong interspecific territorial disputes between brightly colored coral reef fishes are a good example of territorial competition (Sale 1980).

Finally, **encounter** competition results when nonterritorial encounters between foraging individuals result in negative effects on one or both of the interacting individuals. The best examples come from laboratory studies of parasitoids foraging for prey. When two parasitoids encounter each other, they may interact in ways that cause them to stop foraging or to leave a site where there may be more prey (Hassell 1978). The net result is that time and energy that could be used for reproduction is lost or diverted to other nonreproductive tasks. Schoener would also include cases where an encounter between individuals results in injury or death, as when one species attacks or consumes the other. This definition includes situations that have previously been described as cases of competition, such as interactions between the species of *Tribolium* beetles that live in stored grain products (see Park 1962). The main interactions among *Tribolium* involve interspecific consumption of eggs, larvae, and pupae (Park et al. 1965). Including cases of mutual predation as examples of competition potentially blurs the important distinctions between competition and predation and runs the risk of including all predator-prey interactions as just another kind of competition.

Regardless of the mechanism involved, species often compete asymmetrically, in the sense that one species exerts considerably stronger per capita effects than another. Some of the earlier experimental evidence cited in support of strongly asymmetric interactions probably confounded asymmetric per capita effects with initial differences in the densities of manipulated species (Lawton and Hassell 1981). An unequal response to the removal of interspecific competitors may reflect different per capita effects of removed species or different initial densities of species of similar per capita competitive ability. Underwood (1986) has outlined the kinds of careful experimental designs that are required to separate differences in per capita competitive effects from differences in density. Such approaches are feasible only where it is possible to exercise tight control over the densities of competitors.

Extreme cases of asymmetric competition, in which one species has a strong negative effect on a second species and the second species has no detectable negative effect on the first, are sometimes called **amensalisms** (Burkholder 1952). In most experimental settings, it is unclear whether the complete absence of a reciprocal effect is real or just a statistical artifact of the small sample sizes associated with most field experiments.

DESCRIPTIVE MODELS OF COMPETITION

Models of interspecific competition can yield important predictions about the conditions promoting the coexistence or exclusion of competitors. Models are particularly useful tools in situations where laboratory or field experiments are impractical. Models can also be used to generate new hypotheses about the ways that competitors interact.

 Descriptive and mechanistic models of competition.

It is always easy to find fault with models regarding their various departures from the complexities of nature. However, it is important to remember that even relatively simple and seemingly unrealistic models can be very useful, since the ways in which they fail to accurately represent the dynamics of competing species can pinpoint how competition among real species departs importantly from the features abstracted in the models.

Models of interspecific competition can be **descriptive** or **mechanistic.** Descriptive models literally describe how the abundance of one species affects the abundance of another, without specifically including a particular competitive mechanism, such as consumptive depletion of a shared resource, in the model. Instead, competition is represented as a negative function of competitor abundance that slows the rate of increase of the responding species. Mechanistic models explicitly include information about the mechanism responsible for the effects of one species on another. For instance, mechanistic models of consumptive competition would include descriptions of the dynamics of the interacting competitors as well as the dynamics of the resources that are being consumed. In general, theoretical work on interspecific competition has favored the use of relatively simple descriptive models over that of more complex mechanistic ones. There are important trends toward the development of more mechanistic models (e.g., MacArthur 1972; Schoener 1974; Tilman 1982) that will be explored after providing an overview of the descriptive models.

A traditional way to begin exploring models of interspecific competition is to show how simple models for competition among individuals of a single species can be extended to include the effects of two or more species. The logistic equation of Pearl and Reed (1920), which was originally described as a model for human population growth by Pierre-François Verhulst (1838), is a simple descriptive model of how competition limits the growth of populations. A differential equation describes the effects of population size or abundance, N, on population growth rate, dN/dt. The model assumes that a maximum population size—called K, the **carrying capacity**—exists when $dN/dt = 0$. Then

 The logistic equation.

$$dN/dt = rN(1 - N/K) \qquad (2.1a)$$

or, equivalently,

$$dN/dt = rN(K - N)/K \qquad (2.1b)$$

where r is the per capita rate of increase and K is the carrying capacity, or maximum population size, where the population growth rate equals zero. The logistic term, $(1 - N/K)$, has the effect of multiplying the exponential rate of increase, rN, by a factor that decreases toward 0 as N approaches K, thus making the entire population growth rate decrease toward 0 as N nears K. The result is a population with a stable equilibrium population size of $N = K$. Population growth over time follows an approximately sigmoid approach to the carrying capacity, K (Figure 2.1).

The logistic equation can be easily extended to describe competition between two species. Lotka (1925) and Volterra (1926) independently modeled two-species competition using extensions of the logistic equation. Using subscripts to denote values for species 1 and 2,

A model for interspecific competition.

$$dN_1/dt = r_1 N_1 (K_1 - N_1 - \alpha_{12} N_2)/K_1 \qquad (2.2a)$$

$$dN_2/dt = r_2 N_2 (K_2 - N_2 - \alpha_{21} N_1)/K_2 \qquad (2.2b)$$

where all terms are directly analogous to those in the single-species logistic equation, except for terms such as $\alpha_{12}N_2$ in Equation 2.2a. This term uses a competition coefficient, α_{ij}, that effectively translates individuals of species j into species i for the purpose of determining the extent to which those individuals utilize the total carrying capacity available to species i. In other words, the proximity of each population to its carrying capacity depends on both its current population size and the population size of its competitor, weighted by

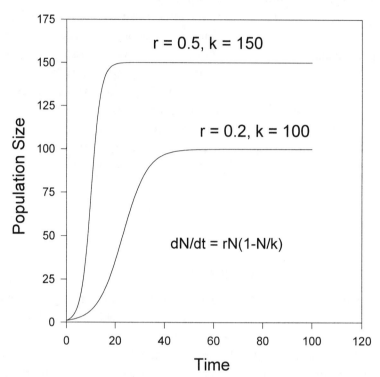

FIGURE 2.1. *Examples of logistic population growth. The two trajectories differ in the exponential rate of increase,* r, *and the carrying capacity,* K, *as shown in the graph.*

the competition coefficient α_{ij}. For example, for the unlikely case of two competitively equivalent species, $\alpha_{ij} = \alpha_{ji} = 1$. When interspecific competitors have a weaker per capita effect than intraspecific competitors, $\alpha_{ij} < 1$ and $\alpha_{ji} < 1$.

These equations yield an important prediction about the conditions leading to the stable coexistence of two competitors. If two species have equal carrying capacities, they will stably coexist 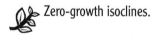 Zero-growth isoclines. only if α_{12} and α_{21} are both less than 1. This result can be shown graphically, using the following argument. A nontrivial equilibrium, one where $dN_1/dt = dN_2/dt = 0$ and where both N_1 and $N_2 > 0$, will occur when

$$(K_1 - N_1 - \alpha_{12}N_2) = 0 \qquad (2.3a)$$

and

$$(K_2 - N_2 - \alpha_{21}N_1) = 0 \qquad (2.3b)$$

When rewritten in the form

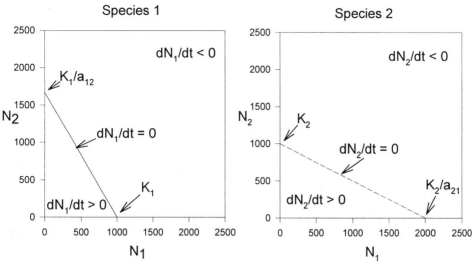

FIGURE 2.2. *Values of population sizes of two species, N_1 and N_2, that result in positive, negative, or zero population growth for species interacting according to Equations 2.2a and 2.2b. The zero-growth isoclines are shown as a solid line for species 1 and a dashed line for species 2. This set of isoclines corresponds to $K_1 = K_2 = 1000$, $r_1 = r_2 = 3.22$, $a_{12} = 0.6$, and $a_{21} = 0.5$.*

$$N_2 = -N_1/\alpha_{12} + K_1/\alpha_{12}$$

and

$$N_1 = -N_2/\alpha_{21} + K_2/\alpha_{21}$$

these equations for two lines are called the **zero-growth isoclines**. They give the values of N_1 and N_2 that yield zero population growth for each species. When plotted on two axes denoting the abundances of N_1 and N_2, the lines can be arranged in four relative positions that correspond to different competitive outcomes and different patterns of dynamics. The area between the origin and each isocline (i.e., the area below the isocline in Figure 2.2) shows the various combinations of N_1 and N_2 where population growth is positive, and the area above each isocline (i.e., the area above the isocline in Figure 2.2) shows conditions where population growth is negative. When both isoclines are plotted on the same graph, the isoclines can be arranged in four relative positions that yield different competitive outcomes. The different outcomes in turn depend on the relative values of K_1, K_2, α_{12}, and α_{21}. Four possible configurations are shown in Figure 2.3, along with the competitive outcomes that they produce.

The four possible competitive situations, defined by the relative positions of the isoclines, are as follows:

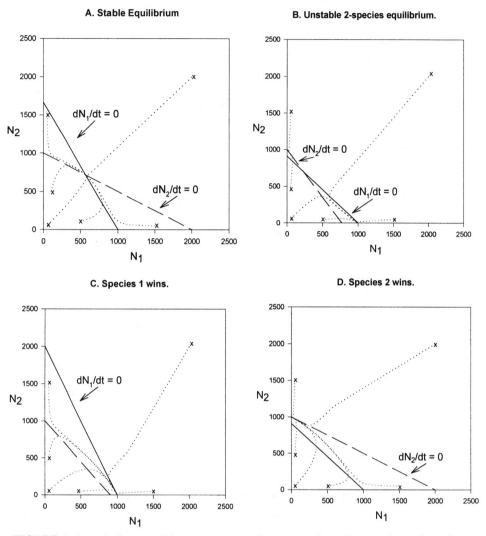

FIGURE 2.3. *The four possible arrangements of zero-growth isoclines in the Lotka-Volterra competition equations, together with population trajectories showing the outcome of interactions starting with different values of* N_1 *and* N_2. *Only the case shown in panel A results in a stable equilibrium where both species persist.*

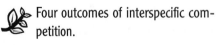 Four outcomes of interspecific competition.

1. An unstable equilibrium, where $K_2 > K_1/\alpha_{12}$ and $K_1 > K_2/\alpha_{21}$.

2. Competitive exclusion of species 1 by species 2, where $K_2 > K_1/\alpha_{12}$ and $K_1 < K_2/\alpha_{21}$.

3. Competitive exclusion of species 2 by species 1, where $K_2 < K_1/\alpha_{12}$ and $K_1 > K_2/\alpha_{21}$.

4. A stable equilibrium, where both species coexist, where $K_1/\alpha_{12} > K_2$ and $K_2/\alpha_{21} > K_1$.

The only situation corresponding to stable coexistence is where $K_1/\alpha_{12} > K_2$ and $K_2/\alpha_{21} > K_1$, as shown in Figure 2.3A. This set of conditions becomes slightly easier to comprehend if we assume for the moment that $K_1 = K_2$. Then the conditions for stable coexistence become $\alpha_{12} < 1$ and $\alpha_{21} < 1$. This is equivalent to saying that the per capita effects of interspecific competition are weaker than the per capita effects of intraspecific competition. Note that if one selects a point in any of the quadrants of the graph in Figure 2.3A, the direction of change in both populations always tends toward the equilibrium point.

Competitive exclusion occurs in cases 2 and 3, where the zero-growth isocline for one species falls completely below the isocline for the other species. The species with the lower isocline loses because its population growth goes from zero to negative under conditions in which the other species can still increase. An unstable equilibrium occurs in case 1 because populations tend toward an equilibrium point when either above or below both isoclines, but tend away from the equilibrium point when between the isoclines. This case will result in the extinction of one species or the other, but the outcome depends on the initial values of the abundance of each species. This is an example of a kind of **priority effect,** where initial conditions determine the outcome of an interaction. We will return to the topic of priority effects in Chapter 9.

A multispecies extension of the two-species model of interspecific competition can be arrived at by modifying the two-species model described above. The model consists of n differential equations, one for the dynamics of each of the n species. The equation for the ith species would be

A multispecies model of interspecific competition.

$$\frac{dN_i}{dt} = N_i r_i \left(k_i - \sum_{j=1}^{n} \alpha_{ij} N_j \right) / k_i \qquad (2.4)$$

where the effect of species i on itself is $a_{ii} = 1$. This approach assumes that the pairwise competition coefficients, the α_{ij}'s, are a fixed property of the interaction between a pair of species and do not depend on the other species present in the system. This assumption means that the joint or aggregate effect of several species on another can be represented by simply summing up the pairwise effects of each competitor species on the species of interest. The assumption of additive competitive effects requires that the αij's do not change as the community becomes more or less complex. Failure to conform to this constraint is sometimes referred to as **nonadditivity,** or the existence of

higher-order interactions. Concerns about the realism of this constraint have inspired several experimental tests that have yielded a mixed bag of results. In some simple systems, such as laboratory communities composed of several species of protists (Vandermeer 1969), interactions are adequately described by an additive model. In other situations, often corresponding to interactions involving somewhat more complex organisms, such as microcrustaceans (Neill 1974), insects (Worthen and Moore 1991), or vertebrates (Morin et al. 1988), interactions can be nonadditive. We will return to this topic in greater detail when we consider experimental studies of competition in Chapter 3.

Conditions required for the stability of multispecies competition systems are considerably more complex than for the two-species case (Strobeck 1973), but the same criteria used to assess the stability of other systems of equations, such as predator-prey models (described in the appendix), still apply. Stability depends in a complex fashion on the rates of increase, carrying capacities, and competition coefficients, unlike the two-species case, which depends only on the carrying capacities and the competition coefficients.

MECHANISTIC MODELS OF COMPETITION

The descriptive Lotka-Volterra model can be transformed into a mechanistic model by expressing the competition coefficients and carrying capacities in terms of rates of utilization and renewal of resources. Specific approaches to modeling consumptive interspecific competition require more complex models (MacArthur 1972; Levine 1976). The models typically treat the competitors as two or more predator species whose dynamics are linked to the dynamics of one or more resources, which can correspond to prey species or other kinds of consumable abiotic resources. Instead of using descriptive competition coefficients to summarize interspecific competition, competition emerges as a result of each species' ability to successfully exploit and deplete resources. In essence, the model describes a simple food web like some of those described in Chapter 6.

Mechanistic models describe interactions between consumers and resources.

One example of this mechanistic approach to models of competition was described by Robert MacArthur (1972) for two predator species feeding on two resource species. The model includes four differential equations that describe the rates of change in the abundances of both predators and resources:

$$dN_1/dt = N_1 C_1 (a_{11} w_1 R_1 + a_{12} w_2 R_2 - T_1) \tag{2.5a}$$

$$dN_2/dt = N_2 C_2 (a_{21} w_1 R_1 + a_{22} w_2 R_2 - T_2) \tag{2.5b}$$

$$dR_1/dt = R_1\left[\frac{r_1}{k_1}(k_1 - R_1) - a_{11}N_1 - a_{21}N_2\right] \qquad (2.5c)$$

$$dR_2/dt = R_2\left[\frac{r_2}{k_2}(k_2 - R_2) - a_{12}N_1 - a_{22}N_2\right] \qquad (2.5d)$$

Here N_1 and N_2 refer to the abundances of the two competing consumer populations, and R_1 and R_2 refer to the abundances of the two resource populations. Each consumer needs to ingest a specific weight of resources to offset the demands imposed by basal metabolism. The consumers acquire resources at rates given by a_{ij}, which can be equated with attack rates; the rates also depend on resource density, the R_j's. Each individual of the resource species is assumed to have a weight of w_j. T_i refers to the weight of resource required to fill the needs imposed by the basal metabolism of consumer species i. Resources in excess of the amount needed for maintenance are converted into new consumers with an efficiency of C_i. The resource species are assumed to grow logistically in the absence of consumers and are assumed not to compete.

Using this framework and some algebra, it is possible to express the competition coefficients and the carrying capacities of the consumers in terms of the following expressions:

$$K_1 = (a_{11}w_1k_1 + a_{12}w_2k_2 - T_1)\Big/\left[(a_{11})^2 w_1k_1/r_1 + (a_{12})^2 w_2k_2/r_2\right] \qquad (2.6a)$$

$$K_2 = (a_{21}w_1k_1 + a_{22}w_2k_2 - T_2)\Big/\left[(a_{21})^2 w_1k_1/r_1 + (a_{22})^2 w_2k_2/r_2\right] \qquad (2.6b)$$

$$\alpha_{12} = (a_{11}a_{21}w_1k_1/r_1 + a_{12}a_{22}w_2k_2/r_2)\Big/\left[(a_{11})^2 w_1k_1/r_1 + (a_{12})^2 w_2k_2/r_2\right] \qquad (2.6c)$$

$$\alpha_{21} = (a_{11}a_{21}w_1k_1/r_1 + a_{12}a_{22}w_2k_2/r_2)\Big/\left[(a_{21})^2 w_1k_1/r_1 + (a_{22})^2 w_2k_2/r_2\right] \qquad (2.6d)$$

By making some assumptions about the mechanisms of competition involved, the distinctions between descriptive and mechanistic models can thus become rather arbitrary.

Other mechanistic models of competitive interactions use differences in the ability of species to grow at various levels of two or more resources to predict conditions in which those species will coexist (Monod 1950; Tilman 1977, 1982). These models were originally developed to describe competition among bacteria or phytoplankton for nutrients that vary in concentration as a function of nutrient supply rates and uptake rates by competitors. The situation

 Monod models of competition.

modeled is one in which nutrients flow at an initial concentration into a community of fixed volume. Nutrients flow into the community at a fixed rate, and organisms and nutrients (usually at a lower than initial concentration, due to consumption) flow out at the same rate. The situation is analogous to a lake of constant volume, where water carrying nutrients flows in at a given rate, and water carrying organisms and somewhat depleted nutrients flows out. The models are multispecies extensions of the Monod model (Monod 1950). The Monod model looks rather like a predator-prey model with a type II functional response (see Chapter 5). For a single consumer species, N, feeding on a single resource, R, the model is as follows:

$$dN/dt = N\mu_{max}[R/(K_\mu + R)] - ND \qquad (2.7a)$$

$$dR/dt = D(R^0 - R) - N/Y[\mu_{max}(R/(K_\mu + R))] \qquad (2.7b)$$

Here μ_{max} is the maximal per capita rate of increase of the consumer, and K_μ is the level of resource concentration at which the per capita rate of increase is $0.5\mu_{max}$. Y describes the conversion of consumed resources into new consumers; in other words, how much resource is required to create a new consumer. The system is assumed to have a constant volume, with new resources entering at a constant rate, D, and consumers and medium with altered resource levels leaving the system at the same constant rate D. R^0 describes the concentration of resources entering the system, and R is the concentration that results from consumption by the consumers. With a little imagination, it is easy to see that this is really nothing more than a predator-prey equation, where resource consumption is described by a type II functional response, described by the term $N/Y[\mu_{max}(R/(K_\mu + R))]$. The resource is "born" at a rate determined by its initial concentration and rate of flow into the system, DR^0. The resource "dies," or becomes unavailable for consumption, at a rate determined by its rate of consumption, given by the functional response term described above, and its rate of flow out of the system at the concentration, R, set by consumption. Consumers increase at a rate determined by resource concentration and die only as they flow out of the system at a constant rate, D.

The Monod model described above has stable equilibrium values of R and N given by

$$R^* = DK_\mu/(\mu_{max} - D) \qquad (2.8a)$$

$$N^* = Y(R^0 - R^*) \qquad (2.8b)$$

An example of a consumer population growing according to the Monod model is shown in Figure 2.4. Note that this model describes concentrations of the resource and the consumer relative to a particular volume of medium. Also

Equilibrium values of resources and consumers in the Monod model.

note how consumer and resource con centrations change as the equilibrium is approached.

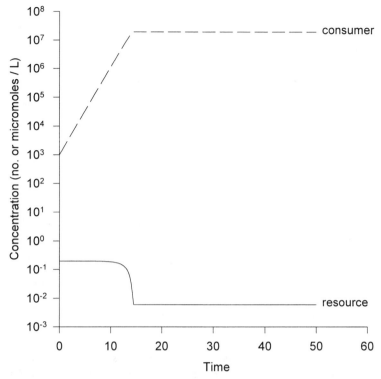

FIGURE 2.4. *Examples of consumer and resource dynamics for a single consumer–single resource Monod model, as in Equations 2.7a and 2.7b. Values of the parameters in this simulation were $\mu_{max} = 0.5$, $K_\mu = 0.01$, D = 0.25, $R^0 = 0.2$, and Y = 1.0×10^8.*

A Monod model for competition between two consumer species really represents predation by two species on some shared resource. A simple example of competition between two species for a single resource would look like the following:

$$dN_1/dt = N_1\mu_{max,1}[R/(K_{\mu,1} + R)] - N_1D \tag{2.9a}$$

$$dN_2/dt = N_2\mu_{max,2}[R/(K_{\mu,2} + R)] - N_2D \tag{2.9b}$$

$$dR/dt = D(R^0 - R) - (N_1/Y_1)[\mu_{max,1}(R/(K_{\mu,1} + R))]$$
$$-(N_2/Y_2)[\mu_{max,2}(R/(K_{\mu,2} + R))] \tag{2.9c}$$

Here the subscripts 1 and 2 refer to consumer species 1 and 2. An important prediction of this model is that one consumer species invariably excludes the other (Figure 2.5). The winner of the competitive interaction can be predicted by determining which consumer species produces the lower value of R^*, the concentration of the resource at equilibrium, in the absence of the other. This

 The R* rule.

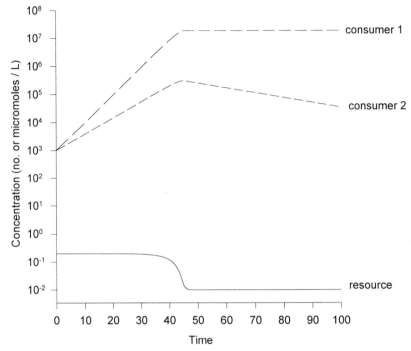

FIGURE 2.5. *Competition in a Monod model for two consumers utilizing the same resource. The species with the lowest* R* *value eventually wins (species 1* R* = 0.006*; species 2* R* = 0.074*). These simulations are based on parameter values for two algae,* Asterionella *and* Cyclotella, *competing for phosphate (Tilman 1977).*

conclusion makes sense, since the species that produces the lowest value of R^* can continue to grow at lower concentrations of the resource than are required to maintain a stable equilibrium in the other species. For two consumer species to coexist, those species must differ in the ways in which they use at least two resources.

An expansion of the Monod model to include two consumers and two resources can be used to predict the conditions leading to coexistence or competitive exclusion. For multispecies competition on multiple resources, the Monod model becomes somewhat more complex. David Tilman's approach (1977, 1982), which is basically a multiconsumer, multiresource variation on the Monod model, requires information about three things: the average mortality rate of each competing species, supply rates of limiting nutrients, and population growth rates as a function of resource supply rates. Mortality rates are assumed to be independent of density and resource supply rates. Population growth rates are assumed to be curvilinear increasing functions of resource supply rates that eventually level off at high resource supply rates due to saturation kinetics of resource uptake mechanisms, as given by the functional responses in the Monod models described above. The corresponding Monod model for consumer species i and resource species j is as follows:

$$dN_i/dt = N_i \min[\mu_{\max,1}(R_j/(K_{\mu,ij} + R_j))] - N_i D \qquad (2.10a)$$

$$dR_j/dt = D(R_i^0 - R_j) - \Sigma N_i/Y_{ij}[\mu_{\max,i}(R_j/(K_{\mu,ij} + R_j))] \qquad (2.10b)$$

The expression $\min[\mu_{\max,1}(R_j/(K_{\mu,ij} + R_j))]$ in Equation 2.10a indicates that the growth rate of each consumer species is limited by the availability of the particular resource that is most limiting, in other words, the resource whose supply rate yields the lowest rate of consumer growth. All other resources are supplied in excess of the demand set by consumer abundance for the limiting resource. This is just a restatement of Liebig's law of the minimum: Where species require several resources to grow, growth rate will be determined by the resource in shortest supply.

For example, for two species that are each limited by a different resource, such that N_1 is limited by R_1 and N_2 is limited by R_2, the model would look as follows:

$$dN_1/dt = N_1 \mu_{\max,1}[R_1/(K_{\mu,11} + R_1)] - N_1 D \qquad (2.11a)$$

$$dN_2/dt = N_2 \mu_{\max,2}[R_2/(K_{\mu,22} + R_2)] - N_2 D \qquad (2.11b)$$

$$dR_1/dt = D(R_1^0 - R_1) - (N_1/Y_{11})[\mu_{\max,1}(R_1/(K_{\mu,11} + R_1))]$$
$$- (N_2/Y_{21})[\mu_{\max,2}(R_1/(K_{\mu,21} + R_1))] \qquad (2.11c)$$

$$dR_2/dt = D(R_2^0 - R_2) - (N_1/Y_{12})[\mu_{\max,2}(R_2/(K_{\mu,12} + R_2))]$$
$$- (N_2/Y_{22})[\mu_{\max,2}(R_2/(K_{\mu,22} + R_2))] \qquad (2.11d)$$

Here, notice that competition occurs through the effect of each species on the consumed resources. Note that the abundance of each competitor does not show up in the equations describing the population dynamics of the consumers. They depend only on resource concentrations, which in turn depend on rates of resource uptake, supply, and outflow.

Tilman's approach has an elegant graphical representation. First, for each competing species and two resources, the minimum concentration of a resource required to balance population growth and mortality is determined. Figure 2.6 shows an example of how this might look. Then for each species, zero-growth isoclines can be plotted as a function of resource concentrations for two potentially limiting resources (see Figure 2.6). For a single species, beginning at some set of resource supply rates, population growth and consumption will change the effective resource concentration (total supply rate minus consumption rate) until population growth stops when it reaches some minimum concentration. The direction of change is given by a

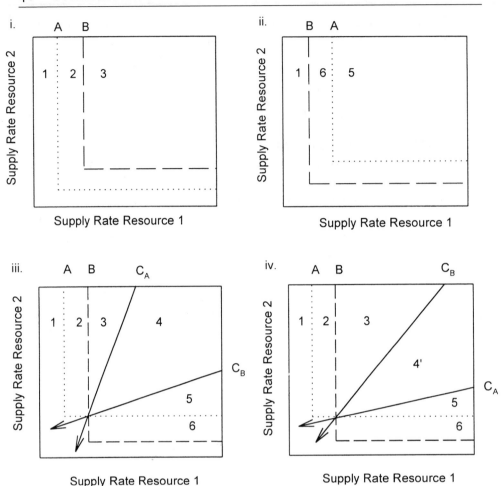

FIGURE 2.6. *(A) Zero-growth isoclines and resource consumption vectors for Monod/Tilman models of competition by two consumers for two resources. The axes describe supply rates of resources. Different outcomes of competition arise from different relative combinations of zero-growth isoclines, which are set by minimum resource supply levels at which populations can persist, and consumption vectors, which describe the relative rates of depletion or uptake of the two resources by each species. Numbered regions in the graphs correspond to different initial values of resource supply rates that yield various competitive outcomes. Case i: In region 1, both species go extinct; in regions 2 and 3, species A predominates and species B goes extinct. Case ii: Species B predominates in regions 5 and 6 and drives species A extinct. Case iii: There is an equilibrium point where the isoclines cross. Both species can stably coexist in region 4. The consumption vectors for each species, labeled C_A and C_B, indicate that each species consumes the resource that limits it at equilibrium at a greater rate than it consumes the nonlimiting resource. Case iv: An unstable equilibrium exists, caused by the reversed relative position of the consumption vectors. In region 4', either species A or species B can exclude the other, depending on initial conditions. (Adapted with permission from Tilman, D.* Resource competition and community structure. © 1982 by Princeton University Press.)

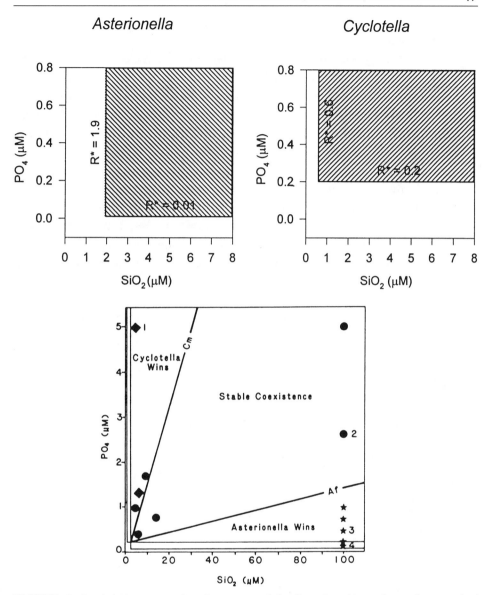

FIGURE 2.6. (B) Top: Examples of zero-growth isoclines for SiO_2 and PO_4 for two algal species, Asterionella and Cyclotella. Within the shaded region each species can increase in population size. Note that the lowest R^* for each species is for a different resource. Bottom: The outcome of competition between these species is described well by the isoclines and consumption vectors. Diamonds = Cyclotella wins; dots = stable coexistence; stars = Asterionella wins. (Adapted with permission from Tilman, D. Resource competition and community structure. © 1982 by Princeton University Press.)

consumption vector, which describes the ratio in which the two resources are consumed by each consumer.

For competition between two species, the outcome depends on the relative positions of the zero-growth isoclines for each species, the consumption vectors, and where the interaction begins in the resource supply space, that is, the initial relative concentrations of the resources. When the isocline for one species falls completely below that of the other, the species with the lowest zero-growth resource supply rate always wins (see Figure 2.6). When the zero-growth isoclines of the two species cross, the situation becomes more complex (Figure 2.7). Tilman's model is mechanistic in the sense that the outcome of competition depends on resource-dependent growth, resource supply rates, resource consumption ratios, and mortality rates of each species. The critical assumptions in the model are that species compete only through the consumption of resources and that resources are indepen-

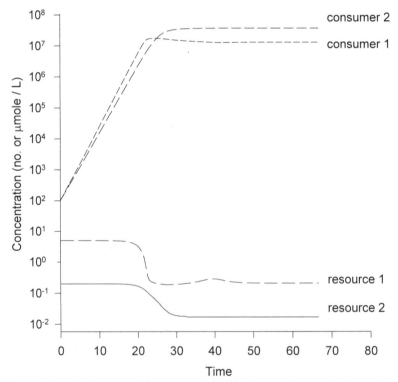

FIGURE 2.7. *Coexistence of two competitors on two resources for Equations 2.11a–d. Parameter values:* $D = 0.5$, $R_1^0 = 0.2$, $R_2^0 = 5$, $\mu_{max,1} = \mu_{max,2} = 1.1$, $K_{\mu,11} = 0.02$, $K_{\mu,22} = 0.25$, $K_{\mu,21} = 1.44$, $K_{\mu,12} = 3.94$, $Y_{11} = 2.18 \times 10^8$, $Y_{22} = 4.20 \times 10^6$, $Y_{12} = 2.51 \times 10^6$, $Y_{21} = 2.59 \times 10^7$. *Parameter values from Tilman (1977) correspond to competition between two algae,* Asterionella *and* Cyclotella, *for phosphate and silica.* Asterionella *has the lower* R^* *for phosphate (0.006 vs. 0.074), and* Cyclotella *has the lower* R^* *for silica (0.424 vs. 1.159).*

dent rather than interactive. The approach seems to work well for phytoplankton growing in chemostats and perhaps, by analogy, for phytoplankton growing in lakes. Its applicability to terrestrial plant communities is much more problematic because of the many other factors, including herbivory, that influence the outcome of interactions among plants.

NEIGHBORHOOD MODELS OF COMPETITION AMONG PLANTS

Plant ecologists were never very comfortable with Lotka-Volterra models of competition, which were designed largely with mobile animals in mind. The sessile nature of plants means that an individual plant generally interacts only with its close neighbors rather than with the entire population of competitors. Consequently, competition among plants might be best portrayed by models that describe how individual plants respond to variation in the abundance of their immediate neighbors. Such **neighborhood models** (Pacala and Silander 1985, 1990; Pacala 1986a,b, 1987) have been developed to describe intra- and interspecific competition among annual plants. The models have been set up in two ways: as computationally intensive computer simulations that keep track of the spatial locations of individual competitors within a given plot of ground, and as analytical models that use assumptions about spatial distributions to capture the essence of spatially constrained competition. The computer simulation models are quite complex because they keep track of the positions of all the individual plants in the modeled population. Analytical models that make use of the average spatial features of interacting plant populations are considerably simpler, yet they yield many of the same predictions as the more complex simulation models.

Competition in sessile organisms depends on proximity.

At their simplest, neighborhood models of competition among plants assume that a focal plant only responds to the number of competitors found within a certain radius, or neighborhood, of the plant (Figure 2.8). Competition is manifested by the ways in which neighbors affect fecundity and survival. In simpler models, survival is independent of the neighborhood density of competitors, but fecundity is not. Figure 2.9 shows how the number of seeds set per plant depends on the number of intraspecific neighbors within a 5-cm radius for the plant *Arabidopsis thaliana*. This kind of relationship is termed a **fecundity predictor**. A fecundity predictor that describes this sort of relation can be modeled as Me^{-cn}, where M is the number of seeds produced by a plant without neighbors, e is the base of the natural logarithms, n is the number of neighbors affecting a particular plant, and c describes the intensity of neighborhood competition.

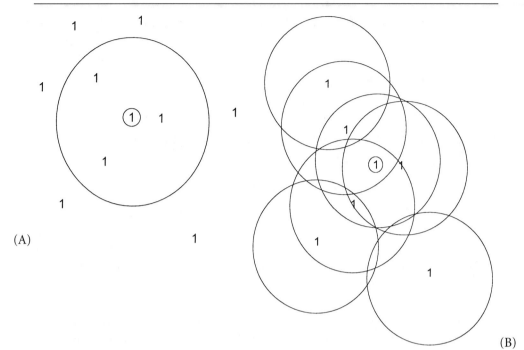

FIGURE 2.8. *(A) A circular neighborhood (large circle) around a focal plant (small circle), containing three competitors of species 1. Plants outside the neighborhood do not affect the focal plant. (B) Overlapping neighborhoods of all the plants in (A). Each plant responds only to the number of plants within its neighborhood.*

For intraspecific competition, Pacala and Silander (1985) have developed the following neighborhood model. It assumes that the number of individuals per neighborhood area in the next generation (N_{t+1}) is based on the number of seeds in the previous generation (N_t), the probability that those seeds will germinate (g), the probability that a germinated seed survives to adulthood and reproduces (P), and the way in which the number of neighbors affects the seed output, or fecundity, of each plant. For simplicity, Pacala and Silander assume that the probabilities of germination and survival do not depend on the density of neighbors. Neighbors only affect reproduction, through their effects on the fecundity predictor described above. The single-species analytical model, assuming a Poisson spatial distribution of seeds and neighbors, is as follows:

$$N_{t+1} = N_t gP \sum_{n=0}^{\infty} \left[\frac{e^{-PN_t g}(PN_t g)^n}{n!} \cdot Me^{-cn} \right] \qquad (2.12)$$

The term in the summation gives the probability that an individual plant will have a given number of neighbors, n, and multiplies that probability by the expected fecundity for a plant with n neighbors, Me^{-cn}, to arrive at an expected

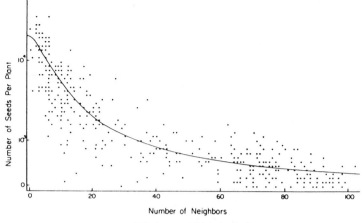

—SFP for *Arabidopsis thaliana.*

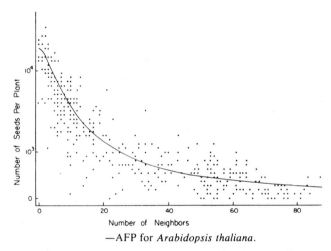

—AFP for *Arabidopsis thaliana.*

FIGURE 2.9. *Neighborhood fecundity predictors for* Arabidopsis thaliana. *The top panel shows a predictor based on numbers of seeds; the bottom panel shows a predictor based on numbers of adult plants. (Reprinted from Pacala and Silander, 1985, with permission of the University of Chicago Press.)*

or average value of fecundity. This value is based on the probability density function of the Poisson distribution. The summation reduces to $Me^{(-gPN_t\gamma)}$, where g is the probability of germination, P is the probability that a germinated seed survives to adulthood, M is the number of seeds produced by a plant without neighbors, and $\gamma = 1 - e^{(-c)}$. The model for intraspecific competition then reduces to

$$N_{t+1} = gPN_t Me^{(-gPN_t\gamma)} \tag{2.13}$$

The single-species model produces increasingly variable dynamics as the average fecundity of a plant without neighbors, M, increases (Figure 2.10).

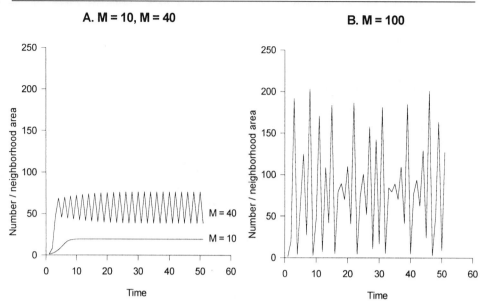

FIGURE 2.10. *Dynamics produced by single-species models of neighborhood competition. Parameter values were g = 1, P = 0.5, c = 0.2, and M = 10, 40, or 100. (Redrawn from Pacala and Silander, 1985, with permission of the University of Chicago Press.)*

This result is precisely what one would expect for a discrete time logistic equation incorporating rates of increase that generate chaotic dynamics (May 1976a).

Pacala (1986a) has developed a similar analytical model for neighborhood competition between two annual plant species without seed dormancy, which can be portrayed by the following pair of difference equations:

$$N_{1,t+1} = g_1 P_1 M_1 N_{1,t} e^{(-g_1 P_1 N_{1,t}\gamma_{11} - g_2 P_2 N_{2,t}\gamma_{12})} \tag{2.14a}$$

$$N_{2,t+1} = g_2 P_2 M_2 N_{2,t} e^{(-g_2 P_2 N_{2,t}\gamma_{22} - g_1 P_1 N_{1,t}\gamma_{21})} \tag{2.14b}$$

where $N_{1,t}$ and $N_{2,t}$ describe the density of seeds of species 1 and 2 at time t, g_i is the probability that a seed of species i will germinate, P_i is the density-independent survivorship of a seedling of species i, M_i is the fecundity of a plant of species 1 when it has no neighbors, and $\gamma_{ij} = 1 - e^{(-c_{ij})}$. Fecundity is related to the number of neighbors via a predictor, or submodel, given by $M_i e^{(-c_{ii}n - c_{ij}m)}$, where n and m are numbers of species i and j within the neighborhood of a focal plant. The c_{ij}'s are analogous to competition coefficients, in that they describe how the density of neighbors depresses fecundity. For randomly distributed plants that follow a Poisson distribution, the average value of the fecundity predictor is $M_i e^{(-g_i P_i N_{i,t}\gamma_{ii} - g_j P_j N_{j,t}\gamma_{ij})}$.

The model can also be expressed concisely in terms of adult plants. The number of adults at time t is $A_{i,t} = P_i g_i N_{i,t}$, in other words, the number of seeds

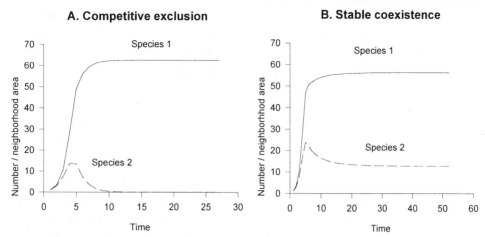

FIGURE 2.11. *Competitive exclusion and coexistence in a two-species neighborhood model of competition among plants. (A) The parameter values were* $g_1 = 0.5$, $P_1 = 0.2$, $M_1 = 35$, $\gamma_{11} = 0.2$, $\gamma_{12} = 0.1$, $g_2 = 0.5$, $P_2 = 0.2$, $M_2 = 30$, $\gamma_{22} = 0.2$, *and* $\gamma_{21} = 0.3$. *(B) All parameter values were the same as (A) except* $\gamma_{21} = 0.3$. *(Model and parameter values are from Pacala, 1986a; figure redrawn with permission from Academic Press, Inc.)*

multiplied by the probability that a seed will germinate multiplied by the probability that the seedling survives to adulthood. Then the model looks like

$$A_{1,t+1} = Q_1 A_{1,t} e^{(-A_{1,t}\gamma_{11} - A_{2,t}\gamma_{12})} \tag{2.15a}$$

$$A_{2,t+1} = Q_2 A_{2,t} e^{(-A_{2,t}\gamma_{22} - A_{1,t}\gamma_{21})} \tag{2.15b}$$

where $Q_i = M_i P_i g_i$, as in the model for seeds described above. Figure 2.11 shows the results obtained for various intensities of intra- and interspecific competition from neighbors. Despite all the complications imposed by the spatial components of the model, it reduces to a framework very similar to the Lotka-Volterra models. The general conclusions are the same: If the intensity of intraspecific competition is greater than the intensity of interspecific competition, then the species will coexist.

COMPETITION, NICHES, AND RESOURCE PARTITIONING

The preceding discussion of various models of interspecific competition shows that the outcome of consumptive competition can depend on how consumers use resources. Because models suggest that coexistence depends on interspecific differences in resource use, studies of how ecologically similar species differ in their use of resources can provide indirect inferences about how potential competitors manage to coexist. By examining patterns of resource use in coexisting species, the minimal differences that are compatible with coexistence can be deduced. Examination of patterns for species that fail to coexist can illuminate what aspects of resource use contribute to apparent competitive exclusion. Such efforts are referred to variously as studies of **niche partitioning**, **resource partitioning**, or **species packing**.

An indirect approach to studies of competition can be particularly attractive when the organisms of interest are large, mobile, long-lived, and therefore difficult to manipulate. Many vertebrates fall into this category. Studies of resource partitioning have been particularly influential in the development of the community ecology of birds (MacArthur 1958), lizards (Schoener 1968; Pianka 1986), and fish (Werner and Hall 1976). The resource partitioning approach is an excellent method for generating hypotheses about competitive interactions among species, and in some settings, it can identify nonrandom patterns of resource use that may be the result of competition. However, overlap in resource use is not sufficient evidence to prove that species compete. Species may overlap in resource use without being resource limited, either because resources are not in short supply or because the species are limited by some other factor, such as predation. Competition also need not be resource based, which raises the possibility that species with very different resource requirements could nonetheless compete via other mechanisms, such as overgrowth or chemical interference.

The Many Meanings of Niche

Grinnell (1914) is usually credited with the first use of the word **niche** in an explicitly ecological context. His usage, "no two species of birds or mammals will be found to occupy precisely the same niche," implies, albeit indirectly, that a competitive relationship between the species is important in affecting the ways in which species make their livings. Later codification of this idea as the **competitive exclusion principle** (Hardin 1960), which states that complete competitors cannot coexist, once again indirectly emphasizes the importance of differences among species that make them less than complete competitors.

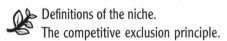
Definitions of the niche.
The competitive exclusion principle.

Other uses of the term *niche* carry subtly different meanings. Elton (1927) used *niche* to describe "what place a species occupies in a community." His meaning is essentially a description of the functional role of a species within the community. Hutchinson (1957) used *niche* to describe the range of physical and biological conditions, including limiting resources, needed for a species to maintain a stable or increasing population size. Hutchinson's definition is that of an *n*-dimensional hypervolume, where the *n* dimensions correspond to independent physical or biological variables that affect the abundance of that species (Figure 2.12). Values of the variables should be linearly orderable, such as temperature, pH, prey size, or perch height. Then for

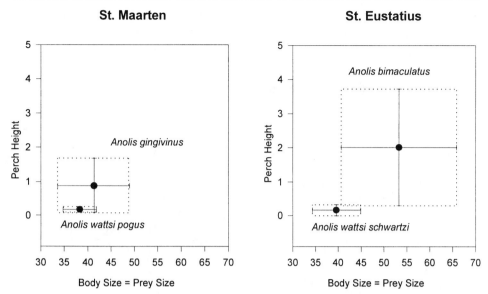

FIGURE 2.12. *Hutchinsonian niches of species pairs of* Anolis *lizards on two islands in the Lesser Antilles. Niche dimensions are defined by perch height and body size, where body size is a convenient index of average prey size. Realized niches for each species are defined by the mean ± one standard deviation for each measure. (Data from Pacala and Roughgarden, 1982.)*

two niche axes, the **fundamental** or **pre-interactive niche** of a species corresponds to those values of the two variables where the species can persist in the absence of competitors. To the extent that a species is found under conditions that are more restricted than those specifying its fundamental niche, its **realized** or **postinteractive niche** is a measure of the potential impact of other species in limiting the range of conditions successfully exploited by that species. One factor that might lead to a reduced realized niche is competitive exclusion by another species that shares part of the same fundamental niche. If many species occupy the same general region of the n-dimensional hypervolume, reductions in the size of a realized niche may reflect the aggregate effects of many small pairwise overlaps in the fundamental niches of species. No single species may account for a large effect, but collectively, the impact of many species may be severe. This notion is known as the **diffuse competition** exerted by an array of species.

Attempts to describe the Hutchinsonian niche for various species invariably show that similar species tend to differ in some aspects of their lifestyles. These differences, in turn, are often presented as an explanation for how those otherwise similar species manage to coexist. The classic example is Robert MacArthur's (1958) description of the foraging differences observed for a set of five small warbler species that nest in the forests of New Hampshire. All the birds are roughly the same size and shape, and all make their livings by glean-ing insects from the foliage of trees. Careful fieldwork revealed that the species differ subtly in the ways in which they feed in the forest canopy (Figure 2.13). Some concentrate their efforts high in the trees, whereas others spend more time near the forest floor. Similarly, some species feed near the core of the

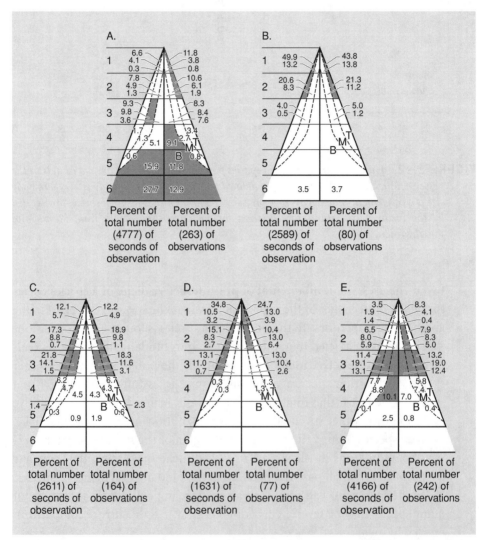

FIGURE 2.13. *Different patterns of foraging activity, which in turn represent different resource utilization niches, of five species of coexisting warblers. Shaded areas indicate strata within trees where the warblers forage preferentially. Numbers indicate the propor-tional utilization of each microhabitat determined by direct observation. (A) Myrtle warbler. (B) Cape May warbler. (C) Black-throated green warbler. (D) Blackburnian warbler. (E) Bay-breasted warbler. (Adapted from MacArthur, 1958, with permission of the Ecological Society of America.)*

tree, adjacent to the trunk, whereas others selectively forage out among the tips of the branches.

The **resource utilization niche** (MacArthur and Levins 1967; Schoener 1989) is perhaps the most operational approach to the study of how species differ in their myriad requirements. The approach usually focuses on consumable resources or factors that serve as surrogates for those resources, such as different microhabitats. Typically, species are characterized in terms of their similarity in resource use. Important measures used include various estimates of resource (niche) overlap, and breadth of resource use (niche width). One frequently used measure of overlap is Levins's (1968) formula,

 Resource overlap.

$$a_{ij} = \sum_{h=1}^{n} (p_{ih} \cdot p_{jh}) \Big/ (p_{ih})^2 \qquad (2.16)$$

where i and j refer to two different consumer species, and p_{ih} and p_{jh} refer to the fractional contribution of resource h to the total use of n different resources by species i and j. Measures of overlap are often applied to dietary data, where the n resources might be categories like species or size classes of prey used by a set of predators (e.g., Schoener 1974; Pianka 1986). In that case, the p_{ih} values are the fraction of the total diet, by volume, mass, or number, belonging to the resource category, h. Resource overlaps are sometimes considered to be synonymous with competition coefficients, but the assumptions required in making that leap are usually difficult to justify. For instance, competition would have to be purely consumptive and food would have to be in short supply for this approach to be reasonable.

Similarly, Levins (1968) suggested that the breadth of resource use might be quantified by

$$w = \sum_{h=1}^{n} 1 \Big/ (p_h)^2 \qquad (2.17)$$

where there are n different resources and resource h contributes a fraction p_h to the total resource use by a particular species. Specialized species would have small values of niche breadth, whereas generalized species would have larger values.

Field ecologists found it easy to apply these measures of resource utilization overlap and breadth to large numbers of species. The patterns that emerged from a collection of 81 field studies of resource partitioning were summarized by Schoener (1974) with regard to the kinds of resource axes and patterns of overlap seen in various collections of organisms. One aspect of this survey was whether studies showed repeated patterns with regard to niche dimensionality, that is, the numbers and kinds of resource axes required to separate sets of coexisting species (Schoener 1974). The broad kinds of axes considered included types of food, habitat, and activity times. Schoener found that most studies were able to separate species by looking at two or three broad kinds of resource axes, with habitat differences occurring more frequently than food type differences, and food type differences occurring more often than temporal ones. There was a weak trend for communities with larger numbers of species to require more resource axes to separate those species. Schoener also examined whether species tended to be separated along complementary dimensions, that is, whether similarity along one resource dimension would be offset by dissimilarity along other resource dimensions (Figure 2.14). There are numerous examples of such complementarity, involving 1) food type and habitat, 2) food type and activity time, 3) habitat and activity time, 4) different kinds of habitat axes, and 5) different types of food axes.

Despite the tremendous amount of effort put into studies of resource partitioning, the significance of these patterns of resource utilization remains uncertain, since by definition, different species must differ in some way and hence can always be separated by the choice of an appropriate niche axis. For this reason, largely observational approaches to studies of the resource utilization niche gradually declined in popularity because of the uncertain relation between the niche measures that were estimated and the presence and inten-

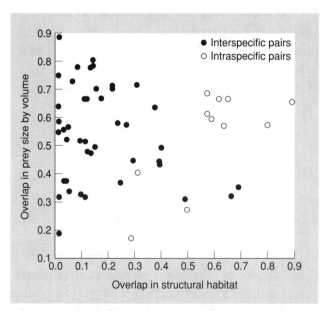

FIGURE 2.14. *An example of niche complementarity for* Anolis *lizards. Species that exhibit high overlap in habitat use tend to have low overlap in food, and vice versa. Intraspecific comparisons show the opposite trend. (Adapted with permission from Schoener, 1974. © 1974 American Association for the Advancement of Science.)*

sity of the interspecific competition. The few attempts that have been made to examine whether resource utilization overlap accurately predicts the intensity of competition have yielded conflicting results (Pacala and Roughgarden 1982; Hairston 1980a) that are detailed in the next chapter.

Other Ways of Thinking About Niches

One shortcoming of the resource utilization niche is its primary emphasis on consumable resources and consumptive competition. One way to return to a more comprehensive Eltonian niche was proposed by Jeffries and Lawton (1984). They suggested that a feature overlooked in purely consumptive approaches to niche metrics was the need for most species to avoid their predators via the utilization of **enemy-free space**. Enemy-free space refers not to a particular physical location but rather to sets of conditions that minimize the impact of predators, rather like the niche refers to sets of conditions that make it possible for a species to persist. The differences among species in the use of such strategies forms another important set of axes in any consideration of how ecologically similar species manage to coexist in a particular place and emphasizes that factors other than competition may be important. Jeffries and Lawton (1984) point out that many of the attributes of species that are used to infer differences in resource use, such as body size and microhabitat, can be just as readily interpreted as strategies for avoiding predators. Species can then be arrayed in a set of axes that define their antipredator adaptations, such as body size, speed of movement, and toxicity, as shown in Figure 2.15. Species may fail to invade a community, either because they lack the requisite antipredator defenses or because they are too similar to other species, and thereby manage to support larger predator populations that have a greater negative effect on prey. This latter possibility, termed **apparent competition** (Holt 1977), is an example of an indirect effect that will be considered in greater detail in Chapter 8.

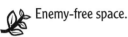 Enemy-free space.

Other approaches to the niche relations of coexisting species use shortcuts to estimate differences in resource utilization or other important aspects of species traits. One popular approach is the notion of the **morphological niche** (e.g., Ricklefs and Travis 1980). The idea is to use differences in the morphology of species as indicators of differences in resource use or other key aspects of lifestyles. Sets of species can be examined to ascertain whether differences in morphology correspond to coexistence. The separation of Galapagos finches by differences in beak morphology provides a simple example of separation of species along a single niche axis, where beak size is a surrogate for the sizes

 Morphology as an indirect measure of resource use.

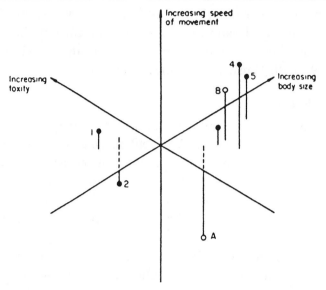

FIGURE 2.15. *A hypothetical example of enemy-free space. Species in a community may differ with respect to toxicity, speed of movement, and body size in ways that influence their susceptibility to predators. (Reprinted from* Biological Journal of the Linnean Society, *Vol. 23, M. J. Jeffries and J. H. Lawton,* Enemy free space and the structure of ecological communities, *pages 269–286, © 1984, by permission of the publisher Academic Press Limited, London.)*

of foods consumed (Figure 2.16; Grant 1986). For cases where more than three morphological measures are of interest and simple graphical presentations are inadequate, statistical techniques like principal component analysis are often used to combine different morphological measures into independent measures of shape, or morphological niche axes (Ricklefs and Travis 1980). Differences among the points corresponding to species in this morphological niche space correspond to differences in size or shape. Dissimilar species are located in different regions of the space (Figure 2.17), whereas morphologically similar species cluster together. A tendency for competitive exclusion within sets of morphologically very similar species should lead to an overdispersion, or spacing out, among coexisting species within the morphological niche space.

    Other comparative studies indicate that some morphological differences among a closely related set of species appear to evolve along independent evolutionary pathways in response to similar ecological selective pressures. Losos et al. (1998) describe patterns of morphological and evolutionary differences among four major *Anolis* ecomorphs, which have distinctive patterns of morphology and habitat use, for lizards living on the Greater Antilles (Cuba, Hispaniola, Jamaica, and Puerto Rico). The set of ecomorphs common to these four islands could have resulted from the colonization of each island by

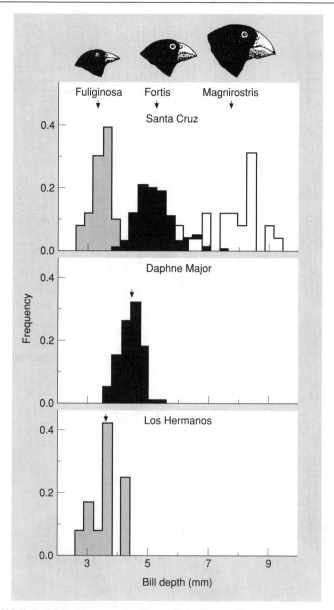

FIGURE 2.16. *An example of morphological differences among species that serve as a surrogate measure for resource use. Here coexisting finch species differ along a single morphological niche dimension, whereas the same species in allopatry are morphologically similar. (Adapted with permission from Grant, 1986. © 1986 by Princeton University Press.)*

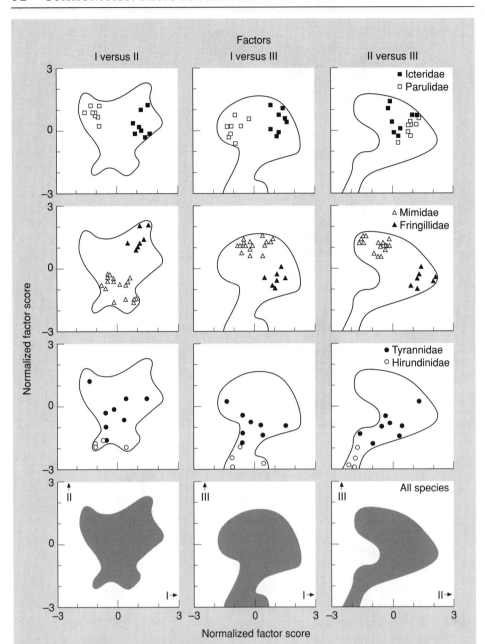

FIGURE 2.17. *Examples of the dispersion of species within morphological niches defined by a principal component analysis of eight morphological characters measured for a set of 83 bird species. (Adapted from Ricklefs and Travis, 1980, with permission of The Auk.)*

the common ancestor of each ecomorph, in which case lizards identified as the same ecomorph on different islands should be closely related. Alternately, the independent evolution of different ecomorphs from different ancestral species on each island would create morphologically similar species that are only distantly related when compared among islands. Reconstruction of the phylogeny of *Anolis* on each island indicates that the ecomorphs are the product of independent pathways of evolutionary change on the four islands. The fact that these patterns evolved repeatedly in these systems strongly suggests that the morphological differences among ecomorphs are nonrandom and tightly linked to ecological differences that promote coexistence.

The morphological niche approach is attractive because it can often be applied to existing collections of organisms in museums. It clearly has its limitations, though. It is usually applied to groups of taxonomically similar organisms, such as a single genus or family of birds or lizards. However, taxonomic similarity is not a prerequisite for strong competition. Species that differ greatly in taxonomic affiliation, and in morphology, may nonetheless compete strongly. The best example of such interactions is the case of competition for seeds in a guild of desert granivores consisting of ants, rodents, and birds. These species are morphologically dissimilar, yet experiments have shown that ants and rodents can compete strongly (Brown and Davidson 1977).

Guild Structure in Niche Space

The existence of guilds, groups of species that use similar resources in similar ways, should be discernible by inspecting whether the niches of guild members cluster together in a larger niche space. Inger and Colwell (1977) and Winemiller and Pianka (1990) have used this approach to attempt to identify nonrandom patterns and clusters in the way that species use resources. Simple inspection of the positions of niches based on a statistical analysis of overlap in microhabitat use by the reptiles and amphibians in a Thai rain forest suggests that guilds, or clusters of species with high overlap, indeed exist (Figure 2.18). However, the existence of these clusters does not imply that the species actually compete.

 Guild structure describes the pattern of niche relations for multiple species.

Another way to look for guilds involves examining patterns of overlap among species as a function of the ranking of species by their amount of niche overlap. This ranking is done by calculating the resource overlaps for all pairs of species in the community and then, for each species, ordering its overlaps with other species from highest to lowest. The species pair with the highest overlap receives a rank of 1, the pair with the next highest overlap is ranked 2, and so on. For each species, a graph of overlap against rank can then

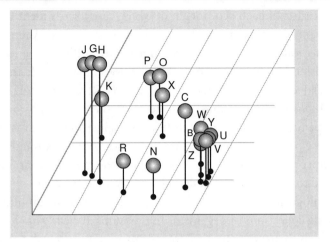

FIGURE 2.18. *Patterns of guild structure in a niche space based on the patterns of overlap in microhabitat use by species of reptiles and amphibians in a Thailand rain forest. Each "balloon" corresponds to the position of a species in niche space. Guilds appear as clusters of species, and different guilds fall in different portions of the space. (Adapted from Inger and Colwell, 1977, with permission of the Ecological Society of America.)*

be plotted. For situations in which guilds exist, overlap among close neighbors is consistently high, since all have similar patterns of resource use. Once beyond the members of that guild, however, overlap drops precipitously, especially if the remaining members of the community comprise another guild that is substantially different in resource use. Figure 2.19 shows the patterns expected in a theoretical community consisting of two guilds of equal size. When real communities of lizards and fish are examined in a similar fashion, a range of patterns are observed (see Figure 2.19).

The models described in this chapter make certain assumptions about how groups of organisms compete. The models also make predictions about the properties of species that favor the coexistence or exclusion of competitors. Experimental studies of interspecific competition considered in the next chapter are variously inspired by models and ideas about the niche relations of species. They range from simple efforts to demonstrate that competition

FIGURE 2.19. *Hypothetical and observed patterns of resource overlap in communities containing guilds of species with similar patterns of resource use. Each line corresponds to overlap between one species and all n other species measured, where the overlaps are ordered from highest (1) to lowest (n). For a hypothetical community consisting of two nonoverlapping guilds—an extreme case—patterns of overlap would look like those shown in (A). Patterns in real communities of lizards and fish are not quite so clear-cut (B). (Adapted from Winemiller and Pianka, 1990, with permission of the Ecological Society of America.)*

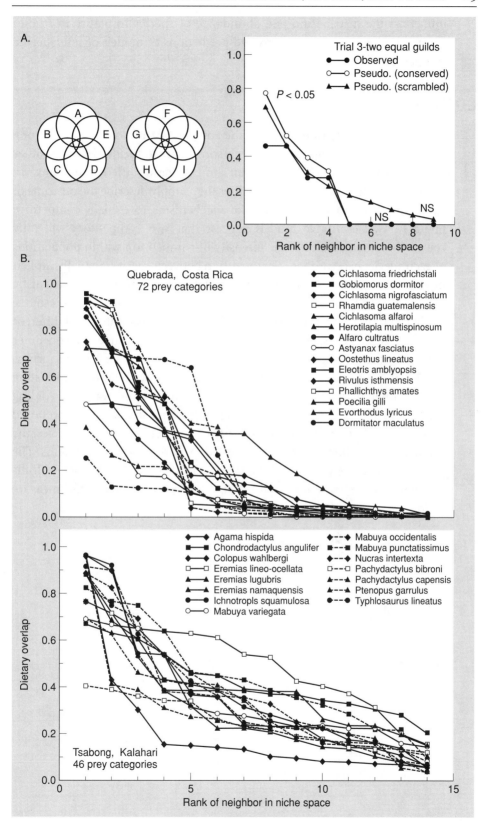

influences the distribution and abundance of species in nature to studies designed to test the assumptions and predictions of models of interspecific competition.

CONCLUSIONS

Ecologists have often emphasized the importance of competition among species in setting community patterns, sometimes to the detriment of other equally likely explanations. Species can compete via six different mechanisms. Models of competition differ in whether they simply describe the outcome of competition, or incorporate specific mechanisms to explain competitive interactions. Descriptive models suggest that two competitors will only coexist in a stable fashion if the intensity of competition within populations of each species exceeds that of competition between species. Spatially explicit models for sessile species arrive at basically the same conclusion. Mechanistic models derived from the Monod model for a single consumer and its resource predict that two species will coexist only when each is limited by a different resource. When two species are limited by the same resource, the consumer that drives the resource to the lowest level excludes the other. Various formulations of the ecological niche provide ways of describing how coexisting species differ in resource use, habitat requirements and ways of avoiding predators. Although measures of overlap in resource use have been proposed as indirect ways to estimate the intensity of competition among species, the validity of this approach remains untested in all but a very few systems. The pattern of resource overlaps displayed by sets of species in abstract multidimensional niche space has been proposed as a way to define and measure guild structure.

CHAPTER 3

Competition: Experiments, Observations, and Null Models

OVERVIEW

This chapter presents case studies of competition among different kinds of organisms in a variety of habitats. Basic experimental designs used to study competition are outlined. Selected case studies highlight special features of competition in different kinds of communities. These case studies also indicate how competition in nature can depart from some of the assumptions of the models discussed in Chapter 2. Surveys of published competition experiments show that competition occurs frequently. Competition also occurs more frequently in some trophic levels than in others. Relatively few studies measure the relative intensity of intraspecific and interspecific competition, despite the fact that theory predicts that interspecific competition should be less intense than intraspecific competition for species to coexist. Similarly, few studies are conducted over enough sites or over sufficient time periods to determine whether competition varies in intensity either spatially or temporally. Null model approaches to the study of community patterns offer an alternative approach to direct studies of competition in communities where experiments are not feasible. Null models compare patterns observed in nature with the expectation of how those patterns would appear if they were generated solely by random, noncompetitive events.

EXPERIMENTAL APPROACHES TO INTERSPECIFIC COMPETITION

The strongest evidence for the importance of competition in nature comes from studies of how species respond to experimental additions or removals of potential competitors (Connell 1975; Hairston 1989). Even then, correct interpretations of the results of experimental manipulations of competitors require that other kinds of noncompetitive interactions between manipulated species can be ruled out. For instance, positive responses to species removals, such as enhanced survival or growth, could happen if the removed species is either a predator or a competitor of the responding species. Sometimes the natural history of interacting species is enough to eliminate noncompetitive interactions from consideration. If the interacting species are both autotrophic plants and therefore cannot interact as predators and prey, then the interpretation of mutually negative interactions as competition is simple. The same is true for interactions among obligate herbivores, or specialized predators with restricted diets that do not include each other. When the interacting species are polyphagous predators that can potentially consume each other, a mixture of competitive and predatory interactions is possible, and the simple interpretation of positive responses to species removals as evidence for either competition or predation becomes more complicated (see Polis et al. 1989).

Ecologists use experiments conducted in different settings to explore the role of competition in community organization. The details of the experimental designs employed also influence the kinds of questions that can be answered about competitive interactions.

 Kinds of ecological experiments.

Field experiments take place in natural settings and usually involve simple removals or additions of potential competitors, coupled with observations of the responses of remaining species. Typical responses include measures of density, growth, reproduction, or survival with or without competitors. Field experiments provide crucial information about where and when particular interactions, including competition, occur in nature. The relatively coarse manipulations (e.g., presence/absence of a competitor) that are possible in most field situations limit the kind of information gained about the details of functional relations in competitive systems (e.g., density or frequency dependence, asymmetry, or additivity). Often, field experiments are restricted to simple additions or removals of species because careful manipulations over a range of competitor densities are impossible to either establish or maintain. Connell's (1961) classic study of competition among barnacles in natural settings is a splendid example of a well-designed field experiment and is reviewed later in this chapter.

Laboratory experiments permit superior control over factors such as the physical environment and species composition, but this control is usually obtained at the expense of reduced community complexity and reduced realism. In seminatural or laboratory situations where greater control over competitor densities is possible, other experimental designs can be used to gain insight into the functional details of competitive interactions. Gause's (1934) experiments on competition among protists in simple laboratory culture vessels are a good example of the advantages and disadvantages of laboratory experiments.

Hybrid experiments often utilize artificial habitats placed in natural settings and have some of the advantages and disadvantages of field and laboratory experiments. For example, experiments using artificial ponds where initial species composition and habitat complexity can be rigorously controlled offer compromises between the advantages and disadvantages of laboratory and field experiments (Morin 1989). Of course, some ecologists feel that field experiments are superior to either laboratory experiments or hybrid experiments, since the latter may either distort or fail to include key features of natural communities (Diamond 1986).

Regardless of the experimental setting, experimental designs involving particular manipulations of the densities of species can test specific hypotheses about the existence and relative intensity of intraspecific and interspecific competition. Examples of some basic designs are sketched in Table 3-1 and described in greater detail below. Minimal experimental designs for **density-dependent competition** usually hold the initial density of a responding

Experimental designs.

species, sometimes called a target species, constant, while varying the density of a competitor (see Table 3-1A). This sort of design is useful to demonstrate that the species of interest actually compete. Most field removals of competitors fit the minimum definition of a design for density dependence, since densities of the manipulated competitor fall at natural and reduced levels. In general, removals are easier to accomplish than additions in field settings. When more than two density levels are used, it is possible to determine whether the per capita effects of competitors increase or decrease disproportionately (nonlinearly) with density. For example, if species compete for discrete units of habitat, say nesting holes for birds, competition might only become intense when densities are sufficiently high so that most sites are occupied. Below that density, per capita effects might be uniformly low. Such nonlinear effects of competitor density would mean, for instance, that the zero-growth isoclines in Lotka-Volterra models of competition are curved rather than straight lines. This fact would in turn influence where the lines intersect, and therefore possibly affect equilibrium densities of competitors.

TABLE 3-1. **Basic kinds of experimental designs useful for testing for intraspecific and interspecific competition.**

	Species 1	Species 2
A. Density Dependence		
Control	N	0
Control	0	N
Competition	N	N
B. Frequency Dependence		
Control	$2N$	0
Competition	N	N
Control	0	$2N$
C. Asymmetry		
Control	N	0
Intraspecific competition	$2N$	0
Interspecific competition	N	N
Intraspecific competition	0	$2N$
Control	0	N

Note: N refers to a density (number or biomass per unit area or volume) of potential competitors.

A different kind of design (see Table 3-1B) is required to show whether the per capita effects of interspecific competitors differ from those of intraspecific competitors. This approach, known as a **replacement series** design (de Wit 1960), is useful in cases where **frequency-dependent competition** is of interest. Replacement series designs are useful when the question is not simply whether competition happens, but whether intraspecific competition differs in intensity from interspecific competition. As shown in Table 3-1B, the combined density of both competitors is held constant, while the relative frequency of each species varies. The design yields information about competition within and between populations of both species, but does not yield useful information about density dependence. If interspecific competitors have weaker per capita effects than intraspecific competitors, mixtures will produce greater yields per individual than will pure stands of a single species. This overyielding phenomenon could result if the species differ sufficiently in resource use so that each species can extract some resources that are unavailable to the other. Replacement series designs are particularly popular in studies of competition among plants (Harper 1977). Plants can be readily planted at constant total densities and different relative frequencies in field or laboratory settings.

A somewhat more elaborate design that combines elements of the basic designs for density and frequency dependence can be used to assess the relative intensities of intraspecific and interspecific competition and to test for **asymmetric competition**. Underwood (1986) suggests that three kinds of competitive asymmetry are of interest: intraspecific asymmetry, interspecific asymmetry, and asymmetry in the relative intensities of intra- and interspecific competition. Intraspecific asymmetry happens when species differ in the intensity of their intraspecific per capita competitive effects. This situation could be assessed by doubling intraspecific density (from N to $2N$) and comparing the relative responses of each species to an identical change in the abundance of intraspecific competitors. Interspecific asymmetry refers to situations in which species differ in the strength of their per capita competitive effects on each other, all else being equal. Inclusion of a mixed-species treatment with a density of N individuals of species 1 and N individuals of species 2 (a total mixed-species density of $2N$) permits comparison of the response of each species to similar increases in the densities of interspecific competitors. Finally, for each species, it is possible to compare the effects of adding N intraspecific competitors or N interspecific competitors. Differences between target species in the relative impact of intraspecific and interspecific competitors describe the third kind of competitive asymmetry. This latter effect is also of interest in determining whether the intensity of competition within species is greater than the intensity of competition between species, as predicted by some simple models considered in Chapter 2.

Other specialized designs are particularly appropriate for sessile species, especially plants. Designs used to assess neighborhood competition (Antonovics and Levin 1980) vary the number of competitors within a particular radius, or neighborhood, of target individuals. This can be done by exploiting natural spatial variation in abundance and assessing competitive effects as a function of the number or biomass of competitors within a particular radius of measured plants (Pacala and Silander 1985, 1990). Alternatively, plants can be thinned to create sets of increasingly greater distances among neighbors to determine the neighborhood distance beyond which effects of competitors become undetectable (Shaw and Antonovics 1986; Shaw 1987). The approach assumes that a competitor's effects will vary with its proximity to a target individual, a reasonable assumption for species that compete for light, water, or nutrients within a fixed area.

EXPERIMENTAL STUDIES OF INTERSPECIFIC COMPETITION

The best way to illustrate particular mechanisms and consequences of interspecific competition is through the description of case studies of competition experiments. There are now so many published experimental studies of competition that a detailed review is beyond the scope of this book. The studies emphasized here are included either because of their historical significance or because they illustrate important mechanisms or concepts. Generalizations about the prevalence and importance of competition in nature require surveys of large numbers of studies conducted in a variety of habitats with a diversity of organisms. Several of those surveys have now been done (Connell 1983; Schoener 1983; Gurevitch et al. 1992; Goldberg and Barton 1992), and their conclusions are reviewed later in this chapter.

Competition in Marine Communities

Animals. Joseph Connell (1961) performed an early influential study of interspecific competition among animals in a natural setting. On the rocky coast of Scotland, two barnacle species typically occupy different locations in the intertidal zone. The smaller of the two species, *Chthamalus stellatus*, generally occurs higher in the intertidal zone than the other species, *Balanus glandula* (Figure 3.1). Such patterns had previously been interpreted as consequences of physiological differences between the species. For example, the observed zonation might occur if *Chthamalus* was more tolerant than *Balanus* of the periodic desiccation experienced at low tide. Differences in larval settlement of both species could also account for the pattern if planktonic larvae of the two species preferentially settled at different heights in the intertidal zone.

 Competition causes vertical zonation in barnacles.

Connell used a combination of observations and clever experiments to show that competition created the observed pattern of intertidal zonation, through preemptive occupation of space and subsequent overgrowth of *Chthamalus* by *Balanus*. Connell observed that larval *Chthamalus* regularly settled in the upper *Balanus* zone, but failed to persist there. He transplanted rocks with *Chthamalus* lower in the intertidal zone into the areas where *Balanus* could be counted on to settle at high densities. Each rock was divided into two parts, one where interactions continued undisturbed and another where *Balanus* were removed from the proximity of each *Chthamalus*. The transplanted *Chthamalus* were rapidly overgrown or crushed by larger, rapidly growing *Balanus*. When transplanted *Chthamalus* were kept free of encroaching *Balanus*, they survived well, indicating that their absence from the *Balanus* zone was caused by interspecific competition rather than by possible physiological constraints (Figure 3.2). Connell's study remains one of the most elegant and compelling examples of the role of interspecific competition in creating patterns within a natural community.

Competition is not limited to sessile marine organisms living on hard substrates. Experiments with bivalve mollusks living in sandy substrates also

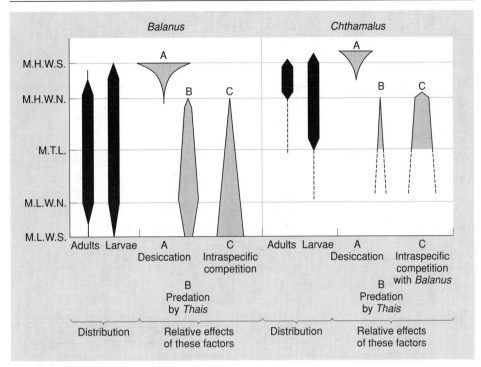

FIGURE 3.1. *Summary of the intertidal zonation of adults and settling larvae of two competing barnacles,* Chthamalus *and* Balanus. *(Adapted from Connell, 1961, with permission of the Ecological Society of America.)*

show that competition occurs. Peterson and Andre (1980) used a combination of caging and transplant experiments to show that a deep-dwelling bivalve, *Sanguinolaria*, competes strongly with two other deep-dwelling species, *Tresus* and *Saxidomus*, but not with a shallow-dwelling species, *Protothaca staminea*. The presence of other species in the same level of the substrate caused an 80% reduction in the growth of *Sanguinolaria*. Peterson and Andre used a clever approach to determine if the mechanism of competition involved a shortage of space or food. They included a treatment consisting of empty shells of competitors, which occupied space but clearly did not feed. The empty shells also depressed growth, but not to the same extent seen in the treatments containing living competitors. The upshot was that competition among the species occupying similar depth strata probably reflects a combination of competition for space and food (Figure 3.3).

Relatively little is known about competition among pelagic marine organisms, which probably reflects the difficulty of conducting experiments with highly mobile animals. Pelagic organisms are difficult to cage or otherwise manipulate, and consequently their competitive interaction remains largely speculative. However, some intriguing observations suggest the possi-

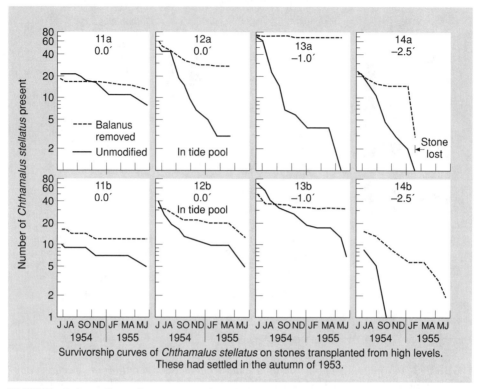

Survivorship curves of *Chthamalus stellatus* on stones transplanted from high levels.
These had settled in the autumn of 1953.

FIGURE 3.2. *Effects of competition from* Balanus *on the survival of* Chthamalus *transplanted into different regions of the lower intertidal zone. In general,* Chthamalus *survives well where competitors are removed (dashed lines), but survives poorly when competing with* Balanus *(solid lines). (Adapted from Connell, 1961, with permission of the Ecological Society of America.)*

bility of complex competitive interactions between the predatory fish and birds that feed on smaller fish in marine systems. Safina (1990) has shown that short-term effects of predatory bluefish (*Pomatomus saltatrix*) on the distribution of smaller prey fish in turn affect the foraging success of two tern species on prey shared by bluefish and terns. The seasonal arrival of schools of foraging bluefish is also correlated with changes in the abundance of foraging terns. Presumably, similar kinds of interactions are possible among other sets of strictly pelagic species, but they appear to have received little study.

Plants. Experimental studies of sessile algae found on hard substrates suggest that competition can influence patterns of species composition. The influence of competition also depends on the extent of herbivory experienced by the algae (Dayton 1975; Lubchenco 1978; Steneck et al. 1991). Working on the coast of Washington, Paul Dayton (1975) found that the dominant alga in the lower inter-

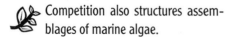 Competition also structures assemblages of marine algae.

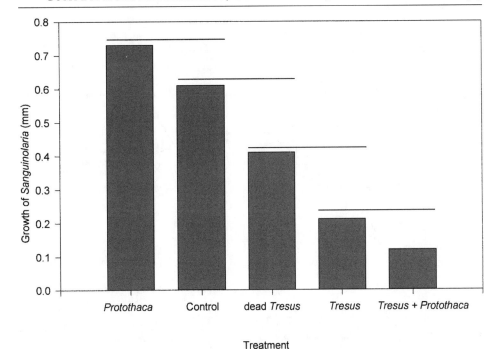

FIGURE 3.3. *Effects of living and dead* Tresus *and living* Protothaca *on the growth rate of* Sanguinolaria. Tresus *lives at the same depth in the sediments as* Sanguinolaria, *whereas* Protothaca *lives in a different stratum closer to the surface. Horizontal lines above the bars indicate sets of means that are not statistically different. (Data from Peterson and Andre, 1980.)*

tidal zone, *Hedophyllum sessile*, had both negative competitive effects and positive effects on different algal species. *Hedophyllum* plays a role analogous to canopy-forming trees in terrestrial forests. It shades out some species of competitors and creates favorable understory conditions for another set of algal species that do not grow well in the absence of a *Hedophyllum* canopy. Experimental removals of *Hedophyllum* caused a group of understory species to decline, while another group of fugitive species rapidly colonized the space from which *Hedophyllum* was removed (Figure 3.4). The fugitive species apparently compete with *Hedophyllum* for light and attachment sites. In contrast, obligate understory species apparently require the modified physical conditions created by *Hedophyllum*, such as reduced light levels and decreased desiccation at low tide.

Competition in Terrestrial Communities

Animals. Nearly 20 years passed after Connell's groundbreaking experiments before convincing experimental demonstrations of competition among terrestrial animals appeared. Most terrestrial ecologists were content to find

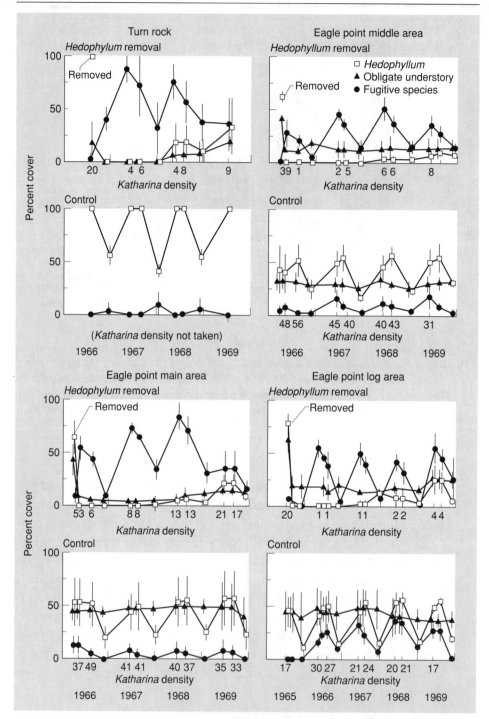

FIGURE 3.4. *Effects of the removal of the dominant canopy alga,* Hedophyllum, *on obligate understory and opportunistic fugitive species of algae. Competition from* Hedophyllum *normally limits the abundance of fugitive species. (Adapted from Dayton, 1975,
with permission of the Ecological Society of America.)*

indirect support for interspecific competition in patterns of field distributions
or resource utilization. For example, Nelson Hairston (1949) noted that two
species of terrestrial salamanders, *Plethodon jordani* and *Plethodon glutinosus*,
differed in their altitudinal zonation in a manner loosely reminiscent of the
pattern shown by Connell's barnacles. The salamanders live in the litter layer
of the forests of the Appalachian Mountains, where they prey on an assortment of small invertebrates. These salamanders do not appear to prey on each
other, although such intraguild predatory interactions occur in other sets of
salamander species (Hairston 1987). *Plethodon jordani* typically lives at higher
elevations than does *P. glutinosus* (Figure 3.5). The amount of altitudinal
overlap in the distributions of the two species differs among isolated populations found on different mountain ranges. Salamander populations overlap
little in the Great Smoky Mountains, but they overlap extensively in the
Balsam Mountains. This pattern suggested that the intensity of competition
among salamanders might be greater in the zone of narrow overlap, whereas
less intense competition might allow greater altitudinal overlap.

Nearly 30 years after he made his original observations on the elevational
distributions of salamanders, Hairston (1980a,b) used a combination of
reciprocal transplant experiments and competitor removals to show that salamanders from the zone of narrow overlap competed more intensely than ones

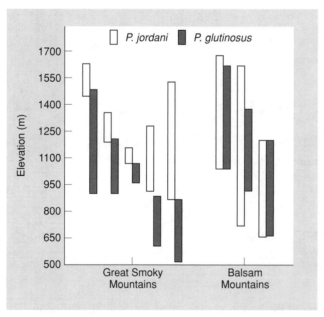

FIGURE 3.5. *Altitudinal distributions of the salamanders* Plethodon jordani *and* Plethodon glutinosus *in the Great Smoky Mountains (low overlap) and the Balsam Mountains
(high overlap). Sets of bars represent replicate transects. (Adapted from Hairston, 1980a,
with permission of the Ecological Society of America.)*

Competition also causes altitudinal zonation in terrestrial salamanders.

from the zone of wide overlap. The salamanders are relatively sedentary, and their abundances can be altered by repeated additions or removals of individuals to or from experimental plots. Hairston established three kinds of plots in each location: *P. jordani* removals, *P. glutinosus* removals, and controls where neither species was removed. Because the salamanders grow and reproduce slowly, the experiment continued for five years, roughly the time required for a newborn salamander to reach maturity. Salamander density in the zone of narrow overlap responded much more strongly to competitor removals than in the zone of wide altitudinal overlap (Figure 3.6). Hairston interpreted this result as support for his earlier inference of an inverse relation between altitudinal overlap and the intensity of competition. Later work in the same system by Nishikawa (1985) showed that the main mechanism of competition among these salamanders was territorial aggression. Salamanders from the zone of narrow overlap were more aggressive than animals from the zone of wide overlap.

Steven Pacala and Jonathan Roughgarden (1982, 1985) studied competition between species pairs of small *Anolis* lizards on two islands in the Lesser Antilles. The lizards feed primarily on arthropods. The lizards differ some-

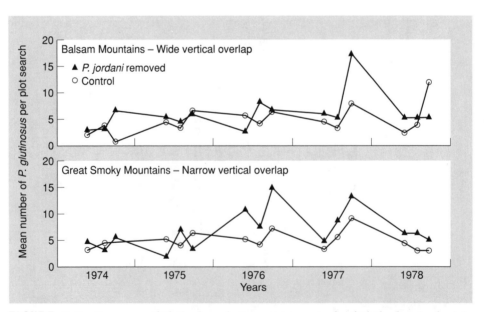

FIGURE 3.6. *Responses of* Plethodon glutinosus *to removals of* Plethodon jordani *in regions of wide and narrow altitudinal overlap.* P. glutinosus *responds sooner, and more consistently, in the zone of narrow overlap. Dashed lines show* P. jordani *removals; solid lines show controls. (Adapted from Hairston, 1980a, with permission of the Ecological Society of America.)*

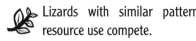 Lizards with similar patterns of resource use compete.

what in body size, which affects the size of prey that can be consumed, and in the locations where they forage in trees and bushes. On St. Maarten, *Anolis wattsi pogus* and *Anolis gingivinus* overlap substantially in average body size and perch height, which implies that they should also be similar in resource use (see Figure 2.12). On another island, St. Eustatius, *Anolis wattsi schwartzi* barely overlaps with *Anolis bimaculatus* in body size and perch height, implying less similarity in resource use than observed for the lizards on St. Maarten.

To test whether different amounts of overlap in resource use corresponded to different intensities of competition, Pacala and Roughgarden established replicated enclosures on both islands, which they stocked with one or both lizard species. Enclosures contained either 100 *Anolis wattsi* plus 60 *A. gingivinus* or 60 *A. gingivinus* alone on St. Maarten, and 100 *Anolis wattsi* plus 60 *A. bimaculatus* or 60 *A. bimaculatus* alone on St. Eustatius. Manipulations of lizard abundance in replicated caged sections of forest on both islands showed statistically detectable interspecific competition (measured as decreases in growth rates) on St. Maarten, where the lizards' resource utilization niches overlapped substantially (Figure 3.7). However, there was no evidence of competition on St. Eustatius, where resource utilization niches barely overlapped. Pacala and Roughgarden interpreted these results as a confirmation of a frequently assumed correspondence between the degree of niche overlap between two species (similarity in resource use as indirectly measured by body size and perch height) and the intensity of ongoing competition. Examination of the stomach contents of lizards removed from the enclosures showed that *A. wattsi* reduced the volume of prey consumed by *A. gingivinus* by two- to threefold. This is consistent with a simple consumptive mechanism of competition.

One other study provides similar support for a direct relation between overlap in resource use and the intensity of interspecific competition. That study focused on competition among coexisting rodent species in deserts of the American Southwest. Munger and Brown (1981) and Heske et al. (1994) used large-scale enclosures and experimental removals to look for competitive interactions within an assemblage of desert rodents. Some of the rodents feed primarily on seeds, whereas others are insectivorous. Removals of a group of larger seed-eating rodents, kangaroo rats in the genus *Dipodomys*, produced significant increases in the abundance of smaller species of seed eaters (Figure 3.8). In contrast, *Dipodomys* removal had no effect on small omnivorous rodents that overlapped little in resource use with the removed species of *Dipodomys*. Effects of competitor removals took several years to become apparent, as in Hairston's experiments with salamanders. In such

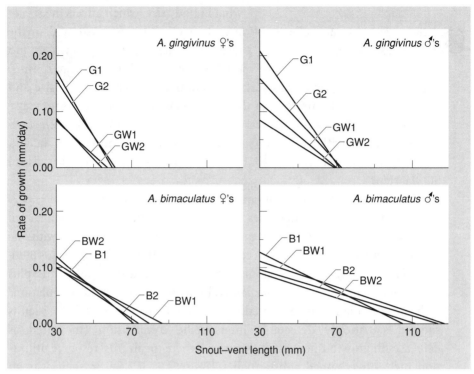

FIGURE 3.7. *Effects of* Anolis wattsi *on size-specific growth rates of* Anolis gingivinus *(high overlap) and* Anolis bimaculatus *(low overlap).* A. wattsi *significantly depresses the growth of* A. gingivinus, *but does not affect the growth of* A. bimaculatus. *Key to treatments: G1, G2,* Anolis gingivinus *alone; B1, B2,* Anolis bimaculatus *alone; GW1, GW2,* Anolis gingivinus *competing with* Anolis wattsi; *BW1, BW2,* Anolis bimaculatus *competing with* Anolis wattsi. *(Adapted with permission from Pacala and Roughgarden, 1982. © 1982 American Association for the Advancement of Science.)*

cases, delayed responses to competitor removals presumably reflect demographic lags caused by the life history characteristics of relatively long-lived species.

On the face of it, Hairston (1980a,b) reached very different conclusions about the relation between overlap and competition than Pacala and Roughgarden (1982, 1985) or Munger and Brown (1981). Hairston's study is consistent with the idea that low overlap is a consequence of competition, whereas Pacala and Roughgarden's results are consistent with the idea that high overlap results in intense competition. These contradictory patterns might be attributed to a number of differences between these studies, including the kinds of niche dimensions considered, the mechanisms of competition, and perhaps the amount of evolutionary time available for species to modify their patterns of resource utilization in response to competitors. The studies suggest that overlap may not have a simple, or single, relation to the intensity

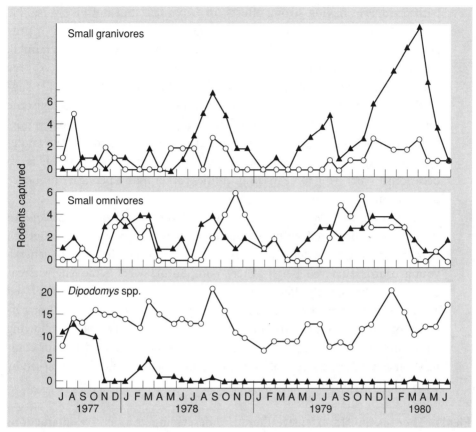

FIGURE 3.8. *Responses of small granivorous rodents and small omnivorous rodents to removals of larger granivorous* Dipodomys *species. Small granivores increase upon the removal of larger granivores, but small omnivores, which have little dietary overlap with large granivores, do not. (Adapted with permission from Munger and Brown, 1981. © 1981 American Association for the Advancement of Science.)*

of competition. Consequently, the utility of overlap measures as a shortcut for the estimation of the intensity of competitive interactions remains contentious.

Studies of uncaged lizard populations in desert habitats suggest that competition may be strongly episodic. In such communities, competition seems to be detectable only in years when resources are in particularly short supply. Arthur Dunham (1980) and David Smith (1981) both published studies describing how lizards found in the deserts of the southwestern United States responded to removals of potential competitors. Dunham observed that the insectivorous lizards *Sceloporus merriami* and *Urosaurus ornatus* each responded favorably to the removal of the other in replicated field plots located in the Big Bend region of Texas. He found evidence of competition in some years, but not in others. The interaction was also highly asymmetric, with *Sceloporus* having strong effects on *Urosaurus*, but not vice versa. He attributed the temporal variation in competition to variable levels of prey abundance (insects) that occurred in response to annual variation in rainfall. Competition was most noticeable in years of low rainfall and reduced prey abundance. Dunham's observations support John Wiens' (1977) earlier suggestion that effects of competition on communities are likely to be episodic rather than constant. According to Wiens, competition would be most important during occasional bottlenecks in resource abundance and virtually undetectable when resources were abundant. In this view, competition is an occasional structuring force in communities rather than a pervasive influence on community patterns.

David Smith (1981), working with a different lizard species, *Sceloporus virgatus*, that occurred with *Urosaurus ornatus* in the Sonoran Desert of Arizona, also found evidence for episodic competition (Table 3-2). In general, young lizards survived better where competitors were removed, whereas old lizards showed no effects. Young females of *S. virgatus* also displayed enhanced growth in the first year following *Urosaurus* removals. Clear evidence of competition occurred only during one of the two years following competitor removals. As in Dunham's study, the strongest evidence for competition came during a year of extreme environmental conditions marked by unusually low rainfall.

Plants. There is an old extensive literature describing competitive interactions among forest trees. The experiments involved usually entail either some sort of trenching to manipulate competition for nutrients and water or removals of the canopy to manipulate competition for light. For example, Chapman (1945) compared the establishment and survival of loblolly pine (*Pinus taeda*) seedlings in clear-cut and shaded portions of a Louisiana forest. The intact forest canopy consisted of a mixture of hardwoods, mostly oaks and gums, and conspecific pines. Seedlings freed from the shade of intraspecific and interspecific competitors survived better and grew faster than seedlings in shaded plots. This susceptibility to competition for light helps to explain why loblolly pine is an early successional species in much of the southeastern United States. It does well in light gaps, such as the ones formed by abandoned agricultural fields, but fails to replace itself in the full shade created by an established forest canopy.

Competition also influences the local distribution of herbaceous plants. Jessica Gurevitch (1986) studied the role of competition in limiting the local distribution of a desert grass. She found that local variation in the distribution of the grass *Stipa neomexicana* was caused by interspecific competition with other grass

 Competition influences the distribution of some terrestrial plants.

TABLE 3-2.   Effects of interspecific competition between *Sceloporus virgatus* and *Urosaurus ornatus* on survival of young and old lizards.

| | | *S. virgatus* Survivorship | | |
Sex	Age	Year	Control	Competitors Removed
Female	Young	1973	0.44	0.28
		1974	0.17	0.42
		1975	0.41	0.56
	Old	1973	0.55	0.50
		1974	0.58	0.65
		1975	0.50	0.56
Male	Young	1973	0.35	0.32
		1974	0.35	0.48
		1975	0.58	0.25
	Old	1973	0.42	0.56
		1974	0.27	0.44
		1975	0.41	0.60

| | | *U. ornatus* Survivorship | | |
Sex	Age	Year	Control	Competitors Removed
Female	Young	1973	0.46	0.46
		1974	0.21	0.48
		1975	0.17	0.08
	Old	1973	0.48	0.41
		1974	0.33	0.27
		1975	0.32	0.38
Male	Young	1973	0.42	0.35
		1974	0.12	0.24
		1975	0.21	0.20
	Old	1973	0.52	0.40
		1974	0.31	0.33
		1975	0.29	0.19

Note: Controls indicate where both species were present; experimentals indicate where the interspecific competitors were removed. Effects of competition were most pronounced on young lizards in 1974.
Reprinted from Smith (1981), with permission of the Ecological Society of America.

species. In Gurevitch's field sites in Arizona, *Stipa* is most common on dry ridge tops where densities of possible competitors are particularly low. Gurevitch found that removals of potential competitors had little impact on *Stipa* survival on the ridges where it was most abundant (Figure 3.9). In contrast, removals of more abundant competitors from midslope and lower slope sites greatly enhanced *Stipa* survival in those locations, where it is usually much less common. The implication is that *Stipa* enjoys a competitive refuge on dry ridge crests, where its important competitors are limited to low abundances by abiotic factors. In contrast, *Stipa* is competitively excluded from wetter downslope sites where other species of competitively superior grasses abound.

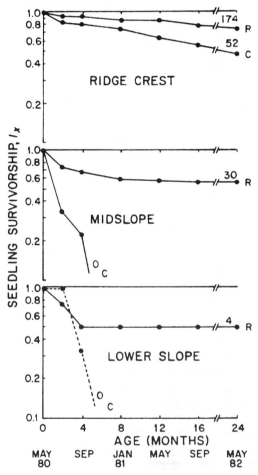

FIGURE 3.9. *Survival of seedlings of the grass* Stipa *in three locations in the presence (C) or absence (R) of competitors. Competition apparently increases in intensity as ones goes from the ridge crest, where* Stipa *is common, to the lower slope, where it is rare. (Reprinted from Gurevitch, 1986, with permission of the Ecological Society of America.)*

In disturbed systems, such as mown fields, competition plays a less striking role in determining the distribution and abundance of plant species. Norma Fowler (1981) studied the potential competitive interactions among an array of herbaceous plants growing in an infrequently mowed field near Duke University in Durham, North Carolina. Previous work by Fowler had shown that the plants in the field fell into two groups distinguished by their seasonal patterns of growth: cool-season, winter-growing species and warm-season, summer-growing species. Removals of either entire groups of species or individual species showed rather sparsely distributed competition among these herbaceous species. Fowler concluded that the effects of competition in this community were diffuse at best and resulted from the cumulative influence of many weak effects exerted by several species. It is also possible that the infrequent mowing of the field simulated a moderate level of herbivory, which in turn moderated some of the competitive interactions within the community.

 Diffuse competition in plant communities.

Competition in Freshwater Communities

Animals. Earl Werner and Donald Hall (1976) explored the effects of competition on diet and habitat utilization by three species of freshwater sunfish: *Lepomis macrochirus,* the bluegill sunfish; *Lepomis gibbosus,* the pumpkinseed sunfish; and *Lepomis cyanellus,* the green sunfish. The fish all use similar habitats when each species is stocked by itself (allopatrically) in comparable ponds. When alone, each *Lepomis* species prefers to forage in vegetation on relatively large invertebrate prey (Figure 3.10). When all three species are placed in the same pond, *L. macrochirus* shifts to feeding in open water on smaller and less energetically rewarding zooplankton, while *L. gibbosus* shifts to feeding on benthic prey. It appears that competition causes a difference in the realized niche of *L. macrochirus*, although the interpretation of these observations is limited by the fact that they were based on unreplicated experiments.

 Competition creates niche shifts in fish.

Bengtsson (1989) used a combination of observations of natural rock pools and experiments conducted in artificial pools to show that interspecific competition influenced the number of *Daphnia* species that regularly coexisted in pools on the Baltic coast of Sweden. Although three *Daphnia* species occur in the system, *Daphnia magna, Daphnia pulex,* and *Daphnia longispina,* usually only one or two species manage to coexist in a given pool. Observations of natural pools show that populations go extinct with increasing frequency in pools

Competition affects the coexistence of *Daphnia* in rock pools.

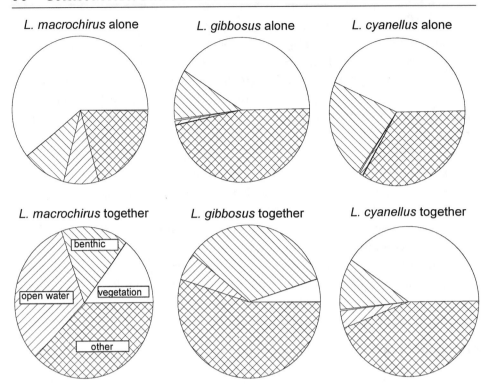

FIGURE 3.10. *Diets, and inferred patterns of habitat use, of the sunfish* Lepomis macrochirus, L. gibbosus, *and* L. cyanellus *when each species is stocked by itself in a pond or when all three species are stocked together and are able to compete. The diet category "other" refers to prey that are not found in specific habitats and therefore cannot be used to infer habitat use. Other categories refer to prey found predominantly in vegetation, open water, or benthic (bottom) habitats. (Redrawn from data in Werner and Hall, 1976.)*

that contain increasing numbers of species (Table 3-3). When Bengtsson added various combinations of one, two, or three *Daphnia* species to artificial rock pools, he found that extinctions were most common in pools containing three species, and somewhat less common in two-species pools (see Table 3-3). He observed no extinctions in single-species pools, reinforcing the conclusion that extinctions were caused by competition with other *Daphnia* species. The number of species that occur in rock pools seems limited by interspecific competition, although three species manage to persist in the larger system of rock pools via a mechanism of competitive hide-and-seek. Bengtsson's experiments are unusual in that they go beyond short-term responses of species, such as differences in growth or survival rates, to document increased extinction rates and actual local losses of species from communities caused by competition.

TABLE 3-3. **(A)** Frequencies of extinctions of *Daphnia* populations in artificial rock pools experimentally stocked with one, two, or three species. **(B)** Frequencies of extinctions in natural rock pools containing one or two species.

A. Artificial Pools

| Pool Volume | Extinctions/Population over 4 Years | | | | | | |
	Daphnia magna (M)	*Daphnia pulex* (P)	*Daphnia longispina* (L)	M + P	M + L	P + L	M + P + L
4 L	0/3	0/2	0/1	M 0/4 P 2/4	M 0/5 L 2/5	P 0/5 L 2/5	M 0/6 P 3/6 L 2/6
12 L	0/1	0/1	0/1	0/2	0/3	P 0/4 L 2/4	M 2/3 P 0/3 L 1/3
50 L	0/1	0/2	0/1	0/1	0/4	0/3	M 0/3 P 3/3 L 1/3
300 L	—	—	—		0/1	—	M 0/3 P 1/3 L 0/3

B. Natural Pools

| Area | Extinction Rate (s.d.)/Population/Year; n = No. of Pools | |
	One-Species Pools	Two-Species Pools
Flatholmen	0.13 (0.037); $n = 82$	0.15 (0.046); $n = 58$
Monster	0.12 (0.038); $n = 74$	0.42 (0.14); $n = 12$
Angskar	0.097 (0.025); $n = 143$	0.17 (0.051); $n = 54$
Tvarminne	0.11 (0.028); $n = 123$	0.16 (0.052); $n = 50$

Reprinted with permission from *Nature* 340: 713–715, J. Bengtsson. © 1989 Macmillan Magazines Limited.

Plants. There are few field studies of interspecific competition among aquatic macrophytes; however, there are important laboratory studies of competition among algae (Tilman 1977) and macrophytes (Clatworthy and Harper 1962). David Tilman's (1977) studies of competition between *Asterionella* and *Cyclotella* are entirely consistent with the Monod/Tilman model of interspecific competition. When grown on a single limiting resource, the species with the lowest R^* wins. Clatworthy and Harper (1962) explored competition among three species of duckweeds in the genus *Lemna* and an aquatic fern,

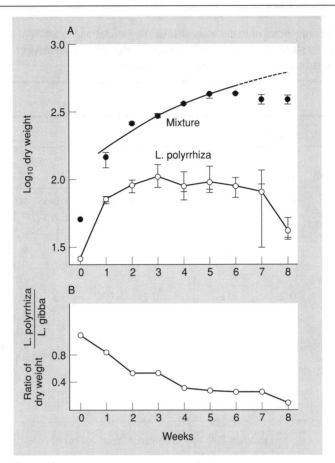

FIGURE 3.11. *Interspecific competition between the aquatic plants* Lemna polyrrhiza *and* Lemna gibba *in laboratory cultures. The combined dry weight of both species levels off, while the fraction of the total weight contributed by* L. polyrrhiza *declines as it is competitively excluded. (Adapted from Clatworthy and Harper, 1962,* J. Experimental Botany *13: 307–324, by permission of the Oxford University Press.)*

Salvinia natans. All four plants have similar growth habits, forming mats of leaves on the water surface. Attempts to predict the outcome of interspecific competition from rates of increase and carrying capacities observed in single-species cultures were not particularly successful. In some cases, such as competition between *Lemna minor* and *Lemna polyrrhiza*, the competitors seemed equally matched and coexisted for long periods. In other situations, such as competition between *Lemna gibba* and *L. polyrrhiza*, *L. gibba* quickly dominated the two-species cultures (Figure 3.11).

Protists. The pioneering laboratory studies of Gause (1934) on competition among ciliates stand out as one of the first attempts to link population dynamics observed in simple communities of competitors with the theoretical framework for competition developed by Lotka (1925) and Volterra (1926). Gause studied populations of *Paramecium aurelia* and *Paramecium caudatum* growing in simple culture tube environments. Both species displayed essentially logistic growth curves when grown in single-species cultures on a diet of the bacterium *Bacillus subtilis*. When placed together, *P. aurelia* competitively excluded *P. caudatum*. Gause was able to estimate the basic parameters (r and K) of the single-species logistic equations for each species, which fit the observed data rather well. He then went on to estimate competition coefficients from the relative sizes (volumes) of the two protist species. Gause assumed that if a given level of food produced a particular volume, or biomass, of each species, then the ratios of the volumes produced under similar conditions of resource availability would provide estimates of the competition coefficients. The estimates obtained (*P. aurelia* = 0.61 of *P. caudatum*, and *P. caudatum* = 1.64 of *P. aurelia*) can be used to simulate competition between the two species. Unfortunately, these estimated competition coefficients predict stable coexistence, whereas exclusion rapidly occurs in these laboratory communities. The competitive abilities of these two species clearly had a more complex basis than the observed difference in their abilities to convert bacteria into consumer biomass.

> Protists provided some of the first experimental evidence for competition.

John Vandermeer (1969) used a similar approach to study whether the results of competition between pairs of ciliate species could be used to predict the outcome of multispecies competition. His estimates of rates of increase, carrying capacities, and competition coefficients provided a remarkably good fit to the observed population dynamics (Figure 3.12). Although simplistic, the Lotka-Volterra model of competition seemed to provide a reasonable description of competition between pairs of protist species in simple laboratory settings. Interactions inferred from pairs of species also predicted the outcome of competition when a set of four species competed together. Vandermeer's study is one of the very few examples to show that the multispecies extension of the Lotka-Volterra equations provides an adequate description of competitive interactions (Figure 3.13). This good fit between theory and data may reflect the relative simplicity of the organisms studied and the simple, artificial laboratory setting where competition occurred. However, protists fit many of the basic assumptions implicit in the Lotka-Volterra equations for competition. Reproduction is continuous, and the small size of the organisms leads to minimal time lags in response to changing environmental conditions. The populations are also relatively unstructured with respect to either size or age classes.

Competition among protists in natural settings is more difficult to study. Gill and Hairston (1972) attempted to determine whether *Paramecium*

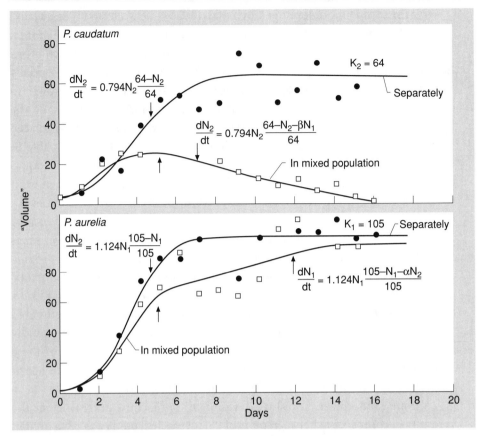

FIGURE 3.12. *The dynamics of competition among the ciliated protists* Paramecium aurelia *and* Paramecium caudatum. *Under these conditions,* P. aurelia *wins after about 16 days. In the absence of interspecific competitors, each species grows logistically to a different carrying capacity. Abundances are expressed in terms of volume, rather than numbers of individuals, because the species differ in size. (Adapted from Gause, 1934, with permission of Dover Publications Inc.)*

aurelia was competitively excluded from a seepage area where other *Paramecium* species occurred. They used plastic tubes pressed into the substrate as experimental enclosures for controlled introductions of *Paramecium* with and without competitors. The failure of *P. aurelia* to persist in all situations where it was introduced led Gill and Hairston to conclude that its absence was probably due to physiological factors. This conclusion is tempered by the fact that it proved difficult to exclude other *Paramecium* species in treatments where they were not supposed to be introduced, so that most tubes initially set up as competitor-free control areas in fact contained large numbers of competitors by the end of the experiment. These complications underscore the difficulty of manipulating small organisms like protists in field experiments.

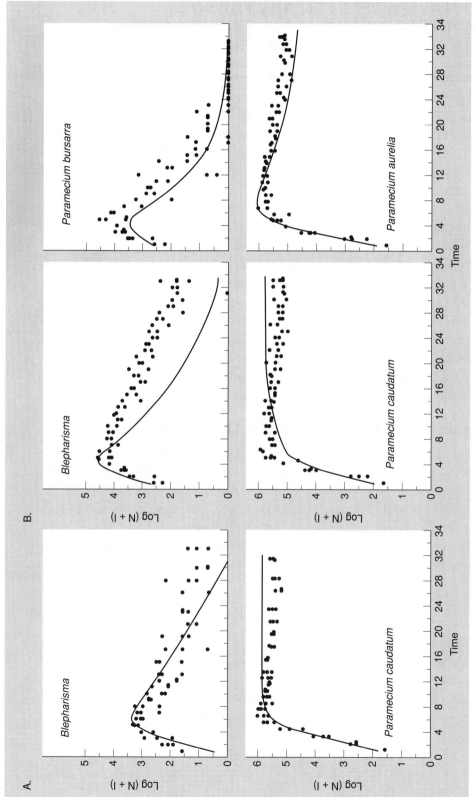

FIGURE 3.13. *Examples of competition between two and four species of protists. (A) Competition between Paramecium caudatum and Blepharisma species. Solid lines show predictions of the Lotka-Volterra model; dots indicate observed abundances in laboratory cultures. (B) Competition between four protist species. Values of the parameters of the competition equations are shown in Table 3-5. (Adapted from Vandermeer, 1969, with permission of the Ecological Society of America.)*

AN OVERVIEW OF PATTERNS FOUND IN SURVEYS OF PUBLISHED EXPERIMENTS ON INTERSPECIFIC COMPETITION

Surveys of published field experiments on interspecific competition show that many taxa compete in a variety of communities (Schoener 1983; Connell 1983; Gurevitch et al. 1992; Goldberg and Barton 1992). The primary goal of the earlier surveys was to assess the frequency of interspecific competition in nature. The general importance of competition became a subject of considerable controversy, partly because the observational studies cited in support of competition's importance were open to alternate interpretations. Later surveys addressed specific issues about the frequency of competition in various trophic levels, taxa, or kinds of communities. A few important generalizations emerge from these surveys, but many important problems remain unresolved because most experiments go no farther than establishing whether competition happens. For example, even in a particularly well-studied group like terrestrial plants (Goldberg and Barton 1992), few studies include the treatments needed to assess whether intraspecific competition is stronger than interspecific competition. Studies that allow comparisons of the intensity of competition over space or time are also infrequent.

 Reviews of published studies confirm that competition is common.

Frequency of Occurrence of Interspecific Competition

The early surveys by Schoener (1983) and Connell (1983) showed that competition occurs frequently. For example, fully 90% (148) of the 164 studies considered by Schoener and 83% (45) of the 54 studies surveyed by Connell demonstrated that competition occurred among some of the species in each study. This figure probably overestimates the frequency of competition at the level of individual species (Connell 1983), which appears to be approximately 43% of the species surveyed.

Asymmetric Competition and Predicting Competitive Ability

The surveys by Connell and Schoener also considered other competitive phenomena, such as the frequency of asymmetric interactions. Competitive interactions are often strongly asymmetric, with interactions between pairs of species typically being very lopsided in intensity. Schoener found that competition was asymmetric in 51 (84%) of 61 studies where the comparison could be made. Connell found that 61% (33) of the 54 competing species pairs displayed some sort of asymmetric competition.

Reasons for asymmetric competition are unclear. One cause of asymmetric competition may be differences in the characteristic sizes of competitors, with larger competitors having greater per capita impacts than smaller ones in competition among plants (Gaudet and Keddy 1988) and among animals (Morin and Johnson 1988). Alternatively, activity levels, which are probably correlated with rates of foraging and resource depletion, may also be a good predictor of competitive abilities in animals (Werner and Anholt 1993). A related notion has been applied to competition among algae or terrestrial plants. Tilman (1982) suggests that a superior competitor is the species that continues to grow at the lowest resource supply rate, in other words, the species that is most efficient at extracting resources at low levels of resource availability. This concept has parallels in the ideas of Lampert and Schober (1980) and Gliwicz (1990), who suggest that competitively superior zooplankton are those species that can continue to grow at the lowest concentration of food. Gliwicz's (1990) data for several species of *Daphnia* support one of the main assumptions of the Brooks and Dodson (1965) size-efficiency hypothesis. Larger species are competitively superior, primarily because they can continue to grow at lower food concentrations than those required for the growth of smaller species. Although all these ideas seem quite reasonable for the case of consumptive competition, similar generalizations for other kinds of competition seem less likely. Asymmetric competition might also occur when organisms compete via very different mechanisms. For instance, one species might chemically inhibit a second species, whereas the second species competes consumptively with the first species.

The identification of a consistent predictor of the competitive impact of one species on another remains an elusive goal for community ecologists. It is apparent that measures of resource overlap (e.g., Levins 1968) can yield ambiguous predictions about competitive interactions. High resource overlap between coexisting species can be taken to mean that species compete weakly, as in the high altitudinal overlap between salamander species in Hairston's study (1980a). In contrast, the high overlap inferred from similarities in body size and perch height between *Anolis* lizards in Pacala and Roughgarden's (1982) study was interpreted in an opposite fashion to suggest a high intensity of ongoing competition. If overlap can be interpreted in such nearly complementary ways, it seems poorly suited as a testable predictor of competitive interactions. Other approaches, such as the R^* criterion associated with Monod/Tilman models of consumptive competition for resources, appear to accurately predict the outcome of competition for single resources in those systems. The applicability of this approach to different kinds of heterotrophic species has been little explored.

Other approaches use an empirical and statistical protocol to determine what life history characteristics are correlated with superior competitive ability. Gaudet and Keddy (1988) found that the best predictor of the competitive impact of an assortment of plant species on a target species was the

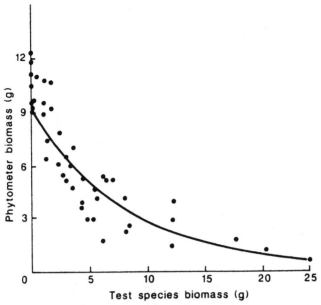

FIGURE 3.14. *Response of a target species (the phytometer,* Lythrum salicaria*) to different species of competitors (test species) that differ in biomass when grown in single-species conditions. Larger competitors have a greater competitive effect, as shown by greater reductions in phytometer biomass (Gaudet and Keddy 1988). (Reprinted with permission from* Nature *334: 242–243, C. L. Gaudet and P. A. Keddy. © 1988 Macmillan Magazines Limited.)*

biomass that the competitors attained when grown alone under comparable conditions (Figure 3.14). This makes sense, since size attained will be a measure of the efficiency of resource utilization under comparable conditions. Gilpin et al. (1986) have used a comparable approach to assess the life history correlates of competitive ability in simple laboratory communities of fruit flies in the genus *Drosophila.* They found that species with higher values of production as larvae and adults were the best competitors, a result very similar to the one obtained by Gaudet and Keddy.

Interphyletic Competition

Strong competitive interactions can occur between taxonomically dissimilar species. Historically, ecologists focused on competition between taxonomically similar species, often coexisting sets of species in the same genus or family. This focus reflected the once prevalent notion that morphological similarity was perhaps related to similarity in resource use and that the species most similar in taxonomy, morphology, and resource use were the species most likely to compete. Experiments have shown that taxonomically

TABLE 3-4.   Responses of granivorous ants and rodents to experimental removals of each other in circular plots 36 m in diameter.

Taxon	Total Rodents Removed	Total Ants Removed	Control	Increase Relative to Control (%)	Removal > Control
Ant colonies	543	—	318	70.8	9/10
Rodents					
Number	—	144	122	18.0	15/25
Biomass (kg)	—	5.12	4.13	24.0	16/25

Note: Removal > controls column indicates numbers of censuses in replicated plots where abundances in removals exceeded abundances in controls. Ants increase in response to rodent removal, and rodents increase in response to ant removal (Brown and Davidson 1977).
Reprinted with permission from Brown and Davidson (1977). © 1977 American Association for the Advancement of Science.

TABLE 3-5.   Values of the parameters of the Lotka-Volterra competition equations estimated for four species of protists studied by Vandermeer (1969).

Species	r	K	α for P. aurelia	α for P. caudatum	α for P. bursaria	α for Blepharisma
P. aurelia	1.05	671	1	1.75	−2.00	−0.65
P. caudatum	1.07	366	0.30	1	0.50	0.60
P. bursaria	0.47	230	0.50	0.85	1	0.50
Blepharisma	0.91	194	0.25	0.60	−0.50	1

Note: The estimates of the competition coefficients, α, describe the effects of species in a given column of the table on the species in a given row.
Source: Data from Vandermeer (1969).

disparate species, such as granivorous rodents and ants (Table 3-4; Brown and Davidson 1977), can compete strongly as long as they exploit a shared resource, in this case seeds. Other studies have shown that anuran tadpoles and aquatic insects, which both feed on periphyton, also can compete (Morin et al. 1988). Such findings seemed novel to terrestrial and freshwater ecologists schooled to expect competition only among taxonomically similar species. In contrast, marine ecologists were accustomed to seeing competition for attachment sites among species in different phyla or kingdoms, and they were unimpressed by the rediscovery of this phenomenon in nonmarine systems.

Nonadditive Competition

Another problem concerns the ability to extrapolate from competition between pairs of species to sets of three or more species. The problem arises from the way in which multispecies competition has been modeled (see Equation 2.4). The multispecies extension of the Lotka-Volterra competition equations assumes that per capita competitive abilities, the α_{ij}'s or competition coefficients, depend only on the pair of species in question and do not vary with the composition of other species in the system. There have been few tests of this assumption. Vandermeer (1969) tested this notion in relatively simple laboratory systems of one to four species of bacterivorous protists (Table 3-5). In these simple systems, pairwise competitive effects are indeed adequate predictors of competition in slightly more complex communities. In other words, aggregate competitive effects are additive and can be estimated by simply summing up the pairwise competitive effects obtained from the product of per capita competitive effects and competitor densities. In other systems, per capita effects of interspecific competitors depend on the identity and density of other species in the system (Neill 1974; Case and Bender 1981; Morin et al. 1988). For instance, Neill (1974) found that his estimates of the α_{ij}'s in a system of up to four competing species of microcrustaceans depended on the mix of species present. This means that the α_{ij}'s were not simply a property of a particular species pair but were also influenced by the ecological context. Here, the ecological context included differences in the number of other species in the system.

If such higher-order interactions commonly appear in a variety of systems, then competitive interactions in complex systems will not be readily predictable from competitive interactions observed in simple systems. This situation means that each particular assemblage of competitors must be studied as a special case to understand the particular competitive interactions that produce a particular community pattern. Reasons for such nonadditive competitive effects, sometimes termed **higher-order interactions** (Vandermeer 1969; Wilbur 1972), are numerous and may include changes in the size, activity, behavior, or other properties of organisms that affect their per capita effects on others (Strauss 1991). Such changes caused by the presence of additional species comprise a kind of indirect effect that is discussed further in Chapter 8.

The Prevalence of Competition on Different Trophic Levels

Different theories make different predictions about the importance of competition among species found on different trophic levels. Hairston, Smith, and Slobodkin (1960), hereafter HSS, predicted that in terrestrial systems, competition should occur among primary producers, top predators, and decomposers more frequently than among herbivores. Menge and Sutherland (1976) predicted a rather different pattern in marine communities, with the importance of competition increasing with trophic level. The reviews by Schoener (1983) and Gurevitch et al. (1992) specifically address this issue, although they rely on different criteria. Schoener (1983) found that relative frequency of competition, either tallied by study or by species, conformed to the expectations of HSS. However, the differences observed among trophic levels were not statistically significant.

Gurevitch et al. (1992) used a different approach, meta-analysis, to assess the relative importance of competitive effects in different trophic levels. Meta-analysis provides a way to combine the results from a large number of independent studies that goes beyond simply tallying the numbers of studies that did or did not find evidence of competition. Meta-analysis effectively pools the results of many studies into a larger and potentially more revealing analysis. As a result, trends that might be dismissed as being of marginal statistical significance in isolated studies may emerge as important patterns if they are consistent across a large number of studies. Gurevitch et al. found the strongest, but not most frequent, effects of competition to occur among herbivores! This result probably has much to do with the fact that many studies included in their analysis focused on aquatic herbivores in artificial communities (hybrid experiments) where predators were excluded.

Variation in Competition over Space and Time

Connell (1983) and Schoener (1983) came to similar conclusions about the frequency of temporal variation in competition among species, but differed in the importance that they ascribed to their findings. Connell found substantial evidence for temporal variation in competition, with 59% of the species that competed showing annual variation in the intensity of competition. A somewhat smaller number of the competing species, 31%, also showed spatial variation in competition. Schoener (1983) found annual variation in competition in 11 (48%) of 23 cases where it might be discerned, a percentage very close to that noted by Connell. Schoener noted that even when temporal variation in competition occurred, that variation took the form of temporal variation in severity rather than temporal variation in the presence or absence of competition.

Relative Intensities of Intraspecific and Interspecific Competition

The relative intensity of intraspecific and interspecific competition is directly relevant to theoretical predictions about conditions that promote coexistence or lead to competitive exclusion. Recall that simple two-species Lotka-Volterra models of competition predict coexistence when intraspecific competition is stronger than interspecific competition, and otherwise predict competitive exclusion. This prediction also follows from ideas about competition and resource utilization, since intraspecific competitors should be more similar than interspecific competitors in resource use. Connell specifically focused on the issue of the relative strengths of intraspecific and interspecific competition in those experiments specifically designed to test for both kinds of interactions. He found that for 14 studies involving 42 species and 123 subexperiments, interspecific competition was stronger than intraspecific competition for 31% (13) of the species and 17% (21) of the subexperiments. Situations where interspecific competition is stronger than intraspecific competition are the cases where exclusion is likely to occur.

NULL MODELS AND STATISTICAL AND OBSERVATIONAL APPROACHES TO THE STUDY OF INTERSPECIFIC COMPETITION

By their very nature, some kinds of organisms, such as birds and large mammals, are too mobile to be successfully manipulated in field experiments. Other organisms, such as trees, are so

 Null models provide a statistical way to test for competition.

long-lived that responses to experimental manipulations of competitors might not be seen within the lifetime of the average ecologist. Nonetheless, these experimentally intractable organisms are conspicuous and intriguing, and their biology provided much of the impetus for the development of ecological theory (MacArthur 1958). How might the role of competition in organizing their patterns of distribution and abundance be best explored?

The traditional approach to studies of competition among experimentally difficult species is to resort to natural experiments (Diamond 1986), which rely on comparisons of the ecology of species in two or more locales that differ in the presence of potential competitors. The key idea is that observed differences in abundance, body size, morphology, or resource use can be attributed to the presence or absence of a competitor. The problem is that such natural experiments lack natural controls, which would ensure that the only important difference between the compared sites is in the abundance of the interspecific competitor. In practice, important differences in other factors, including predators, parasites, resource levels, and the history of community assembly, may be difficult to rule out, making the interpretation of such comparisons problematic at best.

One way to make natural experiments more rigorous is to compare the observed patterns with patterns that might be expected to occur purely by chance in the absence of any biological interactions among species. These expectations are derived from a null or neutral statistical model, which assumes no interactions among species. Null model approaches may be useful in situations where experimental approaches are impossible, but their interpretation often depends critically on the assumptions used to predict patterns that might result from strong interspecific competition, as opposed to the usual null hypothesis of randomness. Randomness can have many meanings, and it has proven difficult for ecologists to agree on the best way to formulate such expectations of random community patterns that might arise purely by chance among noninteracting species. The following examples highlight how the approach has been applied to various problems in the community ecology of potentially competing species.

The first example concerns patterns of species packing and morphological character displacement among assemblages of passerine birds. Ricklefs and Travis (1980) analyzed patterns of morphological dissimilarity among sets of coexisting bird "species." Their approach involved positioning each species within a kind of morphological niche space defined by several morphological measures made on the bird species. The underlying assumption was that morphological differences should correspond to differences in resource use among species.

 Morphological differences among coexisting birds are no greater than random.

Limits to morphological similarity, and similarity in resource use, imposed by interspecific competition within these communities would be expected to be reflected by limits in the proximity of each species and its nearest, and therefore most morphologically similar, neighbor. For each community, an average nearest neighbor distance was computed. This distance was then compared with the nearest neighbor distances calculated within two sets of artificial, null model communities, whose morphological properties should have nothing to do with competition.

The first null model community was constructed by randomly placing points corresponding to artificial birds in the morphological space and then calculating the nearest neighbor distance between those points. That nearest neighbor distance should reflect the morphological difference expected among species purely by chance. The second approach involved selecting real birds from a larger regional pool that were not known to coexist, in other words, a hypothetical community assembled from real species, and then calculating nearest neighbor distances. This approach assumed that the morphological differences observed within a random collection of species would differ from the patterns found in a community where competition limits the permissible morphological (and ecological) similarity between species.

When real communities containing different numbers of species were compared with artificial ones, there was no obvious difference between the

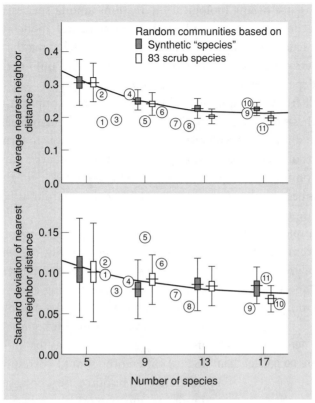

FIGURE 3.15. *Comparisons of the average distance among real bird species in a morphological niche space and the average distance among randomly placed points in the same
space, for communities containing different numbers of real and synthetic species.
(Adapted from Ricklefs and Travis, 1980, with permission of The Auk.)*

average minimum morphological distance between coexisting species in real
communities and "species" in randomly assembled communities (Figure
3.15). The conclusion is that bird species were no more different in morphology than expected by chance. Thus, Ricklefs and Travis concluded that
although coexisting bird species do differ in morphology, those differences do
not provide much evidence for the importance of competition in structuring
these communities.

Strong et al. (1979) applied a similar approach to a classic problem, the
apparent pattern of character displacement in the sizes of bills among coexisting species of Galapagos finches in the genus *Geospiza*. Like Ricklefs and
Travis, Strong et al. found that the differences in morphological features of
sets of coexisting Galapagos finches were no greater than might be expected
by simply assembling sets of birds drawn randomly from within the Galapagos archipelago without regard to their propensity to coexist (Figure 3.16).

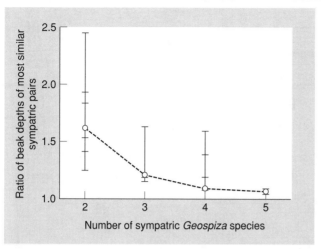

FIGURE 3.16. *Observed ratios of beak depth for pairs of the most similar coexisting Gala-pagos finches, and expectations of the ratio obtained from a null model by randomly pairing populations of finches regardless of their island of origin. (Adapted from Strong et al., 1979, with permission of The Society for the Study of Evolution.)*

Dayan et al. (1990) used a different kind of analysis to assess morpholog-ical displacement within an assemblage of wild cats inhabiting the Middle East. Unlike the previous cases, they found **Differences among coexisting cats** clear evidence of striking, nonrandom dif-**are consistent with competition.** ferences among coexisting species in the size of the canine teeth, which are crucial to the way in which the cats dispatch their prey. The pattern observed suggests an exceptional regularity in the spacing of species along an axis defined by the size of the canine teeth (Figure 3.17). This regularity would be consistent with the hypothesis that the morphology—and ecology—of these potentially com-peting predators is distinctly nonrandom.

Schoener (1984) has also found evidence for exceptional (nonrandom) regularity in the morphological differences among hawks in the genus *Accip-iter*. In this case, and in the case above, the nonrandom morphological pattern is consistent with the hypothesis of a limiting similarity to the sizes of ecolog-ically similar coexisting species. It must be mentioned, however, that the role of interspecific competition in generating these patterns, while quite plausi-ble, is by no means proven.

Other uses of null model approaches to the study of community patterns address patterns of species co-occurrence in discrete units of habitat, such as the distributions of bird species among islands. If species compete strongly, the distributions of those species over a set of habitats might be mutually

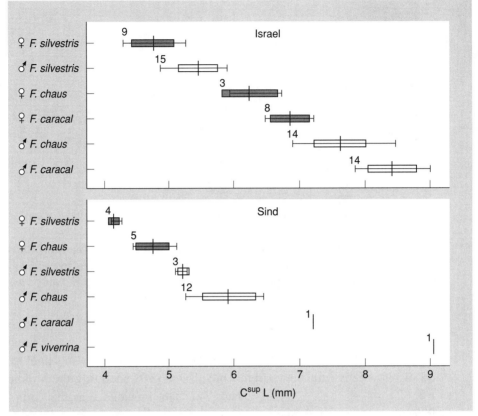

FIGURE 3.17. *Diameters of canine teeth for small species of coexisting cat species in Israel and the Sind. In Israel, canine diameters are significantly nonrandomly spaced (regularly spaced). (Adapted from Dayan et al., 1990, with permission of the University of Chicago Press.)*

exclusive. If only a single pair of species is considered, the analysis is straightforward, and either the species co-occur less frequently than expected or they do not. The problem is that communities contain many possible species pairs, and over a large number of distributional comparisons, some pairs might be expected to display exclusive checkerboard distributions simply by chance. Connor and Simberloff (1978, 1979) suggested this situation to be the case for a reanalysis of data previously used to argue for the existence of exclusive distributions caused by competition among island-dwelling birds (Diamond 1975). The problem with this analysis, and with all null model approaches, is that the assumptions used to produce the null expectation for patterns of co-occurrence are neither simple nor widely agreed upon by other workers (Gilpin and Diamond 1984).

CONCLUSIONS

Experiments provide an essential way to demonstrate that species actually compete in nature or in the laboratory. Different experimental designs can show whether species simple compete, or whether the strength of competition within species differs from that observed between species. A collection of case studies shows that species in marine, aquatic, and terrestrial environments compete via an assortment of mechanisms, sometimes producing striking community patterns that are often manifested as differences in spatial distributions of species. Surveys of published experiments confirm the commonness of interspecific competition, and suggest that the frequency of interspecific competition may vary in interesting ways among different trophic levels. When experiments are impossible, null models provide a statistical alternative against which patterns assumed to be generated by competition can be compared. The unbiased formulation of a null model for a particular competitive pattern is by no means straightforward. Some null model studies fail to detect nonrandom patterns that would support interspecific competition, while others do.

CHAPTER 4

Predation and Communities: Empirical Patterns

OVERVIEW

This chapter describes situations in which predators regulate prey populations and alter the species composition of communities. The main emphasis is on important patterns and processes documented by experimental manipulations of predators in natural or laboratory settings. Some of the mechanisms that generate important community patterns are explored further in the next chapter, which discusses mathematical models of predator-prey interactions. Predators affect community composition in diverse ways. Some predators feed selectively on competitively superior species that would otherwise exclude weaker competitors. One result of selective predation on competitively superior prey is enhancement of the number of prey species that manage to coexist, since predators reduce the interspecific competition among surviving prey. When prey do not compete strongly, or when predators do not selectively attack superior competitors, predation can simply reduce the number of coexisting species. Predators can also drastically affect species composition without changing species richness, by creating communities dominated by species that have particularly effective antipredator strategies. Species with effective antipredator strategies are often poor competitors and are often the first species to be competitively excluded when predators are absent. This complementarity of competitive ability and resistance to predation suggests the existence of a life history trade-off between the abilities of species to compete or resist predation.

PREDATION

Predation is the consumption of all or part of one living organism by another. Predation is operationally defined by a +/− interaction between an individual predator and prey, where the predator benefits from the interaction (+), while the consumed prey does not (−). This +/− interaction may not apply to surviving prey, since survivors can benefit from reductions in the density of conspecific prey, especially if prey are sufficiently abundant to compete for some resource. Predator-prey interactions involve species that reside on many different trophic levels and include the impacts of herbivores on plants, carnivores on herbivores, carnivores on other carnivores, parasites and parasitoids on hosts, and diseases on victims. Many of the examples in this chapter stress that predation can dramatically affect the distribution and abundance of prey species. What is less studied, and less appreciated, is that predation can also profoundly influence patterns of energy flow and nutrient cycling within ecosystems, since it controls the flow of energy and materials from lower to higher levels in food webs.

Published natural history accounts are replete with examples of spectacular antipredator adaptations. Those adaptations provide indirect but compelling evidence for predation's role as a potent agent of natural selection, population regulation, and community organization. Examples of adaptive syndromes attributable to frequent and intense predation include mimicry complexes, crypsis, and aposematic coloration (see Wickler 1968 for a review), chemical defenses (Eisner 1970; Rosenthal and Berenbaum 1992), antipredator behaviors (Harvey and Greenwood 1978), and an assortment of mechanical defenses in both animals (Vermeij 1987) and plants (Crawley 1983). These adaptations are often striking in their sophistication, elegance, and effectiveness. Differences among species that are thought to reflect evolutionary responses to competitors, such as differences in size, morphology, resource use, or habitat utilization (see Chapter 2), seem extremely modest when compared with the evolutionary responses of prey to predators. This kind of indirect evolutionary evidence is just a small part of the larger case for the importance of predation in population regulation and community structure. That case is supported by the many different lines of evidence outlined in this chapter.

EXAMPLES FROM BIOLOGICAL CONTROL

The successful biological control of introduced pests by deliberate introductions of predators provides some particularly striking observational evidence for predation's importance in regulating populations and structuring communities. The four examples briefly outlined below share a common theme. Deliberate or accidental introductions of species into areas beyond the range of their natural enemies can result in explosive episodes of population growth that create ecological and economic problems. Control of such introduced pests is sometimes obtained by the controlled introduction of a specialized predator. Some of the best examples of biological control involve the control of introduced plants by invertebrate herbivores and the control of an introduced herbivore by a viral pathogen.

Two species of prickly pear cactus, *Opuntia inermis* and *Opuntia stricta*, became important pests after their introduction into Australia, where they apparently had no effective natural enemies (Dodd 1959). *Opuntia* populations grew rapidly and transformed many areas of rangeland into impenetrable thickets. Cactus thickets offered very poor grazing for livestock, which provided an economic incentive to control cactus abundance. In an effort to control the cactus outbreak, the herbivorous moth *Cactoblastis cactorum* was introduced into Australia from its native range in Argentina. Larval moths feed only on cactus, and in the process of feeding, burrow through the cactus and make it susceptible to infection by other pathogens. The introduction of *Cactoblastis* caused a spectacular decline in the abundance of cactus. *Opuntia* is now greatly reduced in abundance through many parts of Australia where it was previously a nuisance, and it now coexists with low densities of its natural enemy in a sort of patchy game of hide-and-seek.

Some evidence for the importance of predation comes from examples of biological control.

A second example of the successful biological control of an introduced plant by an herbivore involves St. John's-wort, *Hypericum perforatum*, in California (Huffaker and Kennett 1959). *Hypericum* is another serious rangeland pest, although for somewhat different reasons than *Opuntia*. *Hypericum* contains phototoxic chemicals, which when eaten by cattle cause the cattle to become highly sensitive to sunlight. This sensitivity results in the development of skin lesions that make the cattle unmarketable. To control *Hypericum*, the specialized herbivore *Chrysolina*, a chrysomelid beetle, was introduced into regions infested by *Hypericum*. Larvae of the beetle burrow into the roots and stems of the plant, killing it. The beetle was a spectacularly effective control agent and eliminated the plant from most habitats except for shady sites, where the beetle does poorly and the plant manages to hang on. A visitor to California who was unaware of the history of this ongoing interaction could easily, but wrongly, conclude that *Hypericum* was a specialized shade-loving species, since it seems to now be found mostly in shady sites where *Chrysolina* is an ineffective predator. *Hypericum* is, of course, restricted to those sites by a relatively inconspicuous predator that is now at low abundance throughout the plant's range because of its success in keeping its specialized food plant at low abundance.

A third example of successful biological control involves the small aquatic fern *Salvinia molesta*, which is native to tropical waters in Brazil. It has been widely introduced throughout the tropics, where it can become an important aquatic weed (Room et al. 1981). Lush mats of *Salvinia* can become so dense that they impede boat traffic and prevent subsistence fishing (Mitchell et al. 1980). In other locations, *Salvinia* becomes so abundant that it alters ecosystem processes (Thomas 1981). In some parts of its introduced range, *Salvinia* has been effectively controlled by the introduction of a small weevil, *Cyrtobagus singularis*, that selectively consumes *Salvinia*. The control effort initially showed little promise because the first weevils used in attempts at control were collected from a closely related species, *Salvinia auriculata*. Although *S. auriculata* and *S. molesta* are so similar that they were once thought to be the same species, the natural enemies of *S. auriculata* had little effect on *S. molesta*. After the subtle differences between the two *Salvinia* species were recognized, the weevil *Cyrtobagus singularis* was collected from *S. molesta*, and it proved to be a much more effective control agent. The adult weevil feeds on *Salvinia* buds, and its larvae feed on the plant's roots and rhizome. This example of successful biological control shows just how species-specific some predator-prey interactions can be.

A fourth and final example of biological control underscores that successful examples of biological control are not limited to introduced plants and their herbivores. Introduced animals can also be controlled by their natural enemies. European rabbits, *Oryctolagus cuniculus*, were probably introduced into Australia in 1859 as game animals. In the absence of natural enemies, they proliferated, literally, like rabbits. At the height of their population explosion, they expanded their range into previously rabbit-free territory at the rate of 70 miles per year (Ratcliffe et al. 1952; Ratcliffe 1959). Grazing by the rabbits seriously degraded rangeland, where the rabbits competed with introduced livestock, mostly sheep, that also grazed on plants. Introduction of large mammalian carnivores to control the rabbits would have wreaked havoc with the unique Australian marsupial fauna, and was therefore to be avoided. However, *Oryctolagus* are susceptible to attack by the specialized *Myxoma* virus, which occurs naturally in South American populations of another rabbit species, *Sylvilagus brasiliensis*. *Myxoma* causes a mild, nonlethal infection in *Sylvilagus brasiliensis*, but is highly lethal in populations of *Oryctolagus* that have no recent evolutionary history of exposure to the virus. Following the introduction of the virus, *Oryctolagus* populations crashed. The virus was spread by various blood-feeding insects, including mosquitoes and fleas. The control effort fell short of completely eradicating the rabbits, in part because the virus rapidly evolved to become less virulent but more readily transmitted, and in part because the rabbits became somewhat resistant to the *Myxoma* virus (Fenner 1983).

Many other similar attempts to control introduced insect pests by means of specialized parasites or parasitoids have been somewhat less successful, for reasons that remain unclear. The important lesson learned from successful cases of biological control is that tight regulatory predator-prey interactions can be highly species-specific and can limit prey populations well below the levels of abundance attained in the absence of natural enemies.

IMPACTS OF PREDATORS ON DIFFERENT KINDS OF COMMUNITIES

In his review of ecological experiments conducted in a variety of communities, Nelson Hairston (1989) concluded that most of the generalizations that could be made about the processes that structure communities remained valid only within the confines of certain broad categories of communities, such as terrestrial forests, deserts, and successional, freshwater, and marine habitats. Important differences in habitat structure and physical processes among these types of habitats apparently alter the importance of different processes. In rough accord with Hairston's analysis, the following examples of the impact of predation on community structure are separated into studies conducted in three broad types of habitats: terrestrial, freshwater, and marine. Despite Hairston's cautious reluctance to make generalizations that spanned these major habitat types, there do appear to be some general features of predation that transcend the idiosyncrasies of terrestrial, freshwater, and marine systems. Those general patterns are summarized after the descriptions of selected case studies in each habitat.

Marine Communities

Robert Paine (1966, 1969a,b, 1971, 1974) conducted several very influential experimental studies of the impact of predators on community structure. His key finding was that predators sometimes enhanced the number of species that managed to coexist in a limited area. Predators enhanced species richness by virtue of their ability to prevent competitive exclusion among prey. The important players in his community included about 16 common species of

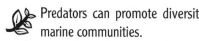 Predators can promote diversity in marine communities.

algae and animals living in close association on the exposed rocky shores of the northern Pacific Coast of the United States. Adequate space for firm attachment against the buffeting waves is an essential resource both for settlement and for subsequent growth of these intertidal organisms. Many of these species are consumed by a large predatory starfish, *Pisaster*, as shown in the food web outlined in Figure 4.1. Several other species, including a large anemone (*Anthopleura*), a sponge (*Haliclona*), a nudibranch (*Anisodoris*), and four common macroalgae, are not consumed by *Pisaster* but still contend for space. At first glance, the coexistence of so many potential competitors for space seems to be at odds with the **competitive exclusion principle** (see Chapter 2), since any competitively superior species should rapidly exclude the remaining weaker competitors. Paine resolved the paradox by showing that intense predation by *Pisaster* effectively prevented competitive exclusion and maintained the number of coexisting species normally observed.

Paine's first experiment involved removing *Pisaster* from one site, leaving *Pisaster* unmanipulated in an adjacent site, and then following the changes that occurred in the abundance of species. After several years, species richness dropped from 15 to 8 species in the manipulated site, and the bivalve mollusk *Mytilus* occupied most of the space. In the area where *Pisaster* continued to forage, species richness remained unchanged. Paine suggested that predation by *Pisaster* normally prevented the competitively dominant species, *Mytilus*, from monopolizing available space. Paine later (1969a) termed *Pisaster* a **keystone species** because of its important role in maintaining the diversity

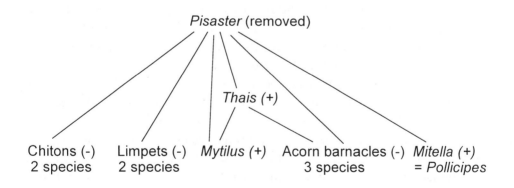

Algae: *Porphyra (o)*, *Endocladia (-)*, *Rhodomela (-)*, *Corallina (-)*

FIGURE 4.1. *Feeding relationships among major species in the rocky intertidal community dominated by* Pisaster ochraceous. *Increases in response to* Pisaster *removal for 3 years are indicated by (+), and decreases by (−). After 10 years, only the response of* Mytilus *remains positive. (Redrawn from Paine, 1966, with permission of the University of Chicago Press.)*

within the community. Subsequent experiments showed that the results of his previously unreplicated experiment were repeatable (Paine 1974). Other experiments conducted with different species in analogous communities on the rocky coast of New Zealand produced similar results (Paine 1971), where the predatory starfish *Stichaster* prevents the bivalve mollusk *Perna* from monopolizing space. These results bolster the generality of Paine's original finding, even though such interactions do not occur on all rocky shores (Underwood et al. 1983).

Jane Lubchenco's (1978) work on another rocky intertidal system on the East Coast of the United States did much to extend the generality of Paine's findings. Her important findings suggested that predation only maintains prey diversity when predators feed selectively on competitively superior prey. The prey assemblage studied by Lubchenco consisted of several species of macroalgae (seaweeds) found in tide pools. The main predator on the algae is the herbivorous snail *Littorina littorea*. Pools with high densities of snails are dominated by a long-lived red alga, *Chondrus crispus*. Pools with low densities of snails are dominated by an assortment of ephemeral algae, predominantly the green alga *Enteromorpha*. Pools with moderate densities of snails display a maximal diversity of algal species. Lubchenco showed that *Littorina* prefers to graze on *Enteromorpha* and exhibits low preference for *Chondrus*. A simple but telling experiment that involved removing *Littorina* from pools where *Chondrus* predominated and transplanting the removed snails to pools originally containing few snails and abundant *Enteromorpha* demonstrated the importance of predation in maintaining this pattern. The changes in algal abundance resulting from the addition or removal of the predator *Littorina* are shown in Figure 4.2. In the absence of grazing by snails, *Enteromorpha* became abundant and outcompeted other algae, including *Chondrus*. In the pool that received the transplanted snails, *Enteromorpha* rapidly declined, although *Chondrus* was slow to invade and take its place.

A survey of unmanipulated pools showed that algal species diversity was greatest in pools containing moderate densities of grazing snails (Figure 4.3). This situation is what would be expected if snails act as a keystone species and reduce the intensity of competition among algal species. A similar survey of the algal species on exposed rocky sites showed a different pattern, with algal diversity declining with increasing snail abundance. The difference between the pattern in sheltered pools and exposed rocky shelves apparently reflects a reversal in competitive ability among the algae. In exposed sites, *Chondrus* is competitively superior to *Enteromorpha*; however, the snails' feeding preferences remain unchanged, and *Enteromorpha* remains the preferred prey. The conclusion here is that for a predator to enhance the prey diversity, it must feed preferentially on the competitively dominant prey species.

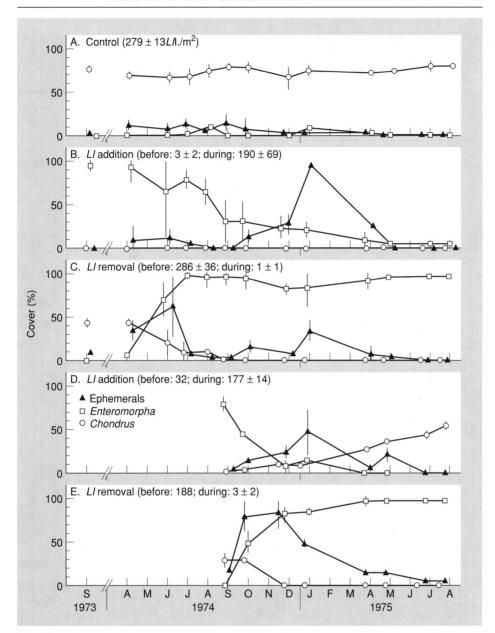

FIGURE 4.2. *Effects of the herbivorous snail* Littorina littorea *on the abundance of algae in tide pools. Control refers to unmanipulated pools (A). Removal pools (C and E) had* Littorina *removed. Removed snails were translocated to addition pools (B and D). Addition of snails results in a decrease in* Enteromorpha, *whereas snail removal has the opposite effect. (Adapted from Lubchenco, 1978, with permission of the University of Chicago Press.)*

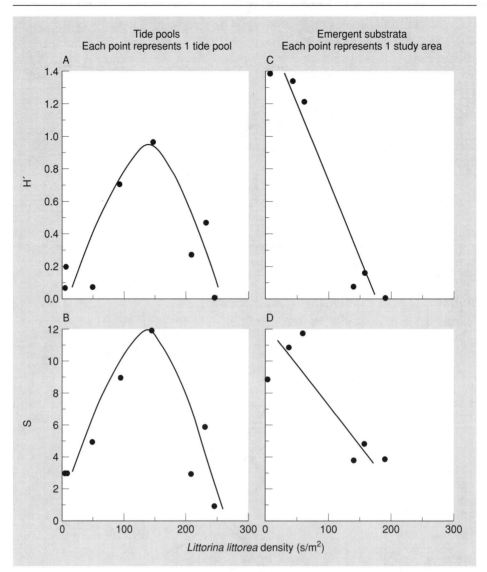

FIGURE 4.3. *Patterns of algal diversity and species richness as a function of predator density in tide pools and emergent substrata. Predators show similar preferences for prey in both sites, but the competitive ability of the prey differs among sites. In tide pools (A and B), the preferred species is competitively dominant. On emergent substrata (C and D), the preferred prey species is competitively inferior. (Adapted from Lubchenco, 1978, with permission of the University of Chicago Press.)*

Work by many others in a variety of rocky intertidal communities has done much to extend the generality of the relation between predation intensity and prey species richness in intertidal communities. The pattern is not universal. Situations exist in which predators have little influence on community structure (Underwood et al. 1983). Those situations seem to be characterized by such low densities of potentially competing prey that competition for space is weak or absent. Where competition among prey is unimportant, predators cannot enhance prey diversity, because the competitive exclusion of some prey species by others seldom occurs.

Terrestrial Communities

The effects of keystone predators observed in some marine communities have parallels in terrestrial systems, especially in systems of plants and their herbivores. Effects similar to those described by Paine and Lubchenco were well known to plant ecologists long before Paine conducted his first experiments. For example, Darwin (1859) observed that mowed lawns supported a greater diversity of plants than sites that were allowed to grow after a period of mowing. Although lawnmowers probably differ in important ways from natural herbivores, other studies show that the effects of natural grazers on plant communities appear to be equally important.

Tansley and Adamson (1925) studied the effects of grazing by rabbits (*Lepus cuniculus*) in floristically rich chalk grasslands in Britain. Naturally grazed sites usually contained 43 to 49 species of herbaceous plants, and little woody vegetation. Six years after rabbits were excluded from two patches of grassland in 1914, the grass *Bromus erectus* had increased to predominate within the plots, and although several species became less frequent, the total number of species present in the ungrazed plots actually increased slightly, to 59 to 66 species (Figure 4.4). After even longer periods of time, woody vegetation became established in the rabbit exclosures, and the plots without herbivores began a gradual successional change toward dominance by woody vegetation (Hope-Simpson 1940). Watt (1940, 1957) observed similar differences in response to rabbit removal from areas of grassland. The implication is that many of the herbaceous species naturally present in grasslands were maintained by rabbit grazing and would rapidly be competitively excluded by a few competitively superior species in the absence of grazing.

Some terrestrial herbivores also maintain plant diversity.

There is an interesting postscript to the story outlined above. Britain experienced widespread declines in rabbits after 1954 due to an epidemic of *Myxoma* virus. When rabbit populations crashed, the kinds of vegetation changes seen in Tansley and Adamson's rabbit exclosures happened on a broad scale throughout the British chalklands (Harper 1977). Indeed, Watt (1981) attributed the establishment of a new pine woodland to the absence of rabbit grazing following the myxomatosis outbreak.

The effects of grazers on plant species diversity do seem to depend on grazer density in a distinctly nonlinear way. At very high grazer densities, the impact of grazers on plant species richness leads to a species-poor community

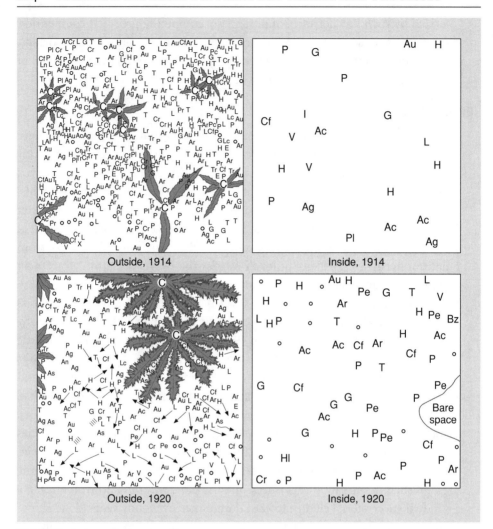

FIGURE 4.4. *Schematic representation of mapped vegetation in plots from which grazing rabbits were excluded (inside) or were present (outside). Different letters correspond to different plant species. (Adapted from Tansley and Adamson, 1925, with permission of Blackwell Science Ltd.)*

consisting mostly of a few grazer-resistant species (Crawley 1983). There are also suggestions that the positive impact of moderate grazer densities on plant species richness could represent a mutualistic association. In African grasslands, plants that are regularly grazed are more productive than ungrazed plants (McNaughton 1979). In turn, other grazers can benefit from this increased productivity by selectively feeding in previously grazed sites (McNaughton 1976). However, where grazers become extraordinarily abundant, they may have catastrophic effects on plant communities. Where African elephants have become crowded into nature reserves with few natural

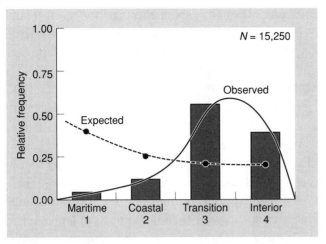

FIGURE 4.5. *Observed versus expected frequency of* Haplopappus squarrosus *across four climatic zones in California. Frequency estimates are based on roadside counts of plants (observed) and flowers (expected). (Adapted from Louda, 1982, with permission of the Ecological Society of America.)*

enemies, deforestation has resulted from their feeding on trees (Connell 1978). Similarly, high densities of domesticated grazers, such as cattle or sheep, can shift the composition of plant communities toward communities of low species richness dominated by unpalatable species.

The impact of herbivores on plant communities is not limited to the effects of large, conspicuous vertebrates. Invertebrates can also have important effects. Svata Louda (1982) showed that large-scale geographic patterns of plant distribution and abundance can also be driven by natural enemies. Louda studied the geographic distribution of *Haplopappus squarrosus*, a small shrub native to a variety of habitats in California. The curious feature about the plant's distribution is that it is least abundant in the places where it flowers most

 Herbivores can influence plant distribution patterns.

luxuriantly, which would seem to be the locations where the plant should reproduce best and become most abundant. Using the number of flowers per plant as an estimate of potential reproductive output, and using this potential reproductive output as an estimate of potential abundance, there is a marked discrepancy between potential and observed abundance (Figure 4.5). Plants are less common than expected in maritime and coastal regions, and more common than expected in transition and interior regions. The discrepancy can be explained by geographic variation in the impact of an assortment of herbivorous insects, including Tephritid flies, Lepidoptera, gall-forming Hymenoptera, and thrips, that attack flowers and seeds. If insects limit *Haplopappus* recruitment in maritime and coastal zones, insect removals in those areas should produce proportionally greater effects on the numbers of seeds, seedlings, and juveniles produced, which is precisely the pattern that Louda found. Figure 4.6 compares the numbers of viable seeds per plant, seedlings per plant, and juvenile plants in plots treated with insecticide (insect removals) and untreated controls. Insect removals increased seed production in all sites, but disproportionately increased seedling and juvenile plant production near the coast. Although this example focuses on just one plant species, the implications for communities are clear.

Studies by Valerie Brown (1985) and Walter Carson (1993) have shown that exclusion of herbivorous insects with insecticides causes large changes in the flowering frequency and species composition of old field communities. The effects are somewhat more subtle than the impacts of larger vertebrates and may take several years to become apparent. For example, Carson (1993) followed the impact of excluding arthropods from old field communities near Ithaca, New York. Arthropods were excluded from some plots by regular spraying of pesticides, while other plots were not sprayed and served as controls where the full effects of arthropod herbivory would be seen. The standard pattern in these communities consists of a nearly monospecific overstory of the goldenrod *Solidago* and a more diverse understory of several herbaceous species. The main impact of arthropod removal was to increase the abundance of the dominant "canopy" species, *Solidago*, which caused a

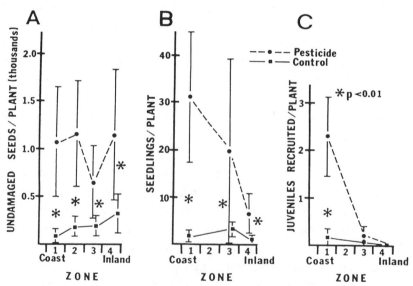

FIGURE 4.6. *Geographic variation in the response of* Haplopappus *to insect removal by pesticides. Dashed lines show responses to insect removal, and solid lines show control values, with statistically significant differences denoted by asterisks. The strongest effects of insects on seedlings per plant and established juveniles occurred in the coastal zone. (Reprinted from Louda, 1982, with permission of the Ecological Society of America.)*

decrease in the species richness of understory herbs (Figure 4.7). The inference is that the observed diversity of understory plants is maintained to some extent by herbivory on the dominant species of *Solidago*.

Striking examples of the effects of predators on terrestrial animals are much less common than examples of plant-herbivore interactions. This scarcity probably reflects the difficulty of doing field experiments with highly mobile organisms that often occur at low densities. The studies that have been done seem to suggest that terrestrial predators reduce numbers of coexisting

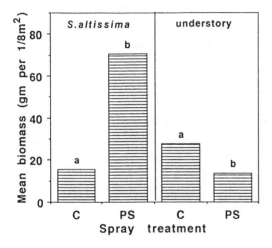

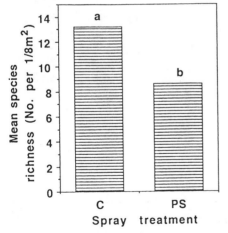

FIGURE 4.7. *Effects of insecticide removal of predators, herbivorous arthropods, on the biomass of the dominant "canopy" species,* Solidago altissima, *and the number of coexisting understory species of herbaceous plants. Treatments labeled C are controls; treatments labeled PS were sprayed with insecticide. (Reprinted from Carson, 1993, with permission of the author.)*

animal species, a rather different result from that described for terrestrial plants and their herbivores. Spiller and Schoener (1989) studied interactions between predatory *Anolis* lizards, preda-

Terrestrial predators can also influence assemblages of arthropods.

tory web-building spiders, and their arthropod prey on small islands in the Bahamas. The interaction between lizards and spiders is best described as **intraguild predation** (Polis et al. 1989) rather than a simple predator-prey interaction because lizards eat some spiders, but lizards also potentially compete with uneaten spiders for small arthropod prey. Spiller and Schoener used enclosures to exclude lizards from some sites and to maintain natural lizard densities in others. They found that lizards decreased both the number of individual spiders and the number of spider species relative to plots without lizards (Figure 4.8). This pattern was consistent with previous observations of a negative correlation between the presence of lizards on islands and the abundance of conspicuous web-building spiders. The pattern also has interesting repercussions for the dominant woody plant on the islands, *Conocarpus erectus*, or buttonwood (Schoener 1988). On islands without lizards, herbivorous insects are abundant and plant leaves carry a dense coat of trichomes that discourage insect attack. On

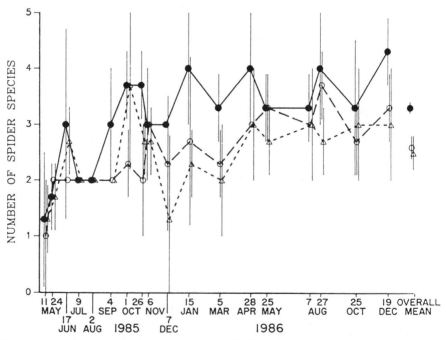

FIGURE 4.8. *Effects of removing predators,* Anolis *lizards, on the species richness of web-building spiders. Solid dots indicate predator removals; open symbols indicate results inside or outside enclosures with natural densities of predators. (Reprinted from Spiller and Schoener, 1989, with permission of the Ecological Society of America.)*

islands with lizards, arthropods are at low abundance and plants display reduced levels of antiherbivore defenses. Similar results obtain for the sea grape, *Coccoloba uvifera*, on the same islands (Spiller and Schoener 1990). This result is an example of what is sometimes called a **tri-trophic level interaction** (Price et al. 1980), or a **trophic cascade** (Paine 1980).

Working in a rather different system, Holmes et al. (1979b) found that birds in temperate forests could act in a fashion similar to tropical lizards in reducing the abundance of terrestrial arthropods. They excluded birds from small patches of temperate forest in New Hampshire using enclosures of bird netting. Most of the arthropods did not respond to bird exclusion, but one group, larval lepidoptera, did increase significantly in plots without birds. This study does not support a broad effect of bird predation on the species composition of terrestrial arthropods, but it remains one of the very few experimental studies of predation in terrestrial animal assemblages. It is remarkable that more studies of this type have not been attempted.

One study of small terrestrial communities reconstructed in the laboratory provides evidence for positive effects of predators on prey species richness. Wade Worthen (1989) found evidence for a kind of keystone predator effect in a common terrestrial community that consists of the array of organisms living on or in forest mushrooms. Shortly after they appear above ground, mushrooms are exploited by a diverse array of invertebrates, including many species of flies and beetles, whose larvae feed on either the mushroom tissue or the yeasts and bacteria that grow on the decaying fruiting body. Important species in the system include three species of fruit flies, *Drosophila*, and a predatory rove beetle, *Ontholestes cingulatus*, that feeds on ovipositing fruit flies and other small arthropods. The beetle perches on mushrooms and captures flies as they attempt to land and oviposit. Worthen showed that mushrooms exposed to three ovipositing fly species in the laboratory primarily yielded a single fly species, *Drosophila tripunctata*, with the other two species, *Drosophila putrida* and *Drosophila falleni*, often being competitively excluded by *D. tripunctata*. When mushrooms were "guarded" or "patrolled" by predatory rove beetles, all three fly species eventually emerged (Figure 4.9). By reducing the initial input of *Drosophila* eggs, either by directly consuming adult flies or by reducing their oviposition, the beetles reduced competition among *Drosophila* maggots and prevented the competitive exclusion of the two competitively inferior species. This stands as one of the very few examples of keystone predation operating in a group of terrestrial animal species.

Freshwater Communities

In contrast to the situation in terrestrial communities, there is an enormous literature on the effects of predation in freshwater communities. The studies

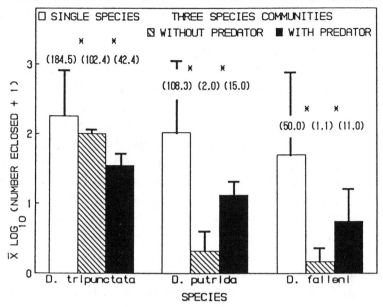

FIGURE 4.9. *Abundances of three species of mycophagous* Drosophila *grown in single-species cultures (open bars, no interspecific competition), in three-species cultures without predators (cross-hatched bars, interspecific competition), and with predators (filled bars, reduced interspecific competition). All species do best when grown without competitors, but predators significantly improve the performance of* D. putrida *and* D. falleni *when competitors are present. (Reprinted from Worthen, 1989, with permission of Blackwell Science Ltd.)*

reviewed below are representative of the main patterns that are seen, but by no means do they represent the complete depth and breadth of what we know.

 Aquatic predators affect the species composition and size structure of zooplankton.

Brooks and Dodson (1965) and Hrbacek et al. (1961) independently proposed that predatory fish influenced the species composition and size structure of freshwater zooplankton. The two critical observations were that large and small zooplankton species tended to have complementary distributions, such that lakes seldom contained an abundance of both large and small species, and that large zooplankton species usually did not coexist with planktivorous fish. Brooks and Dodson proposed the **size-efficiency hypothesis** to explain patterns observed in plankton samples collected from a series of lakes in Connecticut. Lakes with the planktivorous fish *Alosa* contained zooplankton assemblages dominated by species of small body size, such as *Bosmina* (Figure 4.10). Lakes without abundant planktivorous fish contained zooplankton assemblages dominated by different species of much larger average size, including *Daphnia* and the large

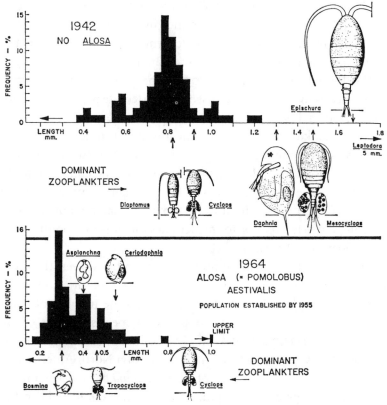

FIGURE 4.10. *Zooplankton species composition in Crystal Lake, Connecticut, before (top) and after (bottom) the introduction of the planktivorous fish* Alosa aestivalis. *(Reprinted with permission from Brooks and Dodson, 1965. © 1965 American Association for the Advancement of Science.)*

copepod *Epischura*. The size-efficiency hypothesis had three key parts. First, larger zooplankton were assumed to be superior competitors for food, phyto-plankton mostly, by virtue of a greater filtering efficiency that was assumed to accompany larger body size. Second, smaller zooplankton with a lower filtering efficiency were assumed to be competitively inferior to larger forms, which might explain their failure to coexist with large-bodied forms. Finally, planktivorous fish were thought to selectively consume large-bodied, compet-itively superior zooplankton, thereby making it possible for small-bodied, less-preferred zooplankton species to persist with fish. Selective predation on larger zooplankton species presumably reflected their greater risk of detection by visually hunting planktivores. Brooks and Dodson presented mostly obser-vational support for their hypothesis, but they did find evidence for a decline in large zooplankton species in one lake after the introduction of planktivo-rous fish.

Dodson (1974) and many others (Zaret 1969; Sprules 1972; Lynch 1979) have shown that part of the size-efficiency hypothesis is correct: Fish, salamanders, and other predatory vertebrates tend to selectively eliminate larger zooplankton species, whereas small zooplankton species often manage to coexist with predators. The reason for the inability of large and small zooplankton species to coexist in the absence of vertebrate predators is more controversial. Dodson (1974) found little evidence for the competitive superiority of larger zooplankton species. Instead he found that some of the largest zooplankton species are predatory copepods that feed on smaller zooplankton species, thus providing a different mechanism for the elimination of small zooplankton species by large ones than the one originally proposed by Brooks and Dodson (1965). Other studies have shown that large zooplankton species can be competitively superior to smaller ones under some conditions (Lynch 1979; Goulden et al. 1982).

The patterns observed in freshwater communities suggest that predators profoundly alter the size structure and species composition of zooplankton assemblages, while exerting little impact on the number of coexisting zooplankton species. Studies of another

Predation by mosquito larva reduces protist diversity in water-filled leaves.

kind of freshwater community, the water-filled leaves of a North American pitcher plant, *Saracennia purpurea*, point to direct negative effects of predators on prey species richness. John Addicott (1974) explored how predation by larvae of the pitcher plant mosquito, *Wyeomia smithii*, affected the number of prey species, mostly protists, that coexisted in pitcher plant leaves. Observations of the numbers of naturally occurring protist species and densities of predatory mosquito larvae indicated that fewer protist species occurred where predators were more abundant (Figure 4.11). Addicott experimentally manipulated predator abundance by stocking leaves with different densities of mosquito larvae. The relation between predator abundance and numbers of coexisting prey species resembled that seen in unmanipulated leaves: Predators tended to depress the number of coexisting prey species. One possible reason predators failed to enhance prey species richness is that the protists may not compete very strongly, even in the absence of predators. Addicott found that in one location without *Wyeomia* there was little evidence of competitive interactions among protist species. Had protists competed strongly there, he expected to find negative correlations in the abundance of species within a large sample of pitcher plant leaves. Instead, the number of negative associations was about what would be expected by chance. The conclusion drawn from observations and experiments is that predators alter community structure by simply deleting increasing numbers of species as predation becomes more intense. Because there is little evidence for competition among prey in the absence of predators, prey

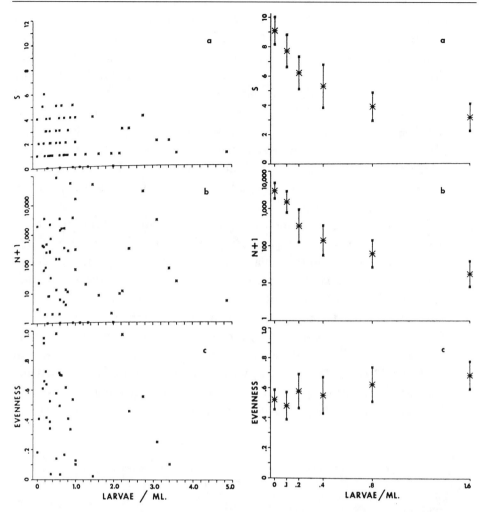

FIGURE 4.11. *Patterns of protist species richness (S), abundance (N + 1), and evenness for naturally occurring (left) and experimentally manipulated (right) densities of the predator* Wyeomia smithii, *a larval mosquito. Predators decrease richness and abundance but have no effect on evenness. (Reprinted from Addicott, 1974, with permission of the Ecological Society of America.)*

species richness would not be expected to increase with increasing predation intensity.

    Effects of aquatic predators are not limited to prey assemblages where small invertebrates predominate. Predators can also influence the species composition of vertebrates. Peter Morin

Predatory salamanders shift the identity of dominant species in groups of larval frogs.

(1983) found that predation by salamanders could shift or reverse patterns of prey species dominance without changing patterns of species richness in an assemblage of six species of frog tadpoles stocked in artificial ponds. The abundance of the predatory salamander *Notophthalmus* was varied over a set of experimental densities that spanned the range observed in natural ponds. Six species of hatchling frog tadpoles were stocked at densities thought to be within the range seen in nature. Artificial ponds made it possible to observe the effects of a range of predator densities on an array of initially identical prey communities. The ponds contained an assortment of alternative prey, including zooplankton and other arthropods. The impact of predators on community structure was measured by determining the abundance of each frog species that successfully completed development and emerged from the ponds as small, recently metamorphosed froglets.

Predators shifted community composition from an assemblage dominated by one apparently competitively dominant species, the spadefoot toad *Scaphiopus holbrooki*, to assemblages dominated by a competitively inferior species, *Pseudacris crucifer* (Figure 4.12). Most communities contained similar numbers of species, but the distribution of individuals among those species was radically altered by predators. Predators actually enhanced the survival of *Pseudacris crucifer*, apparently by reducing the survival of other competitively superior species like *Scaphiopus*. *Pseudacris crucifer* manages to persist with moderate densities of predators by virtue of a suite of antipredator behaviors that include a tendency to remain motionless, so as not to attract the attention of visual predators (Lawler 1989), and shifting patterns of microhabitat use

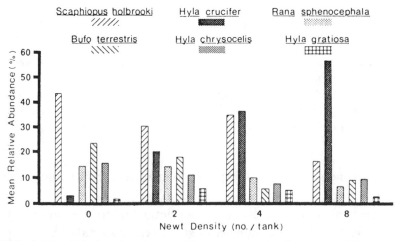

FIGURE 4.12. *Relative abundances of six anuran species that survived to metamorphose from ponds containing different densities of the predatory salamander* Notophthalmus viridescens. *Each mean is based on four replicate communities. Predators shift species composition from dominance by* Scaphiopus *to dominance by* Pseudacris (= H. crucifer). *(Reprinted from Morin, 1983, with permission of the Ecological Society of America.)*

to take advantage of benthic refuges in the presence of predators (Morin 1986).

Predation by fish seems to have similar impacts on the benthic invertebrates that live in ponds (Hall et al. 1970; Crowder and Cooper 1982; Morin 1984a,b). For several groups of organisms, fish drastically reduce prey abundances and shift species composition toward dominance by predator-resistant species. Predator-resistant species are often smaller in size and have well-developed suites of antipredator behaviors.

WHEN IS PREDATION LIKELY TO REGULATE PREY POPULATION SIZE AND COMMUNITY STRUCTURE?

The impacts of predators in successful cases of biological control, taken together with the results of numerous field and laboratory experiments, offer convincing evidence for the ability of predators to vastly alter prey abundances and influence community composition. However, the kinds of effects observed seem highly variable, both among and within the broad classes of habitats considered. It would be desirable to be able to predict whether or not predators will limit prey populations, or whether predators will increase or decrease species richness, from a few easily observed traits. Several conceptual theories, which lack the analytical refinement of the explicitly mathematical models treated in the next chapter, attempt to do this. These theories differ mainly in the frequency with which strong effects of predators are predicted, and they make different predictions about the impacts of predators on species that occupy different trophic levels.

One idea is that predators often have little actual impact on total prey abundances, mostly because consumed prey are those unfortunate individuals that are unable to secure safe territories or refuges from predators (Errington 1946). This idea, sometimes called Errington's Hypothesis, assumes that prey populations are ultimately regulated by competition among prey for predator-free sites. Prey in excess of the number of available safe sites or territories either fall victim to predators or disperse. The idea that predators simply crop the excess prey population clearly does not apply where predators greatly reduce prey abundances, but it may explain situations in which manipulations of predators seem to have little effect on prey numbers (e.g., see Allan 1982).

Hairston, Smith, and Slobodkin (1960) proposed an elegant scenario that continues to stimulate much ecological research. Their argument is based on observation and induction and is often referred to as **HSS**, from the initials of the

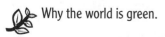 Why the world is green.

authors' last names. The argument is also sometimes known as "**Why the world is green**," for reasons that should become obvious below. The logic runs as follows and was intended to apply only to terrestrial communities, although it has been interpreted much more broadly by others (Sih et al. 1985). First, assume that terrestrial communities can be broadly divided into four compartments: primary producers, herbivores, carnivores, and detritivores (Figure 4.13). What can we say about the relative importance of competition, predation, and natural disasters in regulating populations in these four trophic categories? HSS argued that, in general, natural disasters do not seem to play much of a role in keeping populations low. They then used the apparent presence or absence of food limitation in different groups to argue for or against competition as a regulating force. For example, detritivores are thought to be food limited, since their food, dead plants and animals, accumulates at a negligible rate and is not present in excess of demand. Plants appear to compete for light, water, and nutrients, which suggests that competition also regulates primary producers. Herbivores, however, appear to be surrounded by an excess of food, namely, the plants that make much of the terrestrial world green. This suggests that if herbivores are not food limited, they must be limited by something else. In the absence of regular mortality caused by natural disasters, predators seem to be the only other factor that might limit abundance. Predators, being on the top trophic level, cannot be predator limited themselves, and therefore must be limited either by prey availability (herbivore abundance) or other factors. The end result is that competition for food or resources is thought to be important in regulating abundances of primary producers, top predators, and decomposers, whereas

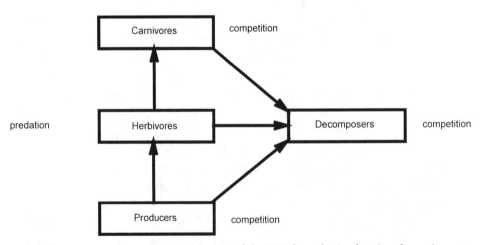

FIGURE 4.13. *Schematic representation of the HSS hypothesis, showing the major compartments in terrestrial food webs, along with the process—competition or predation— that is most likely to regulate populations in each compartment, according to Hairston et al. (1960).*

herbivores appear to be regulated by predators, since the former seldom deplete their food supply. Exceptions that prove the rule are those cases of introduced pest herbivores that defoliate their food plants in the absence of natural enemies of the herbivores, such as the gypsy moth, *Lymantria dispar*, in eastern North America. Other evidence comes from the examples of the biological control of pest plants by introduced herbivores recounted at the beginning of this chapter.

HSS has had its detractors. One important objection is that the excess food supply for herbivores may be apparent rather than real. Because many plants contain toxic compounds that render them unsuitable for food, a large standing crop of plant biomass may not represent a surplus of available food (Murdoch 1966; Ehrlich and Birch 1967). One counterargument to this point is that even relatively toxic plants seem to have at least one herbivore that has evolved the ability to circumvent the plant's armamentarium of chemical defenses (Slobodkin et al. 1967). Others feel that the reticulate nature of many natural food webs makes it unlikely that effects of top predators will simply cascade down through a three-level food chain as envisioned by HSS (Polis and Strong 1996).

Fretwell (1977) extended the reasoning of HSS to situations in which food webs have different numbers of trophic levels. In systems with even numbers of trophic levels (for example, two or four levels), herbivores may not be predator limited and might be able to deplete their food plants (Figure 4.14). A two-level system corresponds to a situation in which herbivorous insect pests are introduced without their natural enemies. A four-level system occurs when a new top predator regulates predators on the third trophic level that would otherwise limit herbivore populations. In systems with odd numbers of trophic levels, plants should be regulated by competition for resources rather than by herbivores, for reasons similar to those suggested by Hairston et al. (1960). Food chain length should increase in a predictable way with productivity, such that longer food chains will occur in more productive environments. The net result of such a relationship between productivity and food chain length would be an alternation of food chains with odd and even numbers of trophic levels along a gradient of productivity, with a corresponding alternation of regulation by competition or predation in the basal trophic levels.

Hairston and Hairston (1993) recently extended the original ideas of HSS to account for some apparent differences between aquatic and terrestrial communities in numbers of trophic levels and the tendency of trophic levels to be regulated by different processes in different systems. They argue that whereas most terrestrial food chains have approximately three trophic levels, lakes and other freshwater ecosystems tend to have four trophic levels (Figure 4.15). This difference in food chain structure results in the conspicuous

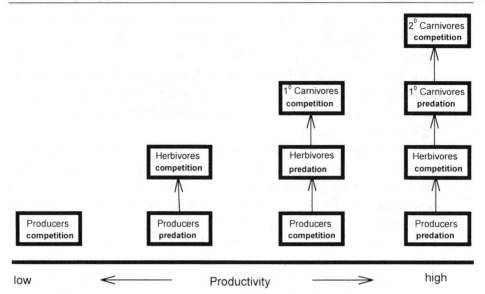

FIGURE 4.14. *Alternation of regulation by competition or predation in each trophic level as a function of an increasing number of trophic levels along a gradient of primary productivity. This scheme follows the ideas presented by Fretwell (1977).*

absence of a large standing crop of producers in lakes compared with many terrestrial systems. The reasons for this difference are potentially complex, and by no means certain, but may reflect the small size of aquatic producers (phytoplankton) relative to their consumers (zooplankton), and the presence of a microbial loop (see Figure 4.15) that redirects energy and nutrients back up into the food chain that would otherwise be lost to detritivores or decomposers. Both factors may contribute to the existence of an extra trophic level in aquatic systems. The end result is that terrestrial communities are green, and aquatic communities are not. This difference in the standing crop of primary producers in aquatic and terrestrial communities can be attributed to the difference in the length of the food chains found in the two habitats.

Bruce Menge and John Sutherland (1976, 1987) used a related approach to predict the relative importance of predation and competition in affecting community patterns. They suggested that the relative importance of predation and competition in rocky intertidal communities varied inversely with increasing trophic status (height in the food web) within a particular

FIGURE 4.15. *Major trophic compartments and patterns of energy flow in terrestrial (top) and freshwater (bottom) communities. Lakes differ from forests by having an additional trophic level and by having a microbial loop, via the bacterioplankton, that feeds back into a single food chain. (Adapted from Hairston and Hairston, 1993, with permission of the University of Chicago Press.)*

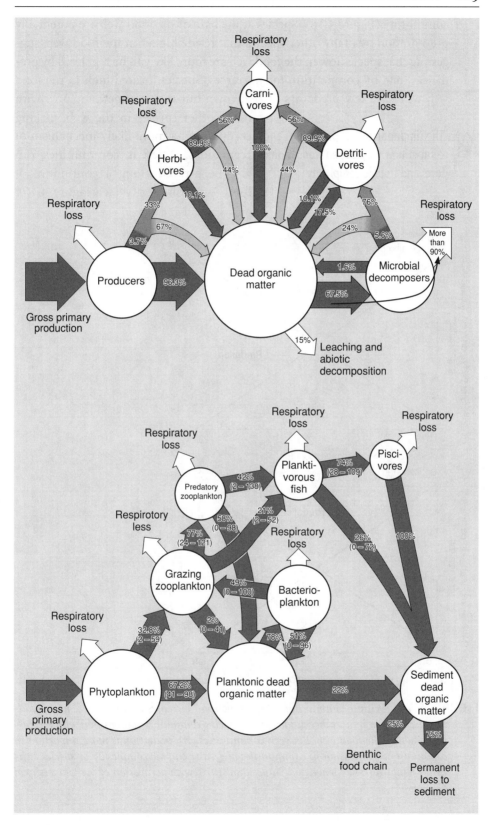

community (Figure 4.16). Species at the base of the food web are potentially preyed upon by many other species that reside higher in the food web, suggesting that species low in the food web are more likely to be regulated by predation than by competition for resources. Species located high in the food web will have few predators of their own but may compete for prey with many other species in the web, making competition a more likely mechanism of regulation. Different communities can also be arrayed along a continuum of food web complexity (see Figure 4.16).

 The importance of predation and competition may depend on trophic level and trophic complexity.

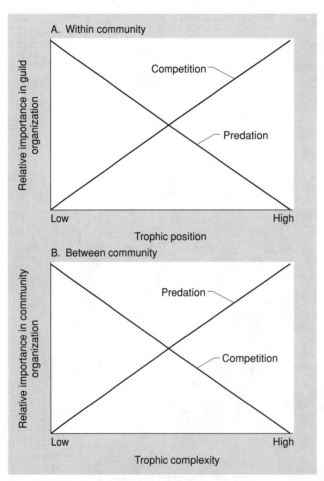

FIGURE 4.16. *Within communities (A), competition increases in importance and predation decreases in importance as factors influencing community composition with increasing height in the food web. Between communities (B), predation is more important in community organization in communities of greater trophic complexity or species richness. (Adapted from Menge and Sutherland, 1976, with permission of the University of Chicago Press.)*

Communities of low complexity and few trophic levels will be predominantly structured by competition (since there are few predators around to limit prey), whereas more complex communities with many trophic levels should be structured by a greater number of predator-prey interactions. In rocky intertidal communities, physical stress, rather than productivity, may play an important role in regulating trophic complexity (Menge and Sutherland 1976; Connell 1975).

OVERVIEWS OF GENERAL PATTERNS BASED ON REVIEWS OF EXPERIMENTAL STUDIES OF PREDATION

Connell (1975) compiled an early and influential review of field experiments that provided evidence for an important role of predation in structuring communities. Like Menge and Sutherland (1976), Connell suggested that the effects of predators on community structure were to some extent determined by the rigors of the physical environment, with predation being less important in physically harsh environments.

Ten years later, after many more experiments on predation in communities had accumulated, Sih et al. (1985) surveyed the ecological literature to assess whether effects of predators on community structure varied systematically with the kind of habitats studied (e.g., terrestrial, freshwater, and marine communities). They also looked for evidence that would support either the HSS or Menge and Sutherland hypotheses concerning the relative importance of competition and predation in different trophic levels. The majority of experimental studies reviewed yielded evidence of important impacts of predation on prey populations, with little difference materializing among studies conducted in different latitudes or habitats (Table 4-1). Effects in which some prey species benefited from the presence of predators materialized in somewhat less than half of the studies that found significant impacts of predators. Predation is a general feature of most of the systems surveyed. When surveys were sorted by the trophic level of the manipulated predator, an interesting pattern emerged (Table 4-2). Striking effects of predators were most common when the predators fed lower in the food chain, a result more consistent with the Menge and Sutherland hypothesis than with HSS.

Hairston (1989) reviewed a somewhat different set of studies than Sih et al. and found general support for the predictions of HSS in terrestrial communities. As mentioned previously, Hairston was reluctant to generalize among different habitats concerning the relative importance of processes that affect species on different trophic levels.

TABLE 4-1.   Summary of the proportion of experimental studies of predation that yielded statistically significant effects, large effects, and unexpected effects on prey abundance. Unexpected effects include positive effects of predators on some prey, as in keystone predation.

Classification	Effect[a,e]	Large[b]	Unexpected[c]
Latitude			
Temperate	95.0 (120)[d]	84.1 (113)	41.4 (116)
Tropical	100.0 (13)	91.7 (12)	30.8 (13)
Polar	100.0 (6)	100.0 (6)	33.3 (6)
System			
Intertidal	94.4 (36)	91.1 (34)	40.0 (35)
Other marine	100.0 (25)	75.0 (24)	32.0 (25)
Lotic	100.0 (17)	70.6 (17)	47.1 (17)
Lentic	100.0 (22)	90.5 (21)	47.6 (21)
Terrestrial	89.7 (39)	75.7 (37)	22.2 (36)
Predator type			
Vertebrate	91.3 (46)	70.0 (40)	42.9 (42)
Arthropod	98.0 (49)	79.2 (48)	30.6 (49)
Other invertebrate	94.2 (52)	88.2 (51)	37.3 (51)
Predator trophic level			
Herbivore	94.9 (59)	86.8 (53)	38.6 (57)
Primary carnivore	96.4 (56)	74.1 (54)	31.5 (54)
Secondary carnivore	100.0 (16)	93.3 (15)	50.0 (16)
Response type			
Community	80.0 (60)	60.9 (46)	19.1 (47)
Population	97.4 (76)	84.9 (73)	50.0 (76)
Individual	93.5 (46)	68.2 (44)	13.6 (44)

[a] Studies with any significant effects as a percentage of all studies.
[b] Studies with large significant effects as a percentage of all studies with significant effects.
[c] Studies with unexpected significant effects as a percentage of all studies with significant effects.
[d] Numbers in parentheses are the number of total studies.
[e] Broken vertical lines indicate contrasts that were made; asterisks indicate comparisons that are significantly different.
Reprinted from Sih et al. (1985), with permission, from the *Annual Review of Ecology and Systematics,* Volume 16. © 1985 by Annual Reviews.

TRADE-OFFS BETWEEN COMPETITIVE ABILITY AND RESISTANCE TO PREDATION

So far, ideas about the conditions leading to keystone predation have stressed differences in the ways in which predators attack competitively superior prey species or differences in the extent to which prey species compete in different

TABLE 4-2. Summary of the proportion of experimental studies of predation that yielded statistically significant effects and large significant effects by predator trophic level and type of system.

System	Effect			Large Effect		
	Herbivore	Primary Carnivore	Secondary Carnivore	Herbivore	Primary Carnivore	Secondary Carnivore
Intertidal	67.7 >[a] (173)[c]	57.8[b] (116)		84.2 > (120)	70.1[b] (67)	
Other marine	59.0 > (144)	40.7[b] (199)		95.1 (82)	68.4[b] (57)	
Lotic		66.7 > (42)	49.3 (134)		72.7 (22)	75.0 (95)
Lentic		73.0 (163)	64.2 (53)		61.3 (106)	55.2 (29)
Terrestrial	63.2 > (174)	53.3 > (45)	28.6 (49)	74.1 (112)	61.1 (36)	

[a] Results of G tests; > indicates the contrasts are significantly different.
[b] All carnivores pooled.
[c] Numbers in parentheses are sample sizes.
Reprinted from Sih et al. (1985), with permission, from the *Annual Review of Ecology and Systematics*, Volume 16. © 1985 by Annual Reviews.

kinds of communities. But do the properties that make some prey competitively superior also render the same prey particularly susceptible to predators, essentially predisposing communities to the phenomenon of keystone predation? There are reasons to suspect that this might be the case.

Figure 4.17 depicts a hypothetical trade-off between the ability of prey species to compete for resources and to avoid or discourage the attacks of predators. The main idea is that species can be ordered along a gradient of their rate of resource acquisition, which in turn affects their rates of growth. Animal species with high rates of resource acquisition typically have high rates of foraging activity, which should in turn attract the attention of visually foraging predators. Similarly, plants with high rates of productivity divert little of their resources into defensive chemicals or structures, and instead invest heavily in highly palatable and relatively undefended structures like leaves and fruits. The precise mechanisms involved in the trade-off between competitive ability and resistance to predation probably vary among broad taxonomic groups. There is some evidence to support a behavioral mechanism for this trade-off in animals (Werner and Anholt 1993). In particular, some animals with higher levels of foraging activity suffer greater levels of

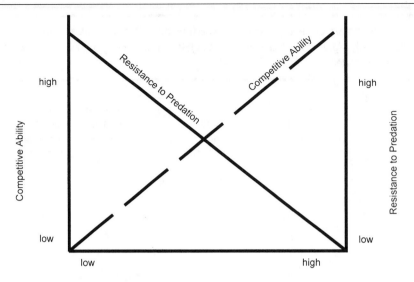

FIGURE 4.17. *Hypothetical trade-offs between competitive ability and resistance to predation along a gradient of the rate of resource acquisition by prey species.*

mortality from predators (Lawler 1989). In intertidal algae, the same structural features that confer resistance to grazers—prostrate growth and structural defenses—predispose crustose algae to shading, overgrowth, and competitive inferiority relative to more vigorously growing but palatable species (Paine 1980, 1984). In terrestrial plants, there is a negative relation between growth rate and the abundance of chemical defenses against herbivores (Coley 1986).

Other features of the predator-prey interaction may also select for preference for competitively dominant prey. Competitively dominant prey are likely to be present in high levels of abundance in previously unexploited patches where they have temporarily escaped from predators. Patches dominated by competitively superior prey should constitute a rich, energetically rewarding, and relatively undefended resource, and predators should experience strong selection to forage selectively in those sites. By virtue of their larger size, or more conspicuous activity, competitively dominant species are also more apparent than others to searching predators (Feeney 1976). The net result is that keystone predation seems to be a logical consequence of the life history traits of predators and competitively unequal prey. In that case, it is unclear why keystone predation, and similar kinds of structuring effects of predators, only materialize in less than half of the field studies of predation that have been reviewed by Sih et al. (1985).

CONCLUSIONS

Predation can influence communities in striking ways. When predators feed selectively on dominant competitors, predation can enhance the diversity of coexisting prey species. if predation does not limit the abundance of superior competitors, or if competition among prey is weak, then predators reduce diversity and create communities dominated by species most resistant to predation. In many freshwater systems, predators have little effect on diversity, but they control the identity of dominant, usually predator-resistant species. Many of these patterns reflect a life history trade-off between competitive ability and antipredator strategies.

CHAPTER 5

Models of Predation in Simple Communities

OVERVIEW

This chapter describes some simple models of predator-prey interactions. Models of predation on single prey species are surveyed to explore the range of predator-prey dynamics that result when models incorporate different assumptions about the way that predators and prey interact. These simple models are then extended to describe the impact of predation on multiple prey species; these extended models are used to explore the conditions in which predation increases or decreases prey diversity. These models of interactions between species on two adjacent trophic levels also provide the background needed to understand models of food webs, considered in Chapter 6, that address interactions among species on several trophic levels.

SIMPLE PREDATOR-PREY MODELS

The simplest models of predator-prey interactions capture the essence of a +/− interaction between species without incorporating very much detailed biology. Their chief advantage is that their dynamic behavior can be analyzed readily. Their disadvantage is that they may be unrealistically simple caricatures of nature, yielding equally unrealistic cari-

 Simple predator-prey models provide a starting point for more realistic models.

catures of the dynamics of natural predator-prey systems. However, even if these relatively simple models turn out to be poor representations of nature, we can still learn much by observing how and where simple models fail to represent the real world. The models can then be modified to make them more reasonable, usually at the cost of reduced analytical tractability.

Lotka (1925) and Volterra (1926) independently formulated a simple predator-prey model that is the point of departure for most other models used to describe the interactions of continuously reproducing populations of predators and prey. Most ecologists would probably agree that this model is a gross oversimplification of nature. Nonetheless, it is a useful starting point for the construction of more realistic, and more complex, models. The model is framed in terms of a pair of differential equations, one describing the rate of change in prey population size (H), the other describing the rate of change in predator population size (P). The equations are as follows:

$$dH/dt = bH - PaH \tag{5.1}$$

and

$$dP/dt = e(PaH) - sP \tag{5.2}$$

where b is the per capita birth rate of the prey; a is a per capita attack rate and aH is the per capita consumption rate, or **functional response**, of predators on a given density of prey; e is the conversion efficiency of consumed prey into new predators; and $-s$ is the rate at which predators die in the absence of prey.

The model makes several simplifying assumptions. Because it is framed in differential equations, all responses to changes in density are assumed to be instantaneous; there are no time lags in the response of predators to prey abundance, or vice versa. Prey are limited only by predators; there is no intraspecific density-dependent competition among prey, and, in the absence of predators, prey simply increase at an exponential rate. Similarly, predators are limited only by prey abundance, and, in the absence of prey, predators die at an exponential rate. Finally, as prey increase in abundance, the functional response, or per capita consumption rate of prey per predator, increases linearly with prey abundance. This is probably a reasonable approximation over the lower range of prey densities, but at higher densities predators would probably become saturated with prey, and the functional response would be expected to level off at some maximal attack rate, say w. The three commonly recognized kinds of functional responses that might occur are shown in Figure 5.1, together with their corresponding formulae. The three types of functional responses were originally noted by Holling (1965), although Solomon (1949) coined the term *functional response*.

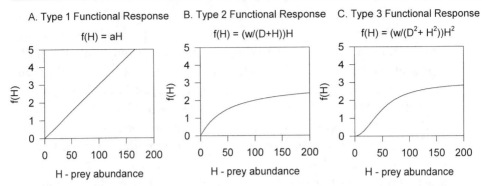

FIGURE 5.1. *Examples of three kinds of functional responses, f(H), which describe how attack rates of predators vary with prey density (see Holling 1965). The type 1 functional response increases at a constant rate, a, as prey density increases. The type 2 functional response increases in a decelerating fashion only up to some maximal rate, w, that is attained at high prey abundances. The type 3 functional response also peaks at a maximal rate, w, but displays a sigmoid approach to that maximal attack rate. D is an empirically determined constant. In these examples, a = 0.03, w = 3, and D = 50; the formula for each functional response is shown with its corresponding graph.*

Models such as the Lotka-Volterra model are usually analyzed with respect to two properties. First, it is of interest to ask whether the model has an **equilibrium**, that is, a set of values of H and P such that $dH/dt = 0$ and $dP/dt = 0$, and H and P are both greater than 0. This state corresponds to a situation in which the predator and prey populations are no longer growing or declining, and neither predator nor prey are extinct. Second, it is of interest to ask whether the equilibrium is **locally stable**. This means, starting at equilibrium values of H and P, if either population changes slightly in size, will it tend to return to its equilibrium value? This corresponds to the tendency for a system to return to a particular equilibrium state rather than to oscillate or go extinct following a change in the size of one or both of the populations. **Global stability** is a more general property that implies that a system will return to the equilibrium point from any set of initial population values. There is a loose, and perhaps too facile, analogy between the prolonged persistence of natural populations and the existence of a locally stable equilibrium in models such as this. Later we will see that under some circumstances model populations without a locally stable equilibrium can persist for a very long time.

 Models can be evaluated to determine whether a stable equilibrium exists.

Several excellent books cover the basics of determining whether a system of equations has a locally stable equilibrium. Good places to start include the books by May (1973), Pimm (1982), Vandermeer (1981), Bulmer (1994), and

Hastings (1997). Other detailed treatments that are accessible to most ecologists include the books by Edelstein-Keshet (1988) and Yodzis (1989). An example of how a stability analysis is done is given in the appendix. Solving sets of differential equations to depict population dynamics is a task best left to standardized computer algorithms, such as the Runge-Kutta algorithm (Johnson and Riess 1982), which are available for most microcomputers. The examples given in this chapter were simulated using the fast Runge-Kutta algorithm available with the MathCad (1994) package for numerical methods. Inspection of these simulations will often indicate whether model populations tend to return to an equilibrium, whether they oscillate, or whether one or both populations fail to persist.

The Lotka-Volterra equations have an equilibrium point given by the conditions $H = s/(ea)$ and $P = b/a$. This point can be determined by setting both equations equal to zero and solving for the values of H and P in terms of the other constants in the model. However, the equilibrium is not locally stable. The populations will instead oscillate if perturbed away from the equilibrium with an amplitude that depends on the size of the departure from equilibrium. The period of the oscillation is determined solely by the parameters of the model. The amplitude of the oscillation depends on how far the initial values of H and P depart from the equilibrium point. Examples of the kinds of dynamics to be expected are shown in Figure 5.2. The model is also not unstable, strictly speaking, since the perturbations do not tend to grow in size over time. This situation is termed **neutral stability** and is a consequence of the peculiar lack of density dependence in the Lotka-Volterra model. Although there are some natural and experimental populations of predators and prey that also display periodic fluctuations (Figure 5.3), it is important to note from the outset that the Lotka-Volterra model is not the only model that produces such oscillatory dynamics. It does, however, capture the essence of the feedback between predator and prey abundance.

The Lotka-Volterra model can be modified in fairly simple ways to make it somewhat more biologically realistic. Reasonable modifications include making the prey populations density dependent and making the predator death rate depend inversely on prey density. Simple inclusion of density dependence in the prey population is enough to shift the behavior of the model from neutral stability to local stability about the equilibrium point. This result is made clear by examining the model described by the following pair of equations:

$$dH/dt = bH(1 - H/K) - PaH \qquad (5.3)$$

and

$$dP/dt = e(PaH) - sP \qquad (5.4)$$

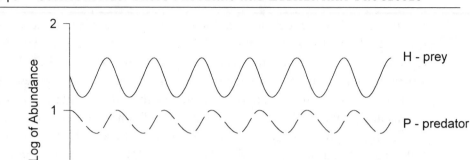

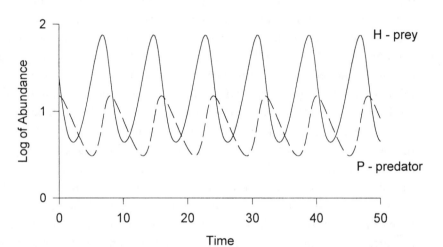

FIGURE 5.2. *Examples of oscillatory dynamics in a Lotka-Volterra predator-prey model. (A) Oscillations in predator and prey abundances that result from using initial values of H = 25 and P = 10 and parameter values* b = 1.5, a = 0.2, e = 0.1, *and* s = 0.5. *The neutrally stable equilibrium is at* H = 25 *and* P = 7.5. *(B) Oscillations of greater amplitude result from using the same model parameters but a different initial value of* P *(P = 15) that is even further from the equilibrium point.*

Using the same set of conditions that produce a sustained oscillation in the Lotka-Volterra equations (see Figure 5.2), this model yields a stable equilibrium that is reached after a series of damped oscillations (Figure 5.4). A stability analysis for this model is described in the appendix. Incorporating a presumably realistic assumption, namely, that prey

Addition of competition among prey stabilizes the Lotka-Volterra equations.

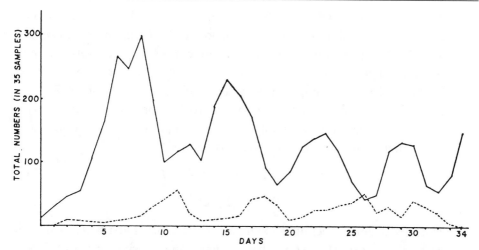

FIGURE 5.3. *An example of predator-prey oscillations in a simple community of protists maintained under laboratory conditions. The predator abundance (*Didinium*) is shown by the dashed line, and prey abundance (*Paramecium*) is shown by the solid line. (Reprinted from Luckinbill, 1974, with permission of the Ecological Society of America.)*

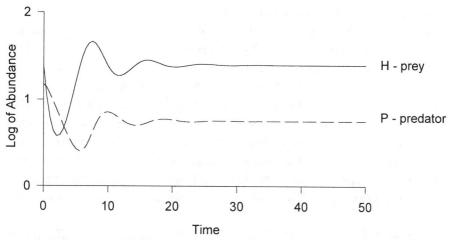

FIGURE 5.4. *Damped oscillatory approach to an equilibrium in a Lotka-Volterra model that incorporates density dependence in the prey population by including a carrying capacity,* k = 100. *The graph shows oscillations of predator and prey numbers that result from using initial values of* H = 25 *and* P = 15 *and model parameters* b = 1.5, a = 0.2, e = 0.1, *and* s = 0.5. *These same parameters produced large sustained oscillations in the Lotka-Volterra model described in Figure 5.2. This model has a stable equilibrium with* H = 25 *and* P = 5.625.

population growth will be limited by competition in a logistic fashion in the absence of predation, makes the behavior of the model more reasonable. Models of a similar form, but which include more trophic levels and more equations, figure prominently in the analysis of possible relations between food chain length and population dynamics (Pimm and Lawton 1977).

Leslie and Gower (1960) described a predator-prey model that incorporated density dependence in both the prey and predator populations. Their model in its most compact form looks as follows:

$$dH/dt = bH - cH^2 - PaH \tag{5.5}$$

$$dP/dt = rP - sP^2/H \tag{5.6}$$

This model looks rather unlike Equations 5.1 and 5.2 at first glance. However, if you define $c = b/K$, assume that r equals some maximal rate of predator increase, assume that the number of prey needed to support a single predator per unit time is j, and let $s = rj$, then by substituting in the above equations and rearranging some terms, you obtain

$$dH/dt = bH(1 - H/K) - PaH \tag{5.7}$$

$$dP/dt = rP[1 - j(P/H)] \tag{5.8}$$

This has the effect of making the prey growth rate density dependent, with a carrying capacity of K, in the absence of predators. It also makes the per capita predator birth rate and the total predator death rate depend on the ratio of predator to prey abundances.

The dynamics portrayed by this model are shown in Figure 5.5. The key feature of the dynamics following a perturbation away from equilibrium is a series of damped oscillations that eventually lead to a return to the original equilibrium. The inclusion of density dependence in the prey population contributes to the stability of the model. Making the predator's death rate depend on the ratio of predator-prey abundance also makes sense, since the predators should not starve when prey are relatively abundant, that is, when P/H is small.

May (1973), Tanner (1975), and Pielou (1977) describe a modification of the Leslie-Gower predator-prey model that includes a nonlinear functional response. This functional response has the effect of decelerating the per capita attack rate of predators on prey such that as prey densities become very large, the functional response attains a maximal value of w. Such effects would occur as the ability of predators to capture and consume prey becomes saturated at high prey densities. The model, called the Holling-Tanner model by Pielou (1977), and discussed by May (1973) and Tanner (1975), illustrates how incorporation of saturation kinetics in the functional response affects predator-prey dynamics. The model looks like the following:

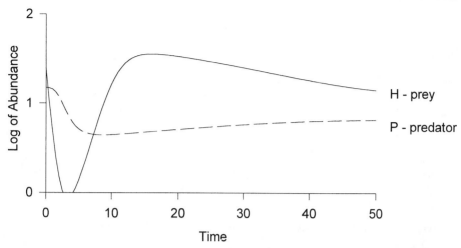

FIGURE 5.5. *Example of the dynamics generated by the Leslie and Gower predator-prey model that includes density dependence in both the predator and prey equations. The dynamics are produced by Equations 5.5 and 5.6, where* b = 1.5, a = 2.0, s = 0.5, e = 0.015, r = ea, c = b/k = 1.5/100, *and initial values are* H = 25 *and* P = 15.

$$dH/dt = bH - cH^2 - (w/(D+H))PH \tag{5.9}$$

$$dP/dt = rP - rj\,P^2/H \tag{5.10}$$

where the term $w/(D+H)$ causes the predator's attack rate to level off at high prey densities. Here w is a maximal attack rate and D is a constant that is determined empirically. As prey density H becomes much larger than D, the per capita attack rate converges on w.

Figure 5.6 shows examples of the range of dynamics produced by the Holling-Tanner model. The model is either stable or locally unstable, in the sense that it shows no tendency to return to the equilibrium point, depending on the choice of parameters used. The model can display another fascinating form of dynamics, however, known as a **stable limit cycle**. Regardless of the initial population sizes, as long as both populations are initially greater than zero, all population trajectories converge on an oscillation whose period and amplitude are determined solely by the coefficients in the model.

 Stable limit cycles are another form of predator-prey oscillations.

The Holling-Tanner model is an example of a very large class of predator-prey models that yield either a stable equilibrium or stable limit cycles. Kolmogorov's theorem (1936) describes the basic conditions that produce this kind of behavior in simple models. May (1973) provides a convenient translation of the ecological conditions specified by Kolmogorov's theorem that yield either a stable equilibrium or a stable limit cycle.

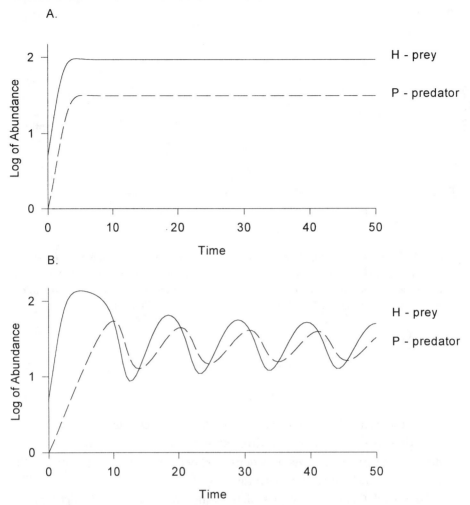

FIGURE 5.6. *(A) An example of a stable equilibrium in the Tanner model of predator-prey interactions. These dynamics correspond to Equations 5.9 and 5.10, where b = 1.5, w = 2.0, j = 3, e = 1, r = ew, k = 150, D = 15, c = b/k = 1.5/150, and initial values are H = 5 and P = 1. (B) Example of a stable limit cycle produced by the Tanner model. These dynamics correspond to Equations 5.9 and 5.10, where b = 1.5, w = 2.0, j = 1, e = 0.25, r = ew, k = 150, D = 15, c = b/k = 1.5/150, and initial values are H = 5 and P = 1.*

One potentially unrealistic assumption of the preceding models is that responses of predators to prey density, and vice versa, are instantaneous. In practice, it seems likely that increases in predator population size will often lag somewhat behind the consumption of prey at a particular density. There are different ways to incorporate time lags into models of predator-prey interactions. One approach is to include explicit time lags in models framed in differential equations. The other approach is to use an alternate mathematical framework, difference equations. For continuous models framed in differential equations, Wangersky and Cunningham (1957) used the following approach:

$$dH/dt = bH_t - P_t a H_t \qquad (5.11)$$

and

$$dP/dt = e(P_{t-T} a H_{t-T}) - sP_t \qquad (5.12)$$

where the subscripts t and $t - T$ refer to the population sizes at time t, now, and $t - T$ time units ago. This has the effect of making current predator reproduction depend on the abundance of predators and prey T units of time in the past. It is sometimes convenient to think of T, the predator time lag, as the amount of time required for consumed prey to be transformed into new predators. Wangersky and Cunningham found that the effect of increasing time lags was increasingly destabilizing, leading to larger oscillations in predator-prey dynamics.

 Time lags tend to destabilize interactions.

The second way to include time lags is to use an altogether different framework for the predator-prey model, namely, a set of difference equations. Difference equations give the values of predator and prey population sizes in one year, or generation, as a function of the population sizes in the previous year or generation. The equations model the population sizes rather than the rates of change in population size treated in differential equations. The simplest predator-prey difference equations are the ones used by Nicholson and Bailey (1935) to model parasitoid-host interactions:

$$H_{t+1} = \lambda H_t \exp(-aP_t) \qquad (5.13)$$

$$P_{t+1} = H_t[1 - \exp(-aP_t)] \qquad (5.14)$$

The model assumes that each prey that survives by evading predators will produce λ offspring, which then become the prey population in the next generation. The probability that prey will evade predators is given by $\exp(-aP_t)$, a function that decreases with increases in either predator abundance, P_t, or a, which is a measure of foraging or searching activity per predator (sometimes called the area of discovery). This probability is derived from the Poisson distribution and is equivalent to the probability that a prey fails to encounter any predators in its lifetime, given that the mean density of predators is aP_t. The predator equation assumes that each attacked prey is transformed into a new predator, as might be the case for a host-parasitoid system in which a single host produces a single parasitoid. The

Time lags can also be represented by models found in difference equations.

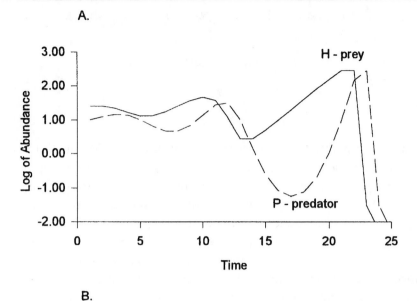

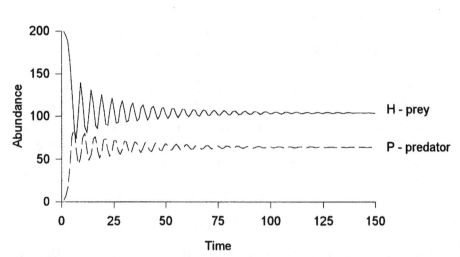

FIGURE 5.7. *Dynamics of predator-prey models based on the Nicholson and Bailey family of discrete time difference equation models. (A) Unstable dynamics that result from using parameter values of* a = 0.068, λ = 2, *and initial values of* H = 25 *and* P = 10 *in Equations 5.13 and 5.14. (B) Stable equilibrium produced by Equations 5.15 and 5.16, which incorporate density dependence in the prey population. Parameter values are* r = 2, k = 200, *and* a = 0.015, *with initial values of* H = 200 *and* P = 2. *(C) Stable limit cycle produced by Equations 5.15 and 5.16 for parameter values* r = 0.693, k = 50, *and* a = 0.068, *with initial values of* H = 50 *and* P = 2. *(D) Chaos produced by Equations 5.15 and 5.16 for parameter values* r = 3, k = 75, *and* a = 0.068, *with initial values of* H = 75 *and* P = 2.

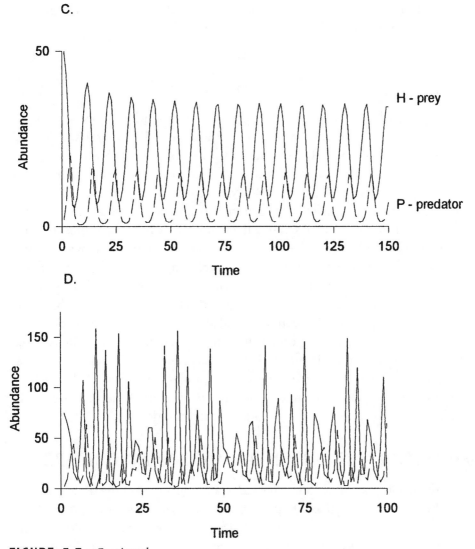

FIGURE 5.7. *Continued*

number of attacked prey is given by the product of prey abundance and the probability that a prey encounters at least one predator in its lifetime (1 minus the probability that it does not encounter a predator in its lifetime).

The dynamics of this simple model are unstable, with fluctuations of increasing magnitude resulting in the eventual extinction of the system (Figure 5.7A). However, almost anything that is done to make the system more realistic, such as inclusion of saturation kinetics in the functional response or inclusion of density dependence in the prey dynamics, will make the system more stable (see Hassell 1978). For instance, the corresponding model that incorporates density-dependent regulation in the prey population is given by the following pair of equations:

$$H_{t+1} = H_t \exp[r(1 - H_t/K) - aP_t] \tag{5.15}$$

$$P_{t+1} = H_t[1 - \exp(-aP_t)] \tag{5.16}$$

This model exhibits a range of different behaviors, including a stable equilibrium, stable limit cycles, and chaotic pseudoperiodic dynamics. Figure 5.7 shows examples of each. Now, however, there exist some choices of parameters that lead to persistent dynamics, a situation that did not obtain for any choice of parameters in the simpler Nicholson and Bailey model.

Although the models described so far seem quite simplistic, they can do a very reasonable job of describing the population dynamics of predators and prey in simple laboratory settings. Gary Harrison (1995) has shown that relatively simple modifications of the Lotka-Volterra predator-prey equations can provide a very good fit to the dynamics of *Didinium* and *Paramecium* described by Luckinbill (1973). The modifications producing the best fit to observed population fluctuations included a term for prey carrying capacity, a nonlinear functional response similar to the one described in the Holling-Tanner model, and a time-lagged numerical response to account for the delay between the consumption of prey and the production of new predators. Relatively simple models can mimic the range of complex dynamics displayed by predators and prey in simple laboratory communities.

MODELS OF PREDATION ON MORE THAN ONE PREY

All the models that we have considered thus far reside at the fringes of community ecology; by including interactions between at least two species they marginally qualify as examples of community dynamics. However, slightly more complex models, which include more species, can be used to explore the conditions in which predators will promote the coexistence of two or more competing prey species.

 Models show when predation can prevent competitive exclusion.

Parrish and Saila (1970) used a simulation model based on differential equations to show that under some circumstances predation can prolong the coexistence of two competing prey species, H_1 and H_2. The inspiration for this model was Paine's (1966) observation that more species coexisted with predators than without predators. Their model had the following form:

$$dH_1/dt = (e_1 - a_{11}H_1 - a_{12}H_2 - a_{13}P)H_1 \tag{5.17}$$

$$dH_2/dt = (e_2 - a_{22}H_2 - a_{21}H_1 - a_{23}P)H_2 \tag{5.18}$$

$$dP/dt = (-e_3 + a_{31}H_1 + a_{32}H_2)P \tag{5.19}$$

The model includes terms for intraspecific and interspecific competition among the prey (the $-a_{11}H_1 - a_{12}H_2$ terms in the equation for prey species 1, for example) and terms for different rates of predation on the two different prey (the $-a_{13}P$ and $-a_{23}P$ terms). Parrish and Saila found that for certain conditions where one prey would competitively exclude the other, predation could prolong the coexistence of the two prey species (Figure 5.8). However, even for the case shown in Figure 5.8, one prey species seems to be slowly on its way to extinction.

Cramer and May (1972) used analytical solutions of the Parrish and Saila model to show that values of parameters that lead to competitive exclusion in the absence of predators can yield stably coexisting prey when predators are present. Figure 5.9 shows the kinds of dynamics observed. Predation can lead to stable coexistence of the two prey either where predators attack the two prey similarly or where the predator feeds selectively on the competitively superior prey species.

Roughgarden and Feldman (1975) used a somewhat more complex model to explore how predation might allow a third prey species to invade a community already containing two other prey species and a predator. The model assumes that competition between species is resource based and that the competition coefficients are a function of the shape (kurtosis) and mean separation (d) between the resource utilization curves of different species:

$$dP/dt = P[(C_1/X)H_1 + (C_2/X)H_2 + (C_3/X)H_3 - s] \tag{5.20}$$

$$dH_1/dt = (bH_1/K)[K - H_1 - a(d)H_2 - a(2d)H_3 - (K/b)C_1P] \tag{5.21}$$

$$dH_2/dt = (bH_2/K)[K - a(d)H_1 - H_2 - a(d)H_3 - (K/b)C_2P] \tag{5.22}$$

$$dH_3/dt = (bH_3/K)[K - a(2d)H_1 - a(d)H_2 - H_3 - (K/b)C_3P] \tag{5.23}$$

The competition function, $a(d)$, describes the competition coefficient between two species as a function of the difference, d, between their resource optima. The function $a(d)$ decreases with increasing values of d, and $a(0) = 1$. X is the constant number of prey required to produce a new predator. C_i are the probabilities of capture of an individual prey by a predator. The predator's probability of death per unit time is s. Roughgarden and Feldman assume equal capture probabilities for H_1 and H_3 in the absence of H_2, denoted by C^*. The equilibrium population sizes of the prey are $H^* = H_1^* = H_3^* = Xs/2C^*$, and for the predator $P^* = (b/C^*)\{1 - [H^*/K(1 + a(2d))]\}$. They then find that the conditions that allow H_2 to invade when H_1 and H_3 are at their equilibrium densities are $k - 2a(d)H^* > 0$, or, equivalently, $(C^*/X)/s > a(d)/k$. Invasion is promoted if either C^* increases or s and X decrease. Invasion is hindered if either $a(d)$ increases or k decreases.

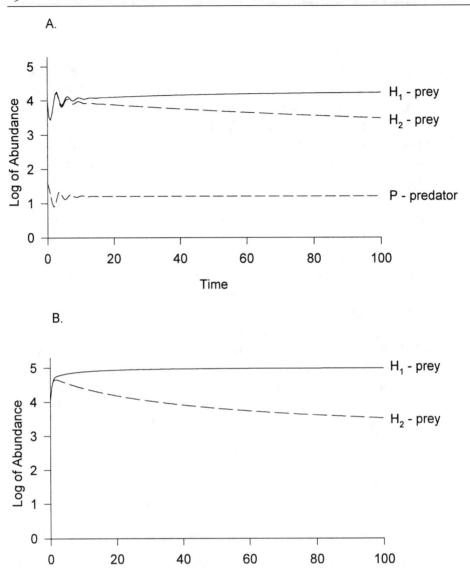

FIGURE 5.8. *An example of prolonged coexistence of two competing prey species resulting from equivalent predation on each prey species. (After Parrish and Saila, 1970.) (A) Dynamics of both competing prey with the predator. (B) Dynamics of the competing prey without the predator.*

Roughgarden and Feldman also used the model to establish three patterns:

1. The minimum niche separation distance, *d*, with the predator is never larger than without the predator. This is another way of saying that

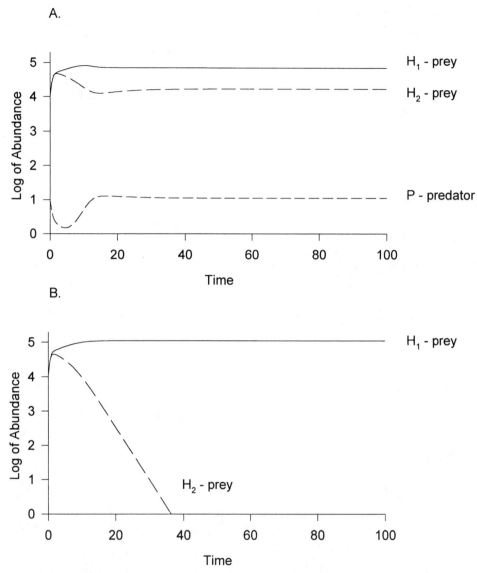

FIGURE 5.9. *Examples of a predator preventing the competitive exclusion of one prey species by another. (A) Dynamics obtained for the parameters provided by Cramer and May (1972) used in the equations of Parrish and Saila (1970). (B) Competitive exclusion in the absence of the predator for the same set of parameter values.*

predators can make it easier, but not more difficult, for a competing prey to invade the community.

2. The minimum niche separation distance depends on the properties of the predator, such that $a(d_{min}) = (C^*k)/Xs$. This means that d decreases as predation becomes more effective.

3. The minimum niche separation distance depends on both the level of predation pressure and the shape, or kurtosis, of the competition function, since $a(d_{min}) = k/2H^*$ and since when predation pressure is high, H^* is low.

Comins and Hassell (1976) describe conditions required for predator-mediated coexistence of two or many (n) competing prey species, using difference equation models. They reach conclusions roughly similar to those in

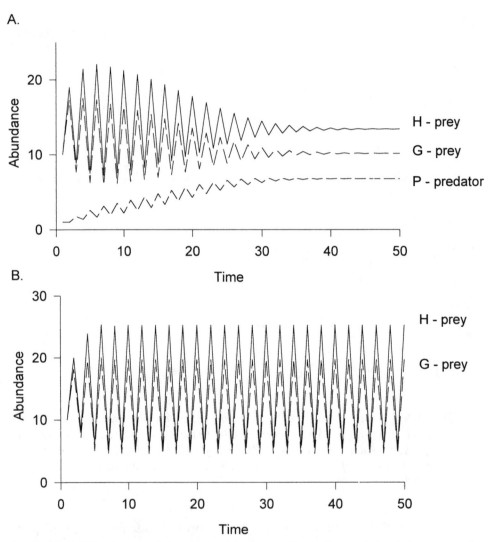

FIGURE 5.10. *An example in which predation stabilizes a limit cycle involving two species of competing prey, H and G. This example is taken from the model of Comins and Hassell (1976), using values of parameters given in Hassell (1978, Figure 7.11b). (A) Oscillations of decreasing amplitude in the presence of a predator. (B) Stable limit cycle dynamics for the same parameter values in the absence of the predator.*

Roughgarden and Feldman (1975). Their model for two prey species, H and G, looks like the following:

$$H_{t+1} = \lambda H_t \exp[(-g(H_t + \alpha G_t) - a_H P_t)] \tag{5.24}$$

$$G_{t+1} = \lambda' G_t \exp[(-g'(G_t + \beta H_t) - a_G P_t)] \tag{5.25}$$

$$P_{t+1} = H_t(1 - \exp(-a_H P_t)) + G_t(1 - \exp(-a_G P_t)) \tag{5.26}$$

The terms $\exp(-g(H_t + \alpha G_t))$ and $\exp(-g'(G_t + \beta H_t))$ describe effects of inter-specific competition between the two prey species H and G. For predation to promote the coexistence of a competitive interaction that is unstable in the absence of the predator, the prey cannot compete so strongly that $\alpha\beta > 1$, and preference for the different prey species must balance any differences in competitive superiority (Figure 5.10). If the model is made more complex by allowing the predators to switch to feeding preferentially on the most abundant prey, predators can sometimes stabilize strong competition among prey where $\alpha\beta > 1$. Similar results obtain when the model is expanded to include greater numbers of prey species.

The models that we have considered above are useful because they have helped us explore some of the conditions that might allow predators to promote the coexistence of prey species. These models all assume that coexistence in natural communities is analogous to the conditions that generate a stable equilibrium in models. As we shall see in a later chapter on nonequilibrium processes, there are other models that produce similar results, namely, the enhanced coexistence of prey species, without relying on the assumption that such coexistence corresponds to a stable equilibrium.

CONCLUSIONS

Relatively simple models of predator-prey dynamics capture the essence of interactions seen in natural and laboratory systems. Predator-prey cycles can arise from simple neutrally stable models, models that produce stable limit cycles, and models with chaotic behavior. Addition of some features, such as competition among prey, can stabilize simple models, while other biologically realistic modifications, including time lags and nonlinear functional responses, tend to destabilize the models. Models of predation on multiple competing prey species show that predators can stabilize unstable competitive interactions, or allow additional prey to invade groups of competitors.

CHAPTER 6

Food Webs

OVERVIEW

This chapter introduces the basic attributes of food webs and reviews general patterns that arise from the examination of large collections of food webs. Simple predator-prey models introduced in the previous chapter are extended to make predictions about the dynamics of species in simple food webs with different structures. These models predict that some features of simple food chains, such as chain length and feeding on multiple trophic levels, may be associated with reduced stability. There are relatively few experimental tests of the predictions that food web theory makes about population dynamics. The available evidence suggests that food chain length may depend in a complex way on both productivity and constraints imposed by population dynamics, since increases or decreases in productivity both lead to decreases in food chain length. Other topics related to food chains and food webs, such as trophic cascades, are discussed in the context of indirect effects in Chapter 8.

FOOD WEB ATTRIBUTES

A **food web** describes the feeding relations among organisms in all or part of a community. Usually those feeding relations are described by a diagram linking the consumers and consumed with lines or arrows, as shown in the examples in Figure 6.1. **Links**, the lines, indicate a predator-prey interaction

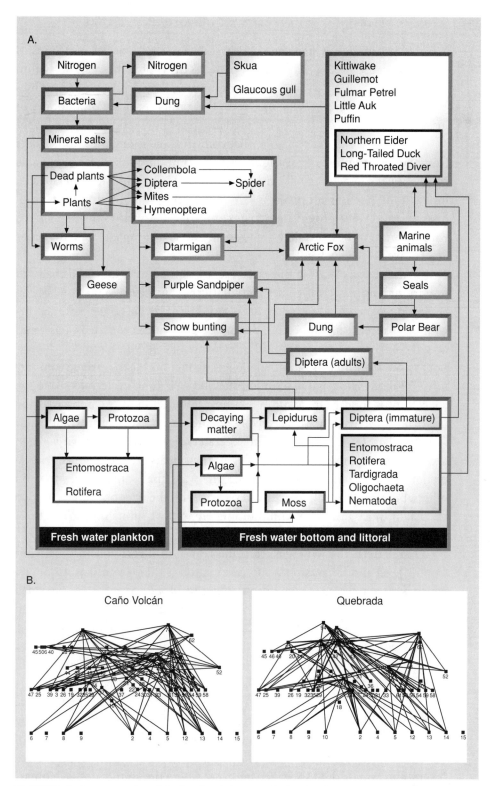

FIGURE 6.1. *Examples of food webs. (A) An early food web, representing the major feeding relations on Bear Island. (Adapted from Summerhayes and Elton, 1923, with permission of Blackwell Science Ltd.) (B) Modern food webs, representing feeding relations within communities dominated by tropical freshwater fish in Venezuela. (Adapted from Winemiller, 1990, with permission of the Ecological Society of America.)*

between two **nodes**, which can correspond to a single species or groups of species. Because food webs focus on patterns of trophic interactions within communities, they describe communities from the rather selective standpoint of predator-prey interactions. To the extent that competition among predators results from the consumption of prey, food webs also outline a subset of the possible competitive interactions within communities. Other kinds of interspecific interactions, such as mutualisms, are not described by food webs. Consequently, food webs provide less than complete descriptions of interactions within communities, but they are probably no less complete than any other descriptive device, such as the niche.

Food webs describe predator-prey relations in community.

Charles Elton (1927) emphasized the use of food webs and food chains as important summaries of community patterns. Figure 6.1 shows one of the earliest published food web diagrams (Summerhayes and Elton 1923), along with more recent, and disarmingly complex, computer-generated webs based on gut content analyses of tropical fish (Winemiller 1990). Elton posed questions about the limits of food chain length that continue to intrigue community ecologists. His original term for the food web, **food cycle**, referred to the collection of food chains within a community. Elton also emphasized the importance of basic patterns involving the sizes of organisms and their feeding relations in food chains. In general, typical predators are larger than their prey, and parasites are smaller than their hosts. This difference reflects obvious biomechanical constraints on the ways that some species feed on others, but these size differences, interacting with the sizes of habitats needed to sustain those predators, could ultimately impose limits on the length of food chains as well.

One pattern that emerges from the common inverse relation between trophic level and organism size noted by Elton is the **pyramid of numbers**, which is often referred to as an **Eltonian pyramid**. The basic idea is that small organisms at the base of the food chain are more numerous than their larger predators, and so on up through the remainder of the food chain. There are, of course, obvious exceptions to this generalization, especially where large primary producers (e.g., trees) are fed upon by much smaller and more numerous herbivores (e.g., aphids or other insects). Similar pyramids can be envisioned for biomass or productivity (measured in units of grams of carbon accumulating per unit area per unit time) for each trophic level. Inverted pyramids of numbers or biomass, where the abundance or biomass of a lower trophic level is less than in an adjacent higher trophic level, can also occur. This inversion can happen when primary producers are highly produc-

Eltonian pyramids of numbers, biomass, and energy.

can have inverted numbers or biomass but not energy!

tive, reproduce rapidly, and are rapidly cropped by consumers. This is sometimes the case in relatively clear oligotrophic lakes, where herbivorous zooplankton reduce phytoplankton to very low levels of abundance or biomass, whereas high turnover rates of phytoplankton can support a large standing biomass of consumers. However, it is thermodynamically impossible to have an inverted pyramid of productivity, since the rate of energy or biomass accumulation in higher trophic levels cannot exceed that in lower levels, which are the sole source of energy for consumers on higher trophic levels.

Raymond Lindeman (1942) made another important contribution to the study of food webs by introducing the idea of **ecological efficiency**, a measure of the fraction of energy entering one trophic level that is passed on to the next higher trophic level. Energy transfer between trophic levels is often rather inefficient, on the order of 5% to 15%.

Energy transfer between trophic levels is inefficient.

Ecological efficiency

This inefficiency of energy transfer between trophic levels provides one possible explanation for the limited length of food chains, since rather little energy remains after passing through four or five trophic levels. This idea is central to the notion that food chains may ultimately be limited in length by the interaction between primary productivity (the rate at which energy is fixed in primary producers as organic carbon) and the inefficiency of energy flow between trophic levels in food chains (Slobodkin 1960).

Despite the early recognition of the importance of food webs, most ecologists viewed webs as little more than descriptive devices. Then, in the 1970s, ecologists using two very different quantitative approaches revitalized the study of food web patterns. Joel Cohen (1978) focused interest on the statistical properties of food webs by showing that comparisons of many webs seemed to point to the existence of repeated properties, some of which are detailed below. Since the publication of Cohen's book, the collection of known food webs, which vary greatly in the taxonomic resolution of the feeding relations that they describe, has grown considerably. At about the same time, but using a very different approach based on Lotka-Volterra models of population dynamics in simple food chains, Robert May (1972, 1973) and Stuart Pimm and John Lawton (1977, 1978) raised interest in the consequences of food web structure for population dynamics. Their models explored whether differences in the structure of food chains and food webs affect the stability of populations.

Models

Most descriptions of food webs are very incomplete, often lumping or aggregating many species into single trophic categories, or nodes, which are sometimes called *tropho-species* to distinguish them from biological species.

Three categories of food webs.

Before discussing the major patterns, it is important to first understand the terms and ideas used to describe aspects of the webs. Food webs are sometimes separated into three categories: **source webs**, **sink webs**, and **community webs** (Figure 6.2). Source webs describe the feeding relations among species that arise from a single initial food source, say a single plant species. Sink webs describe all of the feeding relations that lead to sets of species consumed by a single top predator, the sink. Community webs, at least in theory, describe the entire set of feeding relations in a particular community, although this ideal goal is never realized in practice because of the extraordinary complexity of most communities.

Three types of food webs.

The following terms and concepts describe some rather abstract features of food webs that form the basis for most comparative studies. It is worth keeping in mind that these abstractions are simply a way of quantifying some of the fascinatingly complex interactions within large collections of predators and prey.

Important attributes of food webs.

- **Trophic position.** The nodes or species in the webs are distinguished by whether they are **basal species**, **intermediate species**, or **top predators**. Basal species feed on no other species, but are fed upon by others. Intermediate species feed on other species and are themselves the prey of other species. Top predators have no predators themselves, but prey on intermediate or basal species. These notions refer to the feeding relations drawn in the webs, rather than to strict biological reality. For instance, it is arguable whether true top predators really exist, since the species depicted as top predators in food web diagrams are in fact attacked by various parasites and pathogens that usually are not included in food web diagrams.

- **Links** are simply the lines that link consumers and the consumed. **Undirected links** represent a binary (all or none) property of interactions between a pair of species. If a species occurs in the diet of a predator, they are joined by an undirected link in a food web diagram. **Directed links**

FIGURE 6.2. *Source, sink, and community food webs. (A) Source web, based on the species known to feed on pine, from Richards (1926). (B) A sink food web, based on Paine's (1966) survey of feeding by* Pisaster. *(C) A community food web for Morgan's Creek, Kentucky, from Minshall (1967).*

A.

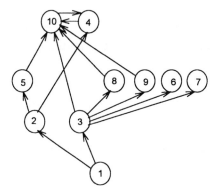

1. Pine, 2. Lepidoptera, 3. Aphids, 4. digger wasps, 5. ichneumon wasps
6. Hemiptera, 7. ants, 8. syrphid flies, 9. coccinellid beetles, 10. spiders

B.

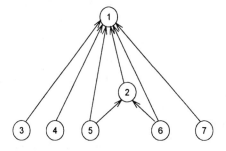

1. *Pisaster,* 2. *Thais,* 3. Chitons, 4. Limpets, 5. bivalves, 6. acorn barnacles

7. *Mitella*

C.

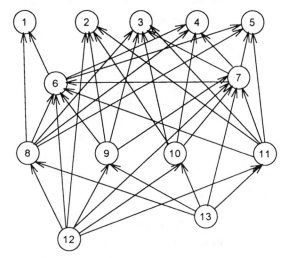

1. *Phagocata,* 2. Decapoda - *Orconectes, Cambarus,* 3. Plecoptera - *Isoperia, Isogenus*
4. Megaloptera - *Nigronia , Sialis,* 5. Pisces - *Rhinichthys, Semotilus,* 6. *Gammarus,*
7. Trichoptera - *Diplectrona, Rhyacophila,* 8. *Asellus,* 9. Ephemeroptera - 5 species,
10. Trichoptera - *Neophylax, Glossosoma,* 11. Tendipididae, *Simulium,* 12. Detritus, 13. Diatoms

are usually represented by arrows, which describe the net effect of each species on the other. Ignoring intraspecific effects, each pair of species can be joined by up to two directed links. When quantitative data on diet composition are available, as in Winemiller (1990), it is possible to use different thresholds to establish linkage; for example, species are linked only if one constitutes greater than some fixed percentage of the diet of another.

- **Connectance** is a way of describing how many of the possible links in a food web are present. One formula for connectance, based on undirected links, is

$$c = L/[S(S-1)/2] \qquad (6.1)$$

where L is the number of undirected links and S is the number of species (nodes). This formula is based on the notion that in a web consisting of S species there are $S(S-1)/2$ possible undirected links, excluding any cannibalistic links. Highly connected systems contain many links for a given number of species. Another notion of directed connectance is the probability for any pair of species selected at random that a species will have a positive or negative effect on the other (May 1973).

- **Linkage density,** L/S, refers to the average number of feeding links per species. It is a function of connectance and the number of species in the web.

- **Compartmentation** refers to the extent to which a food web contains relatively isolated subwebs that are richly connected within subwebs but which have few connections between subwebs. One formula used as an index of compartmentation is

Subwebs

$$C_1 = \frac{1}{s(s-1)} \cdot \sum_{i=1}^{s} \sum_{j=1}^{s} p_{ij} \qquad (6.2)$$

for i not equal to j, where p_{ij} is the number of species that interact with both species i and species j divided by the number of species that interact with either species i or species j, and s is the number of species in the web (see Pimm and Lawton 1980; Winemiller 1990).

- **Trophic level** refers to the number of links + 1 between a basal species and the species of interest. For all but basal species, or species in linear food chains, the notion of a trophic level becomes rather uncertain because the number of links traced from a basal species to a species higher in the food web may vary with the path taken. One way of dealing with this problem is to represent the trophic level of a species as the average of the number of links + 1 counted to arrive at that species from different basal starting points in the web (Winemiller 1990).

■ **Omnivory** occurs when species feed on prey located on more than one trophic level. It is easiest to identify when considering simple food chains or pairs of food chains (Figure 6.3). <u>**Same-chain omnivory**</u> occurs when a species in a particular food chain feeds on trophic levels in addition to the one immediately below its own trophic level (see Figure 6.3). One example is the protist *Blepharisma,* which can feed on bacteria (the basal level) as well as on other protist species (the intermediate level) that consume bacteria. <u>**Different-chain omnivory**</u> occurs when a species feeds at different levels in multiple food chains. <u>**Life history omnivory**</u> occurs when different life history stages or size classes of an organism feed on two different trophic levels. An example would be the herbivorous larvae of frogs, which transform into insectivorous adult frogs after metamorphosis.

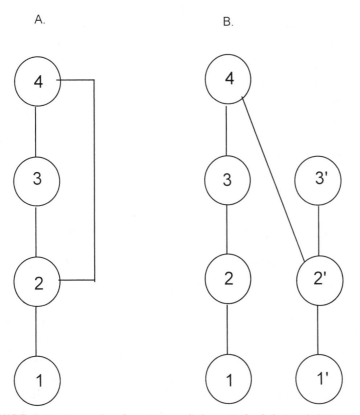

FIGURE 6.3. *Examples of omnivorous linkages in food chains. (A) Same-chain omnivory, in which one species (4) feeds on two levels (2, 3) in the same food chain. (B) Different-chain omnivory, in which a species (4) feeds on different levels (3, 2′) in two connected food chains.*

- **Cycles** and **loops** occur if species have reciprocal feeding relations. A cycle occurs if each of a pair of species eats the other. The top predators in the food web shown in Figure 6.2A are an example of a cycle, where wasps eat spiders, and spiders eat wasps. A loop occurs if species 1 eats species 2, species 2 eats species 3, and species 3 then eats species 1. Cycles and loops generally occur where species have a range of size or age classes and where large individuals of each species are capable of eating smaller individuals of the other.

- **Rigid circuit** properties have to do with the way that overlaps in the prey consumed by predators can be described. For any food web, one can draw a predator overlap graph such that predator species that have at least one prey in common are linked by a line segment (Figure 6.4). If every series of three predators completes a triangle of line segments, the predator overlap graph is said to have the rigid circuit property.

- **Intervality** is a property that is related to the rigid circuit nature of predator overlap graphs. If a food web is interval, overlaps between predators can be represented by a series of overlapping line segments, as indicated in Figure 6.4. If line segments cannot be so placed, such that a segment must be broken to represent prey overlaps, the web is not interval. This admittedly esoteric property of food web graphs has a possible link to the dimensionality of the niche space required to represent feeding overlaps among species. Cohen (1978) has argued that if food webs are interval, then the niche space required to represent overlapping feeding relations is unidimensional, for example, a series of overlapping line segments arranged along a line.

PATTERNS IN COLLECTIONS OF FOOD WEBS

Cohen (1977, 1978) was the first to suggest that even coarsely drawn diagrams of food webs yielded some repeatable, and therefore interesting, patterns. The ecological significance of these and other patterns remains controversial, since many ecologists have serious reservations about the accuracy and completeness of food web descriptions (Paine 1988). Many published descriptions of food webs are simply descriptive devices created to illustrate subsets of important interactions within communities and were never intended to serve as complete descriptions of trophic linkages. For example, Paine's (1966) *Pisaster* sink web only describes interactions between seven nodes, but the community contains at least 300 macroscopic species (Paine 1980)!

Patterns emerge from comparisons of published food webs.

A.

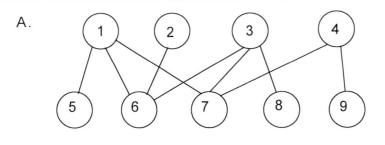

B.

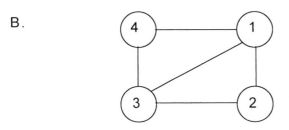

C.

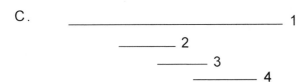

FIGURE 6.4. *Examples of the rigid circuit property of a simple hypothetical food web. (A) The food web. (B) The predator overlap graph, in which line segments connect predators that share at least one prey species. Predators that share no prey species are not directly connected by line segments. (C) An interval graph, showing that overlaps in diet for predators can be represented by overlapping line segments arranged in a single dimension.*

Lawton and Warren (1988), Lawton (1989), and Pimm et al. (1991) have summarized the broad patterns emerging from collections of food webs. The 10 important patterns summarized by Lawton and Warren (1988) are outlined below. Some of these patterns have become more equivocal with the advent of increasingly detailed food web descriptions. Despite the controversy surrounding the significance of these patterns, they are described here to illustrate the kinds of properties that studies of food webs address.

1. Many collections of food webs have constant ratios of predator to prey species, or ratios of basal to intermediate to top predator species. Cohen (1978) found that his collections of community webs yielded ratios of numbers of predators to prey of about 4:3 (Figure 6.5). At first glance

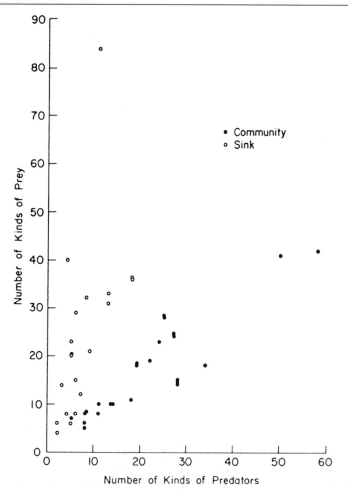

FIGURE 6.5. *Relations between the numbers of predator nodes and prey nodes in collections of community food webs. The linear relation suggests a ratio of approximately four predator nodes to three prey nodes in the community webs shown by the filled circles. (From Cohen, 1978. © 1978 by Princeton University Press. Reprinted by permission of Princeton University Press.)*

this seems odd, since it suggests that a larger number of predator species are being supported by a fewer number of prey species. It is less disconcerting when you consider that most prey "species" in this analysis are in fact highly aggregated collections of taxa—things like "insects" or "plants." Later analyses extended the constancy of proportions to basal, intermediate, and top predators (Briand and Cohen 1984; Cohen and Briand 1984). Subsequent analyses of more detailed food webs have examined the effect that aggregating species into tropho-species has on food web patterns (Sugihara et al. 1989; Martinez 1991). Sugihara et al. found that additional aggregation of already aggregated webs did little to change these patterns. Martinez (1991) found that aggregating a very finely resolved food web had little effect on ratios of predator to prey nodes, but did influence the ratio of top predators to total species, effectively overestimating the ratio of top predators to total species in highly aggregated webs.

2. Cohen's second major conclusion was that, more often than not, food webs tended to be interval in nature. There is no neat linkage between this property of food webs and any single biological process. As noted above, intervality is consistent with the notion that overlaps among predators in the prey that they consume can be represented by a series of line segments arranged in a single dimension. This may be the same thing as saying that a single niche dimension is sufficient to describe the feeding relations within a collection of predators. A descriptive model, called the **cascade model** (Cohen and Newman 1985; Cohen et al. 1985; Cohen et al. 1986), can produce webs that are interval, although the biological mechanism involved in generating these patterns remains uncertain. The cascade model assumes that a constant linkage density exists, and also assumes that species can be ordered into a hierarchy such that species low in the hierarchy can be consumed by ones higher in the ordering of species. This kind of ordering might result if predators must be larger than their prey, or if parasites must be smaller than their hosts.

3. Three-species loops are infrequent (Lawlor 1978). Close inspection of very detailed webs has shown that two-species cycles and three-species loops can arise in systems with size-dependent or stage-dependent predator-prey interactions (Polis 1991). In such systems, the roles of predators and prey can reverse with reversals in the relative sizes of interacting species, as larger organisms generally eat smaller ones.

4. Early analyses suggested that the number of links per species, **linkage density**, was constant across collections of food webs in which the nodes consisted of highly aggregated sets of species (Cohen and Newman 1985; Cohen et al. 1986). If this is the case, then connectance should decline hyperbolically with increasing species richness, according to the relationship given in Equation 6.1. Analysis of other, more detailed, food webs in which nodes correspond to less aggregated groups shows instead that connectance is constant over a fairly broad range of species richness (Figure 6.6; see Martinez 1992).

Connectance fairly constant over broad range of sp richness

5. The average proportions of links between basal, intermediate, and top species also seem relatively constant (Briand and Cohen 1984). This pattern may be no more than a simple consequence of constant linkage

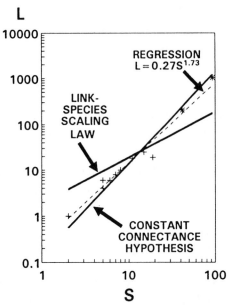

FIGURE 6.6. *Alternate patterns suggested by the constant connectance hypothesis and constant links per species hypothesis. L and S refer to numbers of links and species per web. For highly resolved webs based on less-aggregated trophic nodes, the constant connectance hypothesis provides a better description. (Reprinted from Martinez, 1992, with permission of the University of Chicago Press.)*

density and the constant proportions of species in basal, intermediate, and top positions.

6. Food chains are relatively short, usually containing no more than five or six species (Elton 1927; Hutchinson 1959; Pimm and Lawton 1977; Pimm 1982). This pattern is partly due to the low taxonomic resolution of many webs, as food chains tend to increase in length in more detailed webs (Martinez 1991). Both energetic (Lindeman 1942; Slobodkin 1960) and population dynamic (Pimm and Lawton 1977) hypotheses have been proposed to account for this pattern. These ideas are described in greater detail below.

7. Omnivory appears to be relatively infrequent in some systems (Pimm and Lawton 1978), but this may be a consequence of inadequate description rather than biological reality. In more recent detailed descriptions of some food webs (Sprules and Bowerman 1988; Polis 1991; Martinez 1992), omnivory is common. Omnivory also seems common in webs rich in insects and parasitoids, or in decomposers.

8. Connectance and estimated interaction strength appear to vary between webs in relatively constant and variable environments (Briand 1983).

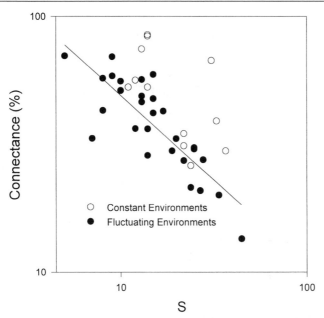

FIGURE 6.7. *For a given level of species richness, s, food webs in constant environments (open circles) have higher levels of connectance than webs in fluctuating environments (filled circles). (Redrawn from Briand, 1983, with permission of the Ecological Society of America.)*

Webs in variable environments appear to be less connected than ones in more constant environments (Figure 6.7). If one assumes that an inverse relation between connectance and per capita interaction strength exists (from May 1973; see the discussion of stability and complexity below), then species in more variable environments also interact more strongly.

9. Webs do not seem to be strongly compartmented or subdivided (Pimm and Lawton 1980). Some exceptions occur in situations in which webs describe communities that span discrete habitat boundaries, but even then, subwebs tend to be interconnected.

10. Food chains in two-dimensional habitats, such as grasslands, seem to be shorter than those in three-dimensional habitats, such as lakes, open oceans, or forests with a well-developed canopy structure (Briand and Cohen 1987).

EXPLANATIONS FOR FOOD WEB PATTERNS

Explanations for food web patterns draw heavily on two kinds of models: dynamic models based on extensions of the Lotka-Volterra predator-prey

Possible causes of some food web patterns.

models, and static models, such as the cascade model of Cohen et al. (1985), that make no specific reference between population dynamics and food web patterns. Dynamic models attempt to explain food web patterns on the basis of food web configurations that promote stable equilibrium population dynamics, which presumably allow populations to persist for long periods of time, as opposed to configurations that are unstable and that presumably fail to persist for very long. The models used to predict these patterns are based on relatively simple Lotka-Volterra models that have been extended to include more than two species (May 1973; Pimm and Lawton 1977, 1978).

For a system of n species, the differential equation for the dynamics of species i looks like

$$dX_i/dt = X_i(b_i + \sum a_{i,j}X_j) = F_i \qquad (6.3)$$

where b_i is the per capita population growth rate of species i, $a_{i,j}$ is the per capita effect of species j on species i, including intraspecific effects when $i = j$, and X_i is the abundance of the species i, in a system of n species. The stability of these systems depends on the properties of the Jacobian matrix (see the appendix), which consists of the matrix of partial derivatives $\partial F_i/\partial X_i$ evaluated at the equilibrium densities of the n species, the X_i^*. Models of simple food chains can be constructed by choosing the elements of the Jacobian matrix from an appropriate range of values. Different food chain configurations can be modeled by setting entries to zero, positive, or negative values, as shown in Figure 6.8. The return time of the system, which is approximately the time required for the system to return to equilibrium following a perturbation, is roughly $1/\lambda_{max}$, the reciprocal of the largest negative eigenvalue of the Jacobian matrix. This approach allows comparisons of the stability and return times for simple-model food webs of different configurations.

Stuart Pimm and John Lawton (1977) used this approach to assess the dynamics of systems of four "species" arranged in food chains of different

 Models suggest that long food chains may be unstable.

length. The assumptions included were that basal species were self-limiting (negative a_{ii}'s for basal species), whereas other species were limited only by their food supply and their predators. For each food chain configuration, numerical entries in the appropriate Jacobian matrix were selected at random from a uniform distribution of values of the appropriate sign and magnitude. This process was repeated 2000 times, a process called Monte Carlo simulation, to produce frequency distributions of return times and to estimate the frequency of stable and unstable webs. One result, shown in Figure 6.9, is that all of the food chains consisting of four

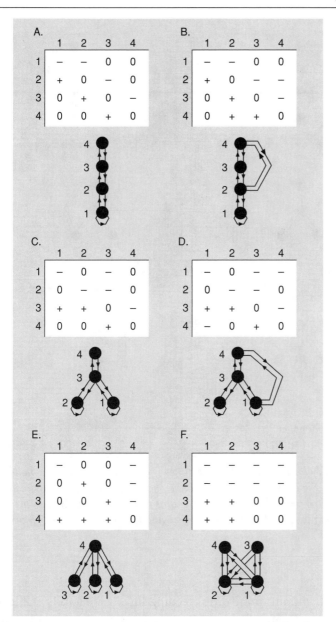

FIGURE 6.8. *Schematic Jacobian matrices and corresponding food chains, showing systems simulated by Pimm and Lawton in their studies of the dynamics of model food chains. Numbers identify species located in particular trophic positions. Positive and negative signs in the interaction matrices correspond to directed links in the food chains. Negative signs on the diagonal correspond to intraspecific density dependence. (Adapted with permission from* Nature *268: 329–331, S. L. Pimm and J. H. Lawton. © 1977 Macmillan Magazines Limited.)*

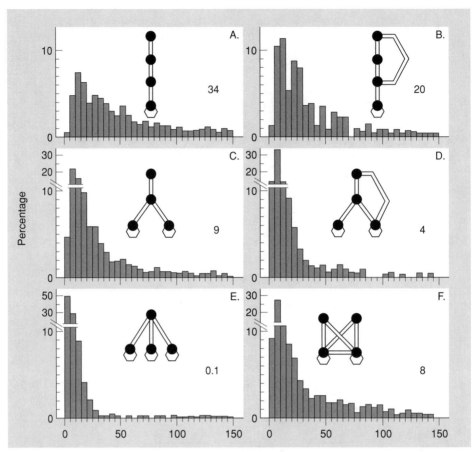

FIGURE 6.9. *Frequency distributions of return times (horizontal axis) for the model food chains described in Figure 6.8. Longer food chains, shown in panel A, have a greater frequency of long return times than the shorter chains in panels C and E. (Adapted with permission from Nature 268: 329–331, S. L. Pimm and J. H. Lawton. © 1977 Macmillan Magazines Limited.)*

species arranged without omnivorous feeding links were locally stable, but return times were substantially longer in longer chains. Longer return times suggest that populations in longer chains would require longer periods of time to return to equilibrium values following a perturbation. Pimm and Lawton equated these prolonged return times in longer chains with reduced stability, in the sense that they would recover more slowly after perturbation. An example of that property is shown for a pair of two-level food chains in Figure 6.10, which are contrived to differ in their return times. If perturbations are large or frequent, populations in systems with long return times might be more prone to extinction.

Recent work suggests that the greater stability of the shorter food chains

A.

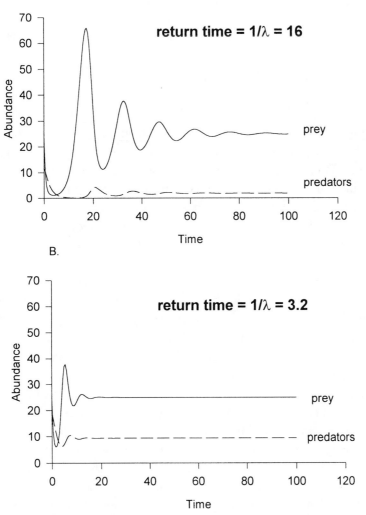

FIGURE 6.10. *Examples of dynamics produced by differences in return times in simple two-level food chains described by Equations 5.3 and 5.4. Differences in return times are generated by giving the prey a higher rate of increase (2.5 versus 0.5) in the system with the shorter return time. Simulations of both systems begin at the same displacement from equilibrium.*

modeled by Pimm and Lawton (1977) may be an artifact of the way that density-dependent population regulation was assumed to operate in model chains. Sterner et al. (1997) pointed out that the shorter food chains modeled by Pimm and Lawton had greater numbers of species on the basal trophic level with density-dependent self-regulation. Consequently, the greater stabil-

Artifact of models

ity may have been a consequence of a greater frequency of density-dependent self-regulation and not of food chain length per se.

The second aspect of food chain architecture considered by Pimm and Lawton (1978) was the effect of same-chain omnivory on population dynamics within these relatively simple four-species food chains. As before, omnivory could be modeled by including appropriate entries in the Jacobian matrix and then evaluating the eigenvalues of the Jacobian. Omnivory had an even more striking effect on dynamics than did food chain length. Fully 78% of the longer chains with an omnivorous link were unstable. Of the remaining 22% that were stable, return times were on average shorter than in comparable food chains without omnivores. The conclusion was that omnivorous systems should be rare, given the unstable behavior of their dynamics. However, those relatively few stable systems that contained omnivores should be more stable (in the sense of having shorter return times) than comparable food chains without omnivores.

Omnivory may also destabilize food chains.

Omnivory effects

Robert May (1972, 1973) used a similar approach to compare the stability of webs differing in species richness, connectance, and the intensity of interactions between species. Rather than using webs of a particular predetermined structure, May constructed randomly connected food webs consisting of s model species. Each of the species was assumed to display intraspecific density-dependent regulation, which is modeled by placing -1's down the diagonal of the Jacobian matrix from upper left to lower right. Interactions between species are modeled by selecting off-diagonal elements of the Jacobian matrix at random and then filling the entries with positive or negative values from a normal distribution with a mean of zero and variance i. The larger the value of i, the larger a nonzero value describing the strength of an interaction is likely to be. In this model, connectance, c, is the probability that an off-diagonal element will be nonzero. May explored the relative contributions of s, species richness, c, connectance, and i, which he termed interaction strength, to the stability of these model systems. His main result was that as s becomes arbitrarily large, to a reasonable approximation, the system will be stable if $i(sc)^{1/2} < 1$. This means that increases in i, s, or c will tend to be destabilizing in randomly connected model food webs. Counter to the conventional wisdom of most field ecologists, (e.g., Elton 1958), increases in the complexity of a system involving increases in either the number of species (n) or the richness of trophic connections (c) should create greater instability in that system. One reason for this is that in increas-

Random food webs may be less stable as complexity increases . . .

ingly complex systems, there are more ways for things to go badly wrong, in the sense that there are more opportunities for unstable interactions to arise.

Other theoretical ecologists have suggested that May's conclusions depend critically on the way in which he constructed his models and that different models lead to rather different conclusions. Donald DeAngelis (1975) found that stability increased with increasing values of connectance, c, under conditions where 1) predators consumed only a small fraction of prey biomass, that is, predators had only modest effects on prey abundance, 2) predators in higher trophic levels were strongly self-regulated, and 3) there was a bias toward what DeAngelis called donor dependence in interactions. Donor dependence implies that for a situation in which species j is eaten by species i, $\partial F_{ij}/\partial X_j > \partial F_{ij}/\partial X_i$, or in other words, the predator's dynamics are more strongly affected by changes in prey abundance than by predator abundance.

Lawrence Lawlor (1978) also questioned whether May's model food webs were biologically realistic, since randomly connected food webs are likely to contain problematic features such as three-species feeding loops. The probability that a randomly constructed web will contain no three-species loops, given that it contains n species and has a connectance of c, is $[1 - 2(c/2)6]^{(s!/(s-3)!3!)}$. This probability becomes vanishingly small as s and c increase to the levels used in May's original study. The upshot is that for many values of n and c, May's approach produces webs that have a high probability of containing three-species feeding loops. Although Lawlor argued that this was an unrealistic feature of May's approach, recent detailed studies of complex natural food webs show that three-species feeding loops do in fact occur (Polis 1991).

> ... but random webs differ from real ones in many ways.

There is one other idea relating stability to complexity, but it differs from the ideas discussed above in focusing on the stability of a top predator rather than on the stability of the entire food web that contains the predator. Robert MacArthur (1955) argued that predators feeding on multiple prey species are more likely to weather crashes in the abundance of a single prey species than are specialized predators that depend entirely on a single prey species for their food (Figure 6.11). The idea is fairly simple and involves the notion that the existence of more than one pathway of energy flow to a predator should buffer the predator against fluctuations in prey abundance, as long as fluctuations in prey abundance are not positively correlated over time (i.e., fluctuations are not simultaneous and in the same direction).

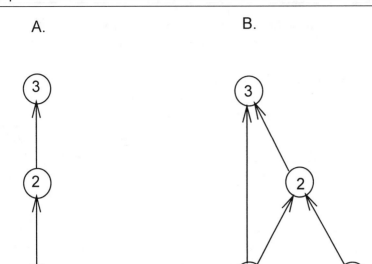

FIGURE 6.11. *Examples of single and multiple trophic pathways in specialized and generalized predators. (A) In the simple food chain, extinction of either species 1 or 2 will lead to the extinction of species 3. (B) In the more complex chain, alternate pathways of energy flow exist, such that some energy will reach species 3 if species 1 is lost and 2 remains, or vice versa.*

EXPERIMENTAL TESTS OF FOOD WEB THEORY

Causes of Food Chain Length

Most explorations of the possible causes of patterns in food webs rely heavily on models because the dynamics of species in natural food webs are difficult to study. Long-lived species require equally long-term studies to separate apparent dynamics from artifacts imposed by life history traits (Frank 1968; Connell and Sousa 1983). For example, very long-lived organisms, such as trees, might appear to be stable simply because their dynamics occur on a different timescale than do those of shorter-lived organisms, such as bacteria. To avoid such artifacts, temporal changes in population sizes must be scaled against the generation time of the organisms in question. It is also very difficult to collect information about the dynamics of complex multispecies systems in which species operate on very different timescales. Consequently, experimental studies of links between food web attributes and the population dynamics of their component species tend to focus on simple systems containing organisms with short generation times. There is also the nontrivial problem of actually determining the feeding relations in a natural food web. Determining the major feeding links in a single food web can consume years of dedicated effort (e.g., see Polis 1991; Winemilller 1990), even without making an attempt to observe population dynamics! Despite all these problems, there have been some experimental tests of food chain hypotheses performed with organisms having short generation times.

If food chain length is determined primarily by the inefficiency of energy flow between trophic levels, experimental manipulations of productivity should affect the lengths of food chains. Pimm and Kitching (1987) and Jenkins et al. (1992) have tested the effects of variation in productivity on the relatively simple food webs that develop in water-filled tree holes in tropical Australia. The longest food chains in naturally occurring tree holes have been resolved to four trophic levels: 1) detritus, primarily leaf litter that falls into the tree holes and forms the basal trophic level and main source of energy that supports the food chain, 2) larval mosquitoes and chironomid midges, 3) larvae of a predatory midge, *Anatopynia*, and 4) predatory tadpoles of the frog *Lechriodus fletcheri*. Another nice feature of this system is that small plastic containers that retain water can be

 Experiments show that food chains are longer in more productive environments.

used as artificial tree holes in experimental studies. Typical tree-hole food webs develop when these containers are placed near trees. Pimm and Kitching (1987) manipulated productivity by adding different amounts of litter to a series of artificial tree holes, and observed the food chains that developed. Litter additions bracketed the normal amount observed (903 g/m^2/yr) and included additions of one-half normal, normal, and two times normal amounts of litter. The additions produced slight, but nonsignificant, increases in the abundance of *Anatopynia*, and significant declines in the abundance of *Lechriodus*. Kitching and Pimm concluded that, if anything, increasing productivity decreased food chain length. Subsequent experiments by Jenkins et al. (1992) examined patterns of food web development over a greater range of experimentally manipulated levels of productivity. This time, productivity varied over two orders of magnitude, including levels of detritus input that were natural, 0.1 times natural, and 0.01 times natural. Community development was followed for up to 48 weeks by establishing a total of 15 replicates at each level of productivity and then destructively sampling 3 replicates in each series after 6, 12, 24, 36, and 48 weeks of community development. These results suggested that decreasing productivity resulted in decreases in the number of coexisting species, the number of trophic links, and maximum food chain length (Figure 6.12).

Other studies of protists in simple laboratory microcosms support the notion that dynamics become increasingly unstable with increases in productivity or food chain length. Luckinbill (1974) showed that an apparently unstable interaction between two ciliated protists, the prey *Paramecium* and its predator *Didinium*, became increasingly stable when the amount of food

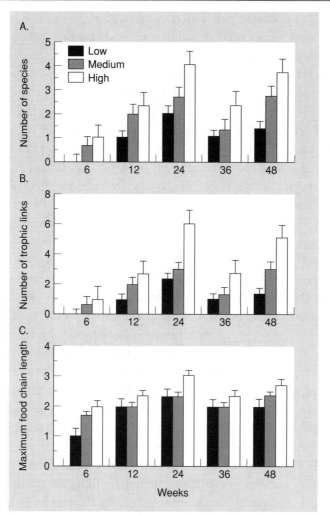

FIGURE 6.12. *Number of species, number of links, and food chain length in tree-hole communities subjected to different levels of nutrient inputs for 48 weeks. Different levels of productivity, denoted high, medium, and low, correspond to 1 times, 0.1 times, and 0.01 times normal levels. (Adapted from Jenkins et al., 1992, with permission from Oikos).*

entering the system was reduced. Luckinbill manipulated food input by adding increasingly dilute suspensions of bacteria, which served as food for *Paramecium*. At the highest food concentration used, 6 ml bacteria per 350 ml total, abundances of *Paramecium* and *Didinium* go through a single strong oscillation that results in extinction after about 6 days. Dilution to 2.0 ml bacteria per 350 ml total yields about five repeated oscillations and persistence for 34 days (Figure 6.13). The relation between persistence and food supply appears nonlinear, with a threshold of greatly enhanced persistence occurring between 4.5 ml bacteria per 350 ml total and 2.5 ml bacteria per 350 ml total.

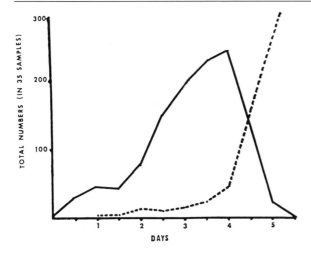

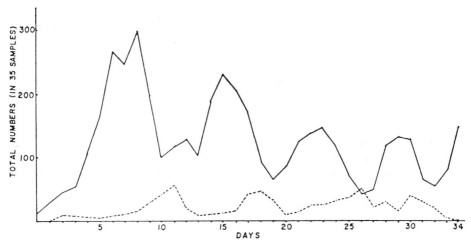

FIGURE 6.13. *Examples of decreasing stability of simple food chains with increasing levels of energy input. Dashed lines show the abundance of the predator,* Didinium. *Solid lines show the abundance of the prey,* Paramecium. *(Top) A single oscillation ending in prey extinction after six days at high nutrient levels. (Bottom) Sustained oscillations for the same species interacting at lower nutrient levels. (Reprinted from Luckinbill, 1974, with permission of the Ecological Society of America.)*

Thus, dynamics become increasingly unstable, and extinction becomes more likely, at higher nutrient levels.

If population dynamics are less stable in long food chains than in short ones, experimental manipulations of food chain length should produce observable differences in population dynamics. Specifically, dynamics should be more variable in longer chains. Sharon Lawler and Peter Morin (1993b) found

> Long food chains display more variable dynamics than shorter chains.

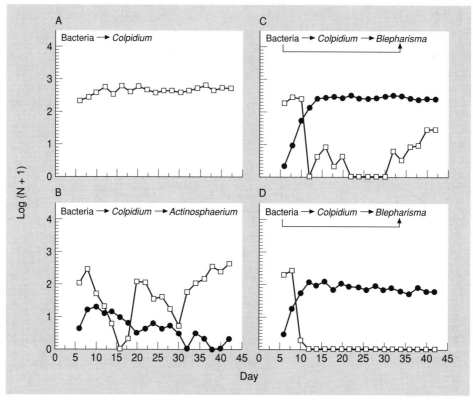

FIGURE 6.14. *Increased temporal variation in population dynamics that accompanies an increase in food chain length by one trophic level. Populations of the same species, Colpidium, in long food chains (B, C, D; open squares) exhibit greater fluctuations in abundance over time compared with their dynamics in shorter food chains (A). (Adapted from Lawler and Morin, 1993b, with permission of the University of Chicago Press.)*

that the population dynamics of protists in simple laboratory food chains become less stable with modest increases in food chain length. They compared the temporal variability of populations of the same bacterivorous protists in short food chains in which bacterivores were the top predators and in slightly longer food chains in which the bacterivores were intermediate species preyed on by another predatory protist. In the majority of cases, an increase in food chain length caused increased temporal variation in abundance (Figure 6.14). Increased temporal variation in abundance would be consistent with longer return times in longer food chains, as in Figure 6.10.

These somewhat conflicting results suggest that productivity influences the length of food chains, but in a curvilinear way (Figure 6.15). Below natural levels of productivity, there is insufficient energy to sustain higher trophic levels, and species may be lost. Above natural productivity levels,

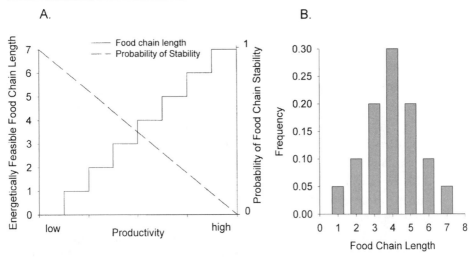

FIGURE 6.15. *(A) Hypothetical relations between productivity and food chain stability or persistence. (B) Effects of productivity and stability on possible distributions of food chain length within or among habitats. At low productivity levels, food chain length is determined primarily by energy availability. Higher levels of productivity make longer food chains energetically possible, but may also decrease the probability that the longer chains will be dynamically stable. This scenario is consistent with observations of decreased food chain length in response to increases or decreases of productivity, if most food chains initially occur at intermediate levels of productivity.*

species may be lost either through direct toxic effects of eutrophication or through the increasingly unstable dynamics that occur in some systems as productivity increases (Rosenzweig 1971; Luckinbill 1974). Some simple predator-prey models become unstable as productivity increases (Rosenzweig 1971). Some simple laboratory predator-prey systems also become increasingly unstable as productivity increases (Luckinbill 1974).

Omnivory, Increasing Trophic Complexity, and Stability

Morin and Lawler (1996) found that omnivorous protists had rather unpredictable effects on their prey and were unable to confirm the hypothesis that omnivores have particularly destabilizing effects on simple laboratory food chains consisting of bacteria and protists. However, they did find that omnivores had consistently larger population sizes than did other nonomnivorous predators under comparable conditions. This conclusion is tempered by the small number of omnivorous species that they examined. One fairly consistent feature of omnivore population dynamics was predicted by MacArthur (1955). Omnivorous protists that can feed on both bacteria and other bacterivorous protists tend to have more stable, less temporally variable dynamics than nonomnivorous, relatively specialized predators that track the fluctuations in a single prey species (Morin and Lawler 1996). Species with more than one prey are less likely to fluctuate greatly in abundance when one of their prey fluctuates in abundance.

Andrew Redfearn and Stuart Pimm (1988) used the comparative method to test MacArthur's hypothesis. They surveyed published accounts of the population dynamics of herbivorous insects that were known to feed on many versus few species of plants. Their results provide some qualified support for MacArthur's hypothesis, in that less-specialized species tend to show reduced fluctuations in population dynamics over time when compared with more-specialized insects that feed on relatively few species.

Sharon Lawler (1993b) also used studies of protists in laboratory microcosms to explore whether more complex food webs were less stable than simple ones. Her simplest systems consisted of four different three-level food chains containing different species of bacterivores and top predators but similar bacteria. Each of these four food chains was known to be stable. These chains were then paired and combined to form eight different communities containing four protist species, or one community containing all eight protist species (Figure 6.16). The main result was that webs containing increasing numbers of species, and increasing possibilities for kinds of predator-prey interactions, exhibited significant increases in the frequency of extinctions of component species. This finding is in general agreement with May's (1972, 1973) original suggestion that increasing complexity in food webs may decrease rather than increase the stability of the system as a whole.

Increasing complexity can decrease food web stability.

Interaction Strength

Paine (1992) has suggested another empirical approach to studies of interactions in natural food webs. His approach focuses on the experimental measurement of interaction strengths for an assortment of predators and their prey. The approach is labor intensive, since it involves measuring how prey respond to replicated removals of various predator species. Paine's operational measure of interaction strength is an index, I, that is calculated using the following expression:

$$I = (D_p - D_0)/(D_p)P \qquad (6.4)$$

where D_p is the density of the prey with a known density of predators, P is the known density of predators, and D_0 is the prey density when predators are removed. Negative values indicate negative per capita effects of predators on prey, but positive effects are possible if predators facilitate certain prey by removing others, as in Paine (1966).

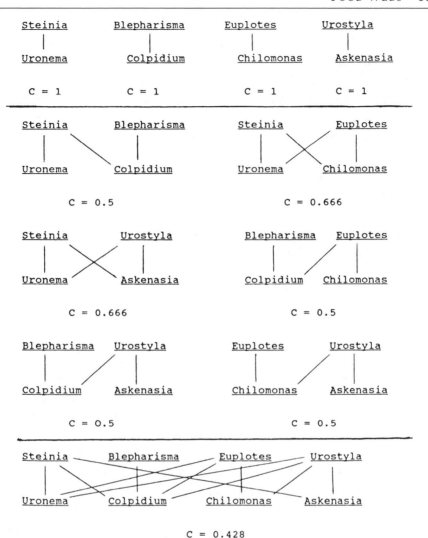

FIGURE 6.16. *More complex food webs produce more frequent extinctions in simple laboratory microcosms. Protist food webs consisted of two, four, or eight protist species, with each web replicated five times and having the values of connectance, c, listed below the web. Only 1 of 40 populations (2.5%) went extinct in the two-species webs, whereas 26 of 120 populations (21.7%) went extinct in the four-species webs, and 11 of 40 populations (27.5%) went extinct in the eight-species webs. (Data from Lawler, 1993b.)*

Application of this approach to an array of seven species of herbivores (predators) known to feed on sporelings (prey), the recently settled juveniles of intertidal brown algae, showed that only two of the seven species had strong significantly negative effects on the prey. The remaining five of the seven interacted either weakly or even positively with the prey. Paine's results suggest that the

> Many interactions in natural food chains are weak.

use of known trophic links, rather than interaction strengths, may badly over-estimate the frequency of important trophic connections in real food webs. It is also important to point out that Paine's measure of interaction strength is very different from the one used by May (1973). Paine's measure potentially includes both direct and indirect effects (see Bender et al. 1984; Yodzis 1988). May's interaction strength involves only direct effects, since it is the value of a partial derivative evaluated at equilibrium for a particular pair of species. The various measures of interaction strength that have been used by ecologists are described and compared in an important paper by Laska and Wootton (1998).

SOME FINAL QUALIFICATIONS CONCERNING EMPIRICAL PATTERNS

Food web research is an active, dynamic, and rapidly changing field. As more and better descriptions of food webs accumulate, some of the original generalizations about food web patterns have become problematic (see Lawton 1989; Pimm et al. 1991). Examples of two current concerns are whether some of the original major patterns seen in collections of food webs are independent of the scale of taxonomic resolution used in depicting the web (termed **scale independence**) and whether the patterns within webs vary significantly within communities over relatively short—seasonal or annual—timescales.

Scale independence refers to whether basic patterns, such as connectance, linkage density, food chain length, or ratios of numbers of taxa in different trophic categories, depend critically on the level of taxonomic resolution employed. The first studies that compared differences involving relatively coarse levels of taxonomic resolution suggested scale invariance (e.g., Briand and Cohen 1984; Sugihara et al. 1989). More recent studies of the effects of aggregating highly resolved webs, in which most nodes in the web correspond to real species or genera, suggest that aggregation may distort some patterns (Martinez 1991, 1992; Polis 1991). Webs with greater taxonomic resolution tend to have greater numbers of omnivores, longer food chains, and roughly constant connectance when compared with webs in which nodes are highly aggregated collections of many biological species.

A second question concerns the degree of temporal variation in food web patterns. Most published food web diagrams depict interactions that are possible, but may include interactions that are infrequent or interactions among seasonally fluctuating species that are seldom simultaneously active in the same community. They are collages, rather than single snapshots, of the interactions

Food web patterns vary over time.

within a community. A few studies have explicitly explored patterns of temporal variation in food web patterns. Kitching (1987) found substantial temporal variation in the composition of his tropical tree-hole communities. Warren (1989) also found substantial temporal variation in the patterns that he observed in an exceptionally well-described pond food web (Figure 6.17). Schoenly and Cohen (1991) also explored patterns of temporal variation in a small collection of webs in which at least some data on temporal variation could be found. The general pattern is that temporal aggregation of food web patterns probably overestimates the actual number of taxa that are interacting at any particular time. By lumping nonsimultaneous interactions, say, interactions between a long-lived predator and short-lived phenologically separated prey, temporal aggregation also overestimates the actual level of connectance in the community at any particular time.

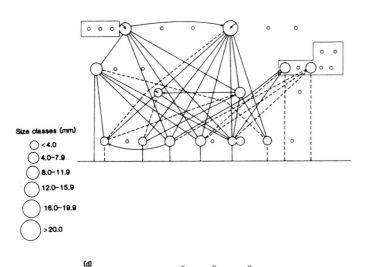

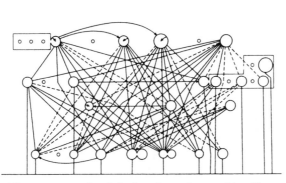

FIGURE 6.17. *The structure of real food webs varies considerably over time. This figure shows temporal variation in the patterns within a food web in a small pond. (Top) Food web in March. (Bottom) Food web in October. (Reprinted from Warren, 1989, with permission from Oikos.)*

CONCLUSIONS

Even if many of the early generalizations about food web patterns eventually fail to survive the careful scrutiny of increasingly detailed data sets, food webs will retain an important role in community ecology. Food webs can identify pathways of potentially important interactions, including indirect effects (Wootton 1994b), and they emphasize that communities are far more complex entities than arbitrary collections of pairwise interactions among species. Experimental tests of food web theory are rare (see Morin and Lawler 1995), and much important work remains to be done in this area.

CHAPTER 7

Mutualisms

OVERVIEW

This chapter completes the description of elementary interactions between species, focusing on positive interactions among species, which include mutualisms, commensalisms, and other kinds of beneficial associations. Mutualisms are reciprocally positive interactions between species. Commensalisms involve unidirectional positive effects, in which one species positively affects another, but the second species has no net effect on the first. The major kinds of mutualisms involve interactions that influence energy supply, nutrition, protection from enemies or harsh environments, and transport of gametes, propagules, or adults. Mutualisms can be facultative or obligate, and simple mathematical models can mimic the full range of mutualisms observed in nature. Case studies of a variety of mutualistic interactions are presented to emphasize the frequently overlooked role of mutualisms in community organization.

KINDS OF MUTUALISMS

Most ecology texts give short shrift to positive interactions among species, emphasizing instead the various negative ways that species can interact as either competitors or predators and prey. The tendency to overlook positive effects of one species on another neglects the potential importance of some of the more fascinating interspecific interactions that can occur in communities.

This oversight is unfortunate given that mutualisms, while often inconspicuous, are common and potentially important forces that influence the structure and function of communities (see Bronstein 1994 and Connor 1995 for reviews). For instance, the mycorrhizal association formed between fungi and the roots of many higher plants can influence seedling establishment and the outcome of competition. Many higher plants are involved in a facultative mutualism with arthropods and vertebrates that pollinate their flowers and disperse their seeds. Although these positive interactions are often emphasized by ecologists who study various forms of plant-animal interactions, their impact on community organization remains little explored.

Boucher et al. (1982) suggest that mutualisms fall into four basic categories: **energetic**, **nutritional**, **protective**, and **transport** associations. Each member of a mutualistic association may benefit from the association in different ways. Mutualisms can include

    Four basic categories of mutualisms.

symbioses, where the two organisms live in close association, but they can also operate without a tight symbiotic association. Energetic mutualisms involve interactions of a primarily trophic nature, in which energy obtained by one mutualist is made available to another. The transfer of photosynthate from endosymbiotic chlorellae to host coral polyps is an example. Nutritional mutualisms involve the transfer of nutrients, such as nitrogen or phosphorus, from one mutualist to another. This appears to be the main role for fungi in mycorrhizal symbioses, although the host plant contributes mainly photosynthate to the fungus. Similar benefits arise from associations between animals and the bacteria and protists that dwell in their guts. Protective mutualisms involve the active or passive defense of one mutualist by another and include a variety of guarding behaviors, such as the protection of domicile plants by the ants that inhabit them. Transport associations usually involve the movement of gametes (pollination) or propagules (dispersal systems), but can also include interactions in which less mobile organisms hitch a ride on others.

Distinctions can be drawn between **obligate mutualisms**, which are co-evolved to the point at which neither member of a mutualistic association can

    Obligate vs. facultative mutualisms.

persist without the other, and **facultative mutualisms**, in which association with the other mutualist is nonessential but nonetheless leads to positive effects on fitness. The same distinction can be made for **commensalisms**, in which only one species experiences a positive effect, and the other species has neither a positive nor negative response.

DIRECT AND INDIRECT MUTUALISMS

Direct mutualisms include most common examples of symbiotic associations among organisms. Flowering plants and their pollinators are an obvious example of a direct interaction, in which the plant gains a reproductive advantage by being pollinated, and the pollinator benefits from the food supply, nectar and/or pollen, provided by the plant. To the extent that the success of each species depends on the other, the relationship may be obligate or facultative. There are probably fewer obligate mutualisms, such as the tight specificity of plant and pollinator that occurs between the orchid *Catasetum maculatum* and the bee *Eulaema tropica* (Janzen 1971a), than diffuse or facultative interactions involving many functionally interchangeable species. Indirect mutualisms involve positive effects between two species that are transmitted through at least one and sometimes more intermediate species. Indirect mutualisms are considered in greater detail in the following chapter on indirect effects.

SIMPLE MODELS OF MUTUALISTIC INTERACTIONS

Models for mutualistic interactions date back nearly as far as models of competition or predation (Gause and Witt 1935). Curiously, these models tend not to be stressed in basic ecology texts, and they have a history of being forgotten and periodically rediscovered (Boucher 1985). May (1976b) and Wolin (1985) provide overviews of the various models that have been used to describe mutualistic interactions between pairs of species. The simplest model for a mutualistic interaction between two species is similar to the basic Lotka-Volterra model for two competing species. The critical difference is that whereas competitors have negative effects on each other that are abstracted as negative values of the competition coefficients, the positive effects of mutualists can be represented as positive interaction coefficients. Using subscripts to denote values for two species, 1 and 2, this model looks like the following:

$$dN_1/dt = r_1 N_1 (K_1 - N_1 + \alpha_{12} N_2)/K_1 \qquad (7.1a)$$

$$dN_2/dt = r_2 N_2 (K_2 - N_2 + \alpha_{21} N_1)/K_2 \qquad (7.1b)$$

where all terms are directly analogous to those in the two-species Lotka-Volterra equations for interspecific competition, except that now the terms $\alpha_{12} N_2$ and $\alpha_{21} N_1$ are positive rather than negative, to indicate the mutualistic nature of the interaction.

This model describes a facultative mutualism when the values of K_1 and K_2 are greater than zero, since either species can grow in the absence of the other. In this model, mutualists offset the negative effects of the other species on the carrying capacity. As shown in Figure 7.1, each mutualistic species attains a higher density in the presence of the other species than when the mutualists occur alone. This model can easily be altered to represent an obligate mutualism by setting the carrying capacities for both species to negative numbers. This changes the observed dynamics drastically, leading either to extinction when initial population sizes fall below a threshold level or to unlimited population growth when the initial population levels exceed that threshold level (May 1976b; Wolin 1985). Since both outcomes seem biologically unrealistic, other models of obligate mutualism have been developed.

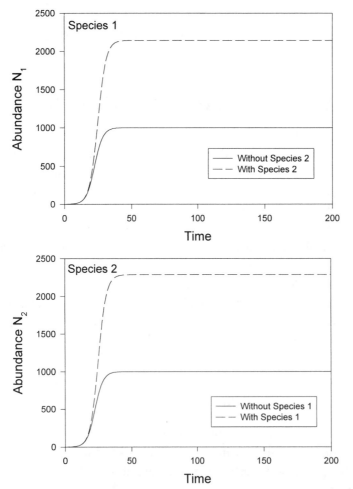

FIGURE 7.1. *Population trajectories for a pair of facultatively mutualistic species with and without the presence of the other member of the species pair. These trajectories result from simulating the model described by Equations 7.1a and 7.1b, using the following parameters:* $r_1 = 3.22$, $r_2 = 3.22$, $a_{12} = 0.5$, $a_{21} = 0.6$, $K_1 = 1000$, $K_2 = 1000$.

Dean (1983) has described one alternate model that can represent a range of interactions running from facultative to obligate mutualism. The model assumes that the carrying capacity of each species is a curvilinear function of mutualist density, up to some maximum value beyond which the presence of mutualists contributes no further increase. The functions describing the relation between mutualist density and carrying capacity are

$$k_1 = K_1\{1 - \exp[-(aN_2 + C_1)/K_1]\} \qquad (7.2a)$$

and

$$k_2 = K_2\{1 - \exp[-(bN_1 + C_2)/K_2]\} \qquad (7.2b)$$

Here K_1 and K_2 are the maximal carrying capacities for each species, which are only obtained at very high densities of mutualists. The lowercase values of k indicate the realized values of carrying capacity set by the abundance of mutualists. The constants C_i determine where the zero-growth isoclines for each species intersect the species abundance axes. The constants a and b influence the curvature of the zero-growth isoclines. The points at which the isoclines intersect the axes, and how their curvature affects the intersection of the isoclines, ultimately influence whether the mutualism is facultative or obligate. The full model then looks like the following:

$$dN_1/dt = r_1 N_1 (1 - N_1/k_1) \qquad (7.3a)$$

$$dN_2/dt = r_2 N_2 (1 - N_2/k_2) \qquad (7.3b)$$

where the lowercase k_i's indicate how the carrying capacity changes with the abundance of mutualists. Substituting the full functions for the k_i's, the equations become

$$dN_1/dt = r_1 N_1 [1 - N_1/\{K_1(1 - \exp[-(aN_2 + C_1)/K_1])\}] \qquad (7.4a)$$

$$dN_2/dt = r_2 N_2 [1 - N_2/\{K_2(1 - \exp[-(bN_1 + C_2)/K_2])\}] \qquad (7.4b)$$

Figure 7.2 shows the range of outcomes that can occur for different values of model parameters. When C_1 and C_2 are both greater than zero, the mutualism is facultative and either species can grow in the absence of the other, although neither species alone attains population densities as great as when both species occur together. When C_1 and C_2 are both equal to zero and ab is greater than zero, an obligate mutualism results. Neither species can grow in the complete absence of the other, but growth at very low densities is possible. When C_1 and C_2 are both less than zero, the mutualism remains obligate, but initial population densities must exceed a lower threshold for population growth to occur. That lower threshold occurs

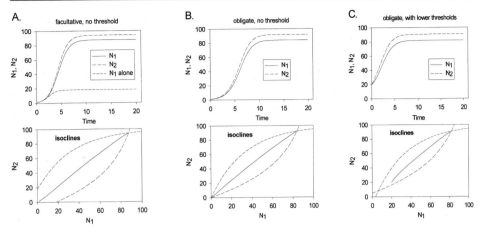

FIGURE 7.2. *Population trajectories for a pair of mutualistic species that differ in whether the mutualism is facultative or obligate, from the model of Dean (1983). The top graphs show population trajectories over time; the bottom graphs show the zero-growth isoclines for each species (dashed lines) and the trajectories of* N_1 *and* N_2 *(solid line) in phase space. (A) Facultative mutualism;* $r_1 = 1.22$, $r_2 = 1.22$, $a = 2$, $b = 3$, $K_1 = 100$, $K_2 = 100$, $C_1 = 20$, $C_2 = 20$. *(B) Obligate mutualism with no lower threshold for population growth;* $r_1 = 1.22$, $r_2 = 1.22$, $a = 2$, $b = 3$, $K_1 = 100$, $K_2 = 100$, $C_1 = 0$, $C_2 = 0$. *(C) Obligate mutualism with a lower threshold for population growth;* $r_1 = 1.22$, $r_2 = 1.22$, $a = 2$, $b = 3$, $K_1 = 100$, $K_2 = 100$, $C_1 = -10$, $C_2 = -10$.

at the point where the zero-growth isoclines for both species cross (see Figure 7.2C).

The models described so far focus on the effects of mutualism on population dynamics. Another kind of model, derived from the economic theory of relative advantage, focuses instead on the conditions under which mutualism will evolve. Schwartz and Hoeksema (1998) consider the evolution of exchange mutualisms, in which each member of the mutualistic association "trades" one kind of resource for another. An example of this sort of interaction is the mycorrhizal association between plants and fungi, in which plants are net providers of carbon fixed in photosynthesis, and fungi are net providers of phosphorus. The model predicts that it will be advantageous for species to trade, or transfer, one resource for another when the species differ in the relative efficiency with which each resource is acquired. The theory predicts that each species should specialize in the resource that it acquires with the greatest relative efficiency and then trade some of that resource for the other resource, which is obtained at lower relative efficiency. The theory works even when one species is more efficient than the other in acquiring both resources, as long as the relative cost of acquiring each resource (measured within each species) differs between species.

EXAMPLES OF OBLIGATE MUTUALISMS

Plant-Pollinator and Plant-Disperser Interactions

Mutualisms involving pollination and seed dispersal involve transport of plant gametes or propagules by animals. Both processes have nonmutualistic alternatives, for example, wind pollination and wind- or water-dispersed seeds. In mutualisms, transport is usually exchanged for a nutritional reward in the form of nectar, pollen, fruit pulp, or lipid-rich elaiosomes. Sometimes the reward is nonnutritive, as in the harvesting of fragrances by male Euglossine bees (Janzen 1971a). The plant-pollinator mutualism is central to the successful reproduction of many flowering plants and their pollinators. The advantages to plants probably involve efficient pollen transfer, avoidance of inbreeding depression, and successful pollination at low plant densities. The pollinators gain an energy-rich food source and, in some special cases, an ensured oviposition site and food supply for their larvae.

Some of the more obvious mutualisms involve obligate relations between flowering plants and their pollinators. Flowering plants in the genus *Yucca* and their specialized insect pollinators, moths in the genus *Tegeticula*, are an example of a particularly tight obligate mutualism (Engelmann 1872; Aker and Udovic 1981). The mutualism is more complicated than a simple plant-pollinator interaction because the adult moth pollinates *Yucca* flowers and also oviposits in developing *Yucca* fruits. By pollinating the flowers, the moth ensures the production of food for its developing larvae. Larvae consume seeds in the developing fruits, but enough seeds escape consumption to produce new *Yucca* plants. The *Yucca* does not self-pollinate, and it appears to depend entirely on pollination by *Tegeticula*. The moth uses an elaborate behavior that involves making a ball of *Yucca* pollen that is deposited on the flower's stigmatic surface. This behavior ensures pollination of the maximal number of ovules and the production of the maximal number of seeds on which the moth larvae feed.

Plant-pollinator interactions need not be as highly species-specific as the *Yucca-Tegeticula* system for mutualism to be important. For instance, there is ample evidence that seed production is limited by pollinator availability in many of the ephemeral spring wild flowers that bloom in the deciduous forests of eastern North America (Motten et al. 1981; Motten 1983). The flowers of several species are visited by a diverse array of small bees and flies. None of the plant-pollinator relations appears to be very specific, but when pollinators are excluded, seed set falls dramatically. Many of the flowers share a broadly similar morphology and coloration, perhaps to facilitate their recognition by pollinators. A few species appear to be involved in a kind of parasitic floral mimicry, since they do not offer nectar rewards to pollinators but are visually similar and bloom at about the same time, thus attracting some pollinators despite the absence of a nutritious reward.

Many plants employ a variety of mechanisms to encourage the dispersal of propagules by animals. These include production of nutritious fruits that contain seeds capable of viably enduring passage though a vertebrate gut. In some cases, the seeds may even require gut passage before being able to germinate (Janzen and Martin 1982). In situations in which the appropriate dispersal agent has become extinct, plant recruitment may be very low or negligible, as appears to be the case for some neotropical plants that may have relied on the now-extinct Pleistocene megafauna for dispersal. A more recent example concerns the absence of seed germination in the tree *Calvaria major*, which is endemic to the island of Mauritius in the Indian Ocean (Temple 1977). Mauritius was home to the extinct dodo, *Raphus cucullatus*, which fed on a variety of fruits and seeds, including the heavy-walled seeds of *Calvaria*. Unless the thick seed walls are abraded, as would happen in the gizzard of a large bird such as the dodo, the seeds, although viable, fail to germinate. The interesting coincidence is that the few remaining *Calvaria* trees on the island are all old, dating back to the time of the last living dodos about 300 years ago. The assumption is that without dodos to process the seeds, no *Calvaria* trees have germinated subsequent to the dodo's extinction, although the few remaining trees continue to produce viable seeds. Consumption of the seeds by another large bird, the turkey, resulted in germinated seeds and the production of the first *Calvaria* seedlings seen in nearly 300 years. Although obligate mutualisms can be beneficial for the species involved, dire consequences await the remaining member of the mutualism if the other becomes extinct.

Dispersal may be advantageous to plants for a number of reasons. Three main ideas have been put forward to explain the advantage that plants gain from seed dispersal (Howe and Smallwood 1982). The **predator escape hypothesis** (Janzen 1971b; Connell 1971) suggests that seeds falling near the parent plant have a higher risk of mortality than do seeds dispersed far from the parent.

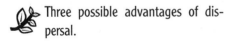
Three possible advantages of dispersal.

The main source of mortality is seed predators that tend to aggregate near the parent plant where seeds are most abundant. The **colonization hypothesis** (e.g., Hubbell 1979) assumes that optimal locations for seedling establishment are constantly shifting in time and space. Consequently, the current location of a parent seed source is a poor predictor of a good site for seedling establishment. The **directed dispersal hypothesis** goes one step farther and assumes that dispersal agents preferentially distribute seeds in sites where germination is more likely. There is some evidence that seeds dispersed and planted by ants benefit from this mechanism.

Some ant-plant associations involve special morphological adaptations of the seeds that promote their dispersal into favorable germination sites by ants (Handel 1978; Beattie 1985). This phenomenon, termed **myrmecochory**, involves the production of special lipid-rich bodies on plant seeds that attract ants that are not specifically seed eaters. These bodies, called **elaiosomes**, are harvested by the ants, and along the way the seeds are moved into particularly favorable sites for germination, the nests of ants. Seeds within ant nests benefit in numerous ways. Ants disperse seeds to new sites, which often minimize interactions with potential seed predators and competitors; favorable levels of moisture and nutrients within ant nests may further enhance the probability of germination and establishment. This strategy works best with carnivorous ants that forage selectively on elaiosomes. Granivorous species of ants that actually consume and kill entire seeds are rarely involved in mymecochory. The advantages of mymecochory to ants are less well understood. The ants clearly gain a food source, and it is thought that elaiosomes may be particularly rich in essential lipids.

There is clear evidence that seeds handled and planted by ants are more likely to survive and grow than are seeds that are not so handled. Hanzawa et al. (1988) compared the survival and reproduction of two groups of seeds of the plant *Corydalis aurea* in the Rocky Mountains of Colorado. One group of seeds with elaiosomes was placed near ant nests, and ants subsequently collected the seeds and planted them after removing the elaiosomes. A second group of seeds was planted by the experimenters (instead of by ants) in the vicinity of ant nests. The seeds and seedlings were then monitored to determine survival and subsequent reproduction. Ants had no effect on seed survival through germination. However, young plants produced by the seeds planted by ants survived better than plants that were not located in ant nests. The seeds planted by ants ultimately produced more offspring, not because of a higher fecundity per plant, but because the plants survived better. These differences in survivorship for ant-handled and control plants are shown in Figure 7.3.

Plant-Defender Interactions

Other spectacular examples of obligate plant-arthropod mutualisms involve interactions between neotropical plants and insects that defend the plants against herbivores or competitors. Janzen (1966) described a fascinating interaction between the swollen thorn acacias, *Acacia* sp. (including *Acacia cornigera*), and obligate acacia-dwelling ants (*Pseudomyrmex* sp.). The ants live within hollow thorns of the acacias. The plants provide two special food sources for the ants: extrafloral nectaries that provide carbohydrates, and nutrient-rich Beltian bodies that form at the tips of leaves (Figure 7.4). In return, the ants vigorously defend the plant, removing other herbivorous

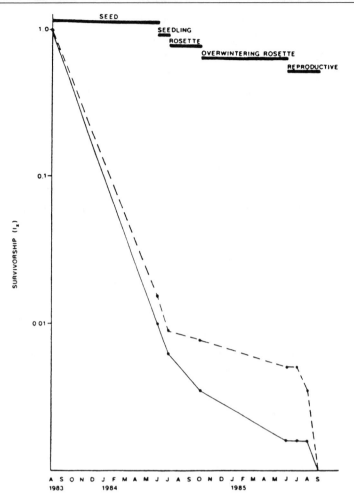

FIGURE 7.3. *Differential survival of plants germinating from seeds handled and planted by ants (dashed line) or handled and planted by experimenters (solid line). Survival of seeds is nearly identical, but survival of plants to reproductive size is considerably greater in ant-handled seeds. (Reprinted from Hanzawa et al., 1988, with permission of the University of Chicago Press.)*

insects from the acacias and even pruning back other plants that might overgrow and shade out the acacia. An interesting commensalism accompanies this ant-plant system (Flaspohler and Laska 1994). Wrens of the species *Campylorhynchus rufinucha* nest preferentially in ant acacias and appear to benefit from the presence of ants that deter possible nest predators. Neither the trees nor the ants appear to gain anything from the presence of the birds.

Other ant-plant associations that occur in temperate forests are not obligate, but they appear to function in a similar fashion. In Michigan, the North American black cherry tree, *Prunus serotina*, has extrafloral nectaries that are particularly active during the first few weeks after bud-break in the early

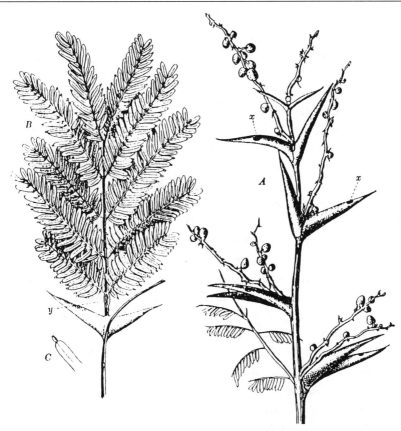

FIGURE 7.4. *Special features of a bull's-horn acacia tree,* Acacia sphaerocephala, *that are used by mutualistic ants of the genus* Pseudomyrmex. *(A) The end of a branch, showing the enlarged hollow thorns that house ant colonies. Ants chew entrance holes, x, in the thorns. (B) The ants feed on nectar provided by extrafloral nectaries, y, found at the bases of leaves. (C) An enlargement of a leaflet tip showing lipid-rich Beltian bodies that grow on the tips. (Reprinted from Wheeler, 1910, with permission of the Columbia University Press.)*

spring (Tilman 1978). The nectaries attract ants, *Formica obscuripes*, to the trees from surrounding woodlands. Unlike the tropical ant-acacia mutualism, the *Formica* do not nest in the trees. The ants prey on insects that might otherwise defoliate the trees, especially larvae of the eastern tent caterpillar *Malacosoma americanum*. Trees benefit from the removal of herbivores, while the ants benefit from an additional food source provided by nectaries on the trees.

ENERGETIC AND NUTRITIONAL MUTUALISMS

Many mutualisms, especially symbiotic associations, involve the transfer of energy or nutrients from one species to another. These associations often involve organisms in different phyla. For example, animals that feed on a diet rich in cellulose contain gut symbionts, either bacteria or protists, that are capable of digesting cellulose. The gut symbionts also produce vitamins and amino acids that are used by their animal hosts. In return, the gut symbionts obtain an abundant and reliable food source and a predator-free environment. These interactions play an important role in carbon cycling in many natural communities, and in some regions may be responsible for the production of large amounts of greenhouse gases, especially methane.

Some mutualisms involving plants and fungi are inconspicuous to the casual observer, but these interactions can be extraordinarily important through their impacts on plant defenses against herbivores and the ability of plants to extract nutrients from soils. Many plants contain symbiotic endophytic fungi that live within the tissues of the plant (Clay 1990). The fungi, rather than being pathogens, produce chemicals that deter the attacks of herbivores on the photosynthetic plants where they dwell. The fungi benefit from the plants, which provide food in the form of photosynthate. Clay et al. (1985) have shown that the fungal endophytes of grasses significantly depress the growth and survival of an insect herbivore, the fall armyworm (*Spodoptera frugiperda*) (Table 7-1). Similar effects are known for the effects of endophytes on grazing mammals (Bacon et al. 1975).

Another inconspicuous but functionally important plant-fungal mutualism involves mycorrhizal associations between fungi and the roots of many plants (Allen 1991). The fungi serve as nutrient pumps, facilitating the uptake of nitrogen and phosphorus by the plants. In return, the plants provide the fungi with carbohydrates produced during photosynthesis. Mycorrhizal asso-

TABLE 7-1. Effects of endophytic fungi on the development of larval fall armyworms (*Sopodoptera frugiperda*) feeding on greenhouse-raised perennial ryegrass (*Lolium perene*). Endophytes significantly reduce early growth and survival, while slightly prolonging the duration of larval development.

Response of Larvae	With Endophytes	Without Endophytes
Weight, day 8	3.30 mg	6.36 mg
Weight, day 12	73.76 mg	125.42 mg
Days to pupation	20.57	19.78
Survival	42%	80%

Reprinted from Clay et al. (1985). Fungal endophytes of grasses and their effects on an insect herbivore. Oecologia 66, page 2, Table 2. © Springer-Verlag.

ciations are ubiquitous. They may be particularly important in allowing plants to become established in habitats where soils have very low nutrient levels. The failure of some introduced plant species to flourish has been attributed to the absence of appropriate mycorrhizal symbionts in novel environments.

Mycorrhizae can also influence the outcome of competition among plants. Grime et al. (1987) have shown that mycorrhizae promote the diversity of herbaceous plants grown in microcosms. Without mycorrhizae, the microcosms were dominated by one species of grass, *Festuca ovina*. With mycorrhizae, a number of subordinate species, primarily forbs, increased in abundance, as measured by yield per plant species (Figure 7.5). The mechanism for enhanced diversity and greater yield of subordinate species is somewhat unexpected. Use of radioactive tracers shows that photosynthate from the dominant species, *Festuca*, is transferred via mycorrhizae to subordinate species. The net result is that the presence of a mycorrhizal network linking

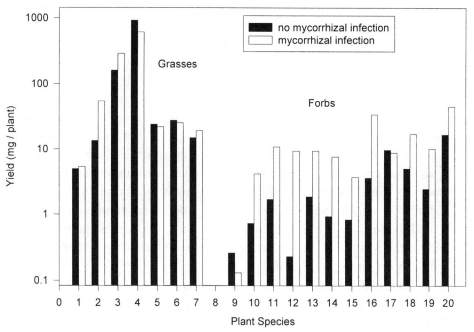

FIGURE 7.5. *The effect of mycorrhizal association on the yield (in mg) of competing plants grown in microcosms, greenhouse flats, for one year. Mycorrhizae increase the yield of several species. The species are keyed by the following numbers: 1 = Anthoxanthum odoratum, 2 = Briza media, 3 = Dactylis glomerata, 4 = Festuca ovina (planted as seedlings), 5 = Festuca ovina (planted as seeds), 6 = Festuca rubra, 7 = Poa pratensis, 9 = Arabis hirsuta, 10 = Campanula rotundifolia, 11 = Centaurea nigra, 12 = Centaurium erythraea, 13 = Galium verum, 14 = Hieracium pilosella, 15 = Leontodon hispidus, 16 = Plantago lanceolata, 17 = Rumex acetosa, 18 = Sanguisorba minor, 19 = Scabiosa columbaria, 20 = Silene nutans. (Data are from Grime et al., 1987.)*

dominant and subordinate species allows the subordinate species to effectively parasitize the competitively dominant grass species by siphoning off a portion of their photosynthate. The implications of such interactions for competition among plants in natural settings certainly merit further study.

A completely different nutritional mutualism occurs in the leaf-cutter ants and the fungi that they culture. Leaf-cutter ants (*Atta, Acromyrmex,* and *Trachymyrmex*) harvest plant material, which is returned to the nest and used as a substrate for the culture of a fungus (*Leucoprinus gongylophora*). The fungus produces special swollen tips on its hyphae that the ants use as food. The fungus lacks proteolytic enzymes, which are provided instead by the ants in the food that they bring to the fungus. Although this association is a conspicuous component of many neotropical communities, small leaf-cutting ants in the genus *Trachymyrmex* can be found as far north as New Jersey in the United States.

A well-known mutualistic association between some fungi and algae has progressed to the point at which the associations are recognized as distinct species, the lichens. Lichens are often important pioneer species in nutrient-poor stressful environments. Indeed it is just such environments that should favor the development of positive interactions among species (Bertness and Callaway 1994).

Other positive interactions among algal species and animals are responsible for the building blocks of some of the more diverse communities that exist. Reef-building corals are associations of cnidarian polyps and endosymbiotic algae called zoochlorellae (Muscatine and Porter 1977). Although the polyps are capable of feeding, much of their nutrition comes from their photosynthetic endosymbionts. Indeed, reef-building corals are limited to fairly shallow marine waters, probably because there is insufficient light for photosynthesis by their endosymbionts at greater depths. In return, the endosymbiotic chlorellae are defended against some of their smaller potential consumers, but not against some of the specialized consumers that feed directly on corals (Porter 1972).

Endosymbiotic chlorellae can have other positive functions that are not strictly energetic. The ciliated protist *Paramecium bursaria* typically contains an algal endosymbiont. The ciliate, while technically a predator on bacteria and other small cells, is bright green when grown under well-lit conditions, due to the abundance of algae in its cytoplasm. Although the algal endosymbiont would seem to confer a competitive advantage under conditions of low prey availability, this appears not to be the case. The interesting advantage conferred by the endosymbiont materializes when *Paramecium bursaria* is under attack by the predator *Didinium nasutum,* another ciliate. Berger (1980) has shown that *Paramecium* with endosymbiotic algae are attacked at a significantly lower rate than are *Paramecium* that have had their chlorellae experimentally removed. Apparently, the chlorellae function as predator deterrents, much in the way that endophytic fungi deter the effects of herbivores on terrestrial plants.

EXAMPLES OF FACULTATIVE MUTUALISMS AND COMMENSALISMS

It is easy to imagine a whole array of facultative mutualisms and commensalisms that might occur in complex communities, in which some species benefit from the presence of others. Just consider the whole range of species that live on or in trees without consuming any part of the plant: Epiphytes, nesting birds, and a whole array of arboreal organisms benefit from the presence of the tree as habitat or substrate. While such arguments seem logically sound, few of these associations have been carefully studied.

Some kinds of facultative associations offer clearly demonstrated benefits to the participants. Anyone who has spent time watching birds is familiar with the phenomenon of **mixed-species flocks**, in which individual birds of several different species move and forage as a loosely associated group. The possible advantages of such associations are at least twofold. First, association with

 Some species benefit form association with others.

a large number of foraging individuals should improve the probability of locating patches of food, since resources discovered by one individual will become apparent to others in the group. Although the presence of more individuals raises the possibility of competition for food once the resource is discovered, that negative effect is presumably offset by the greater ability of the group to locate food. Krebs (1973) has shown that captive birds of the genus *Parus* can learn about the location of food from the foraging behavior of other species in the same setting. A second possible advantage is that a greater number of individuals will be better at avoiding predators. In a group consisting of many individuals, it is more likely that predators will be noticed by some member of the flock or herd. This may be the reason for mixed-species groups of baboons and impala on the African plains (Altmann and Altmann 1970). In response to predators, many birds produce alarm calls that are recognized by other species (Barnard and Thompson 1985). The ability of species to recognize these alarm calls and to respond appropriately suggests that the predator-avoidance component of mixed-species flocks is real.

Other kinds of defenses against predators can result simply from the dilution of palatable prey densities that results from association with unpalatable species. Such **associational defenses** do not require sophisticated behaviors

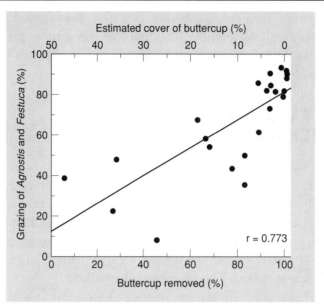

FIGURE 7.6. *Associational defense conferred on the grasses* Agrostis *and* Festuca *by the presence of buttercup,* Ranunculus. *Grazing cattle consume less grass when* Ranunculus *is abundant. (Adapted with permission from Atsatt and O'Dowd, 1976. © 1976 American Association for the Advancement of Science.)*

on the part of the prey species. The action of these defenses relies on the tendency of predators to forage preferentially in patches of highly palatable or highly profitable prey. Where such prey are present at low density, and where they occur in association with unpalatable or unprofitable prey, they are less likely to draw the attention of predators.

One example of associational defenses comes from studies of interactions between terrestrial plants and their herbivores (Atsatt and O'Dowd 1976). Cattle grazing on palatable species of grasses, such as *Festuca* and *Agrostis*, is depressed by the presence of another toxic species, the buttercup *Ranunculus bulbosus*. There is a strong negative relation between *Ranunculus* abundance and grazing on palatable species (Figure 7.6). Other examples of associational defenses come from studies of grazing by various marine herbivores on algae (Hay 1986). Hay found that association of the palatable alga *Hypnea* with the relatively unpalatable alga *Sargassum* resulted in a strong decrease in the loss of algal tissue to grazing by fish under field conditions (Figure 7.7). The result appears to reflect the relatively unpalatable status of *Sargassum* to some herbivores, but it also depends on the palatable alga remaining relatively inconspicuous among the fronds of the unpalatable species. As the palatable species becomes more abundant and more obvious to herbivores, the associational defense breaks down.

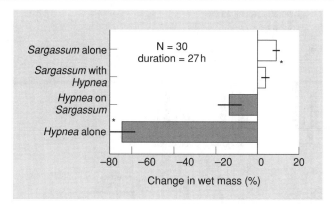

FIGURE 7.7. *Associational defense conferred on the marine alga* Hypnea *by the presence of* Sargassum. *Under field conditions, herbivorous fish consume more* Hypnea, *as shown by the greater loss in mass, when it does not occur with* Sargassum. *(Adapted from Hay, 1986, with permission of the University of Chicago Press.)*

THEORIES ABOUT THE CONDITIONS LEADING TO POSITIVE INTERACTIONS AMONG SPECIES

Bertness and Callaway (1994) have summarized ideas about the conditions leading to positive interactions among sessile species. They suggest that physical stress in the environment plays an important role in determining when positive interactions among species are disproportionately common. In stressful environments, positive interactions are likely to arise in which certain highly stress-tolerant species are able to become established and then subsequently modify the environment to make it less stressful for other species (Bertness 1992; Bertness and Shumway 1993). One example of such

 Mutualisms may be advantageous in stressful or predator-rich settings.

an interaction is the ability of certain highly salt-tolerant plants to invade hypersaline soils in salt marshes. After invasion, plants such as *Spartina patens* and *Distichlis spicatus* shade these sites and alter their physical conditions in ways that tend to decrease soil salinity. As soil salinity decreases, other plant species are able to invade sites from which they were previously excluded. Removal experiments show that other plant species, such as *Juncus gerardi*, benefit from the presence of *Spartina* or *Distichlis* (Figure 7.8). These positive effects only occur under harsh abiotic conditions, particularly high salinity. If salinity is experimentally reduced by watering plots with fresh water, then the interaction shifts from a positive facilitation to a negative competitive interaction. Such interactions have been called **neighborhood habitat amelioration**.

In benign environments, other factors come into play. Attacks by consumers are generally thought to be more frequent in low-stress environments

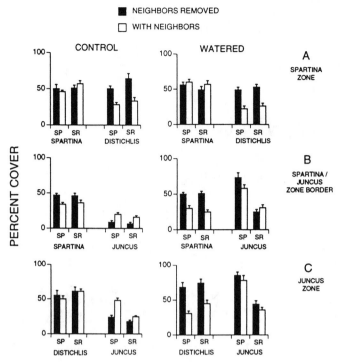

FIGURE 7.8. *Effects of neighboring plants on the percentage of cover of Spartina, Distich-lis, and Juncus in three different zones in a salt marsh in the northeastern United States. In controls with high salinity, neighbors shade plots and reduce salinity, enhancing the abundance of Juncus. In watered plots with reduced salinity, neighbors simply act as competitors and reduce the abundance of Juncus. (Reprinted from Bertness and Shumway, 1993, with permission of the University of Chicago Press.)*

(Menge and Sutherland 1976). This means that prey species experience greater consumer pressure in relatively benign environments, compared with stressful ones. Under high consumer pressure, prey sometimes display what can be termed associational defenses. Associational defenses arise when palatable prey benefit by close association with unpalatable prey (Hay 1986). In the absence of consumers, the prey simply compete, either for space or some other resource.

The frequency of associational defenses and neighborhood habitat amelioration is probably lower at intermediate levels of physical stress. Under such conditions, negative competitive interactions are likely to outnumber positive facilitative interactions. These shifting patterns of positive and negative interactions have been graphically summarized by Bertness and Callaway (1994) in Figure 7.9. This simple conceptual model is one of the first synthetic approaches to predict when and where positive interactions among species are likely to be important. It should apply equally well to obligate and facultative positive interactions among species.

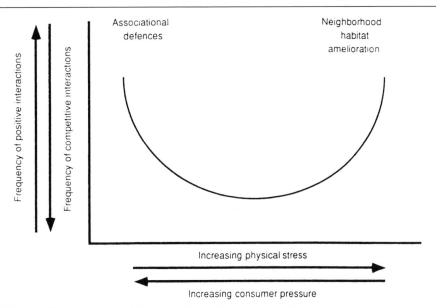

Frequency of positive interactions

Frequency of competitive interactions

Associational defences

Neighborhood habitat amelioration

Increasing physical stress

Increasing consumer pressure

FIGURE 7.9. *A conceptual framework showing how the relative frequencies of positive and negative interactions among species are thought to vary with increasing physical stress and increasing consumer pressure. Physical stress and consumer pressure are thought to be inversely related. (Reprinted from* Trends in Ecology and Evolution, *Vol. 9, Bertness, M. D., and R. Callaway. Positive interactions in communities, pages 191–193. © 1994, with permission from Elsevier Science.)*

CONCLUSIONS: CONSEQUENCES OF MUTUALISM AND COMMENSALISM FOR COMMUNITY DEVELOPMENT

Obligate mutualisms place certain constraints on the ways in which species can successfully colonize developing communities. Obviously, if one member of an obligate association disperses without the other, successful colonization seems unlikely. One way around this problem is to ensure that the mutualistic association disperses as a unit. For instance, queens of fungus-gardening leaf-cutter ants carry an inoculum of fungus with them when they depart from their natal colonies (Holldobler and Wilson 1990). Similarly, the propagules of lichens contain both the algal and fungal members of the mutualistic association. The other problem, already described for the dodo-*Calvaria* system on Mauritius, is that the extinction of one obligate mutualist will result in the eventual loss of the entire association from the community.

For facultative associations, colonization of a new community is a simpler matter, since one member of the association can arrive and become established without the other. There are limits to how long this may take, and a lone member of a facultative mutualism may be at a significant disadvantage in the absence of its associates. The models described earlier in this chapter show that although facultative mutualists can grow in the absence of their associates, populations will attain smaller sizes.

Recognition of the potential importance of positive interactions is crucial for some aspects of applied community ecology, such as the restoration of functioning communities on degraded sites like mine tailings or landfills. Establishment of a flourishing plant community requires not only the introduction of plant propagules but also mycorrhizal fungi, endophytic fungi, pollinators, and agents of seed dispersal. In general, the importance of mutualisms and commensalisms for most aspects of community structure and function remains very poorly understood. This important area is in need of much careful research.

CHAPTER 8

Indirect Effects

OVERVIEW

Pairwise interactions between species, such as competition, predation, and mutualism, understate the complexity of indirect interactions that can propagate through chains of three or more species in complex communities. Indirect effects describe how the consequences of pairwise direct interactions between species are transmitted to other species either through behavioral modifications, altered spatial distributions, or altered abundances in the food web. Indirect effects are a logical consequence of the fact that interacting species are embedded in larger food webs. Indirect effects can complicate interpretations of community-level experiments, since responses to the additions or removals of species can result from direct and indirect effects. However, knowledge of the potential pathways of indirect interactions can be used to generate testable hypotheses that can illuminate which indirect interactions probably account for a particular response.

INDIRECT EFFECTS occur when the influence of one species, the donor, is transmitted through a second species, the transmitter, to a third species, the receiver (Abrams 1987). The observed effect is indirect if the donor influences the receiver through an intermediate species. Complex interactions including both direct and indirect effects are possible, and perhaps likely. Indirect effects can involve changes in a whole host of properties of species.

The most common effects materialize as changes in steady-state abundance, but indirect effects can influence the dynamics, behavior, or even the genetics of the receiver. Indirect interactions potentially occur in any complex community in which chains of three or more interacting species exist, in other words, in all but the simplest communities. Although much of the theory discussed previously in this book has focused on the ways in which pairs of species interact, for example, as abstracted by competition coefficients or functional responses, species are connected in chains or webs of interactions with many other species. As we shall see later in this chapter, indirect interactions, or indirect effects, can complicate the interpretation of ecological experiments, especially when chains of indirectly interacting species are not taken into account. Indirect effects can also complicate the interpretation of experimental introductions or removals of species in complex systems. Despite some of the problems that arise, indirect effects are potentially an integral part of the workings of most complex natural communities. This means that questions about the strength and commonness of indirect effects are of fundamental interest to community ecologists.

There is as yet no generally accepted terminology for the various kinds of indirect effects (Miller and Kerfoot 1987; Strauss 1991; Wootton 1994c).

 Two basic kinds of indirect effects.
Wootton (1993, 1994c) makes a useful distinction between two basic kinds of indirect effects that focuses on whether the transmitting species changes in abundance or changes in its per capita effect on the receiver. An **interaction chain indirect effect** occurs when a species indirectly affects others as a consequence of changes in the abundance of an intermediate transmitter species (Figure 8.1). For example, if species A negatively affects species B, and species C reduces the abundance of species A, then species C will have an indirect positive effect on species B. In contrast, an **interaction modification indirect effect** occurs when a donor species changes the per capita effect of the transmitter on the receiver without changing the abundance of the transmitter (see Figure 8.1). Such effects might arise if species C changes some other attribute of species A—behavior or size, for example—thereby changing the per capita effect of species A on species B. Of course, these two kinds of indirect effects are not mutually exclusive, and mixed effects could occur in which both the abundance and per capita effect of the transmitter change.

The distinction between indirect effects involving interaction chains and interaction modifications can be clarified by a hypothetical example framed in terms of simple models of interacting species. Consider a simple three-level food chain, with a top predator that indirectly affects primary producers in the bottom level through direct effects on the herbivores in the second trophic

A. Interaction Chain Indirect Effect

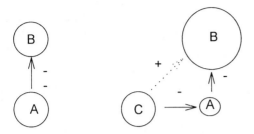

B. Interaction Modification Indirect Effect

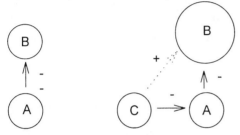

FIGURE 8.1. *Graphical representations of indirect effects involving interaction chains (A) and interaction modifications (B). Direct effects are indicated by solid arrows, with the sign of the interaction indicated by + or −. Indirect interactions are shown by dashed arrows. Population sizes are indicated by the relative sizes of the circles denoting each species.*

level. If the effect is due solely to reduced abundance of the herbivores, then the effect is an interaction chain indirect effect, which could be represented simply by a change in herbivore density in a functional response term for herbivores feeding on producers. However, the top predators might not reduce herbivore abundance and might instead reduce the per capita consumption of producers. Such an interaction modification indirect effect could be represented as a change in the herbivore per capita attack rate within the functional response, instead of a change in herbivore abundance. While this example focuses on predator-prey interactions, similar logic would apply to systems of competitors or mutualists and distinguishes between processes affecting **densities** of interacting species versus **per capita impacts**, such as competition coefficients or attack rates, that relate species densities to the intensity of interspecific interactions.

Some of the interactions already considered in previous chapters are examples of indirect effects. These interactions, and some additional ones,

208 COMMUNITIES: BASIC PATTERNS AND ELEMENTARY PROCESSES

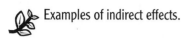 Examples of indirect effects. are outlined graphically in Figure 8.2 to illustrate why they are indirect. For example, purely **consumptive competition** between two species, when the contested resource is a third species, is a kind of indirect effect (see Figure 8.2). In contrast, direct chemical competition between two species involving chemical inhibition of one species by another clearly would not be an indirect interaction. The outcome of **keystone predation**, in which a predator enhances the abundance of one or more inferior competitors by reducing the abundance of a superior competitor, is another kind of indirect effect. The positive effect of the predator on inferior competitors is mediated through its negative impact on an intermediate species, the superior competitor.

Other kinds of indirect interactions are of sufficient interest that they have acquired specific names. These interactions are described in greater detail below and include **apparent competition**, **indirect mutualism**, and **indirect commensalism**. The ideas of Hairston et al. (1960) and Fretwell

A. Apparent Competition

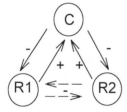

B. Consumptive Competition

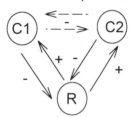

C. Indirect Mutualism

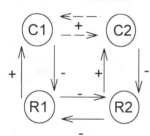

D. Trophic Cascade

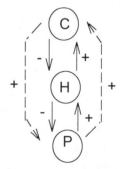

FIGURE 8.2. *Some of the major kinds of indirect effects. Direct effects between pairs of species are indicated by solid arrows, whereas indirect effects are shown as dashed arrows. C indicates consumer species, and R indicates resource species in the interactions diagrammed in (A), (B), and (C). In (D), P indicates primary producers, H indicates herbivores, and C indicates top consumers or predators.*

(1977) about the relative importance of competition and predation in regulating populations of species in different trophic levels describe another kind of indirect effect, called a **trophic cascade**. Here the abundance of primary producers is thought to be indirectly regulated by top predators in food chains with three or more trophic levels. In short chains with only two levels, the regulation would be a direct effect. Other interaction modification indirect effects, called **higher-order interactions**, refer to changes in the ways in which pairs of species interact that are caused by the presence of other species. Although very complex indirect effects involving feedback through large numbers of species can be imagined and modeled, the majority of well-studied indirect effects typically involve chains of only three or four interacting species.

APPARENT COMPETITION

Robert Holt (1977) first described the conditions that might promote an indirect effect called **apparent competition**, in which the presence of multiple noncompeting prey species elevates predator abundance above levels maintained by single prey species, which increases predation pressure on multiprey

 "Apparent" competition.

assemblages (see Figure 8.2). Apparent competition can occur where two noncompeting species share a predator on some higher trophic level. In the absence of predators, each prey species is regulated by purely intraspecific density-dependent mechanisms, and neither prey species competes, directly or indirectly, with the other. The scenario assumes that predator abundance depends on total prey abundance, so that where more species of noncompeting prey coexist, they should support more predators than in situations where only a single prey species occurs. Predators consume prey at a rate that increases with predator abundance. This can lead to a situation in which both prey species occur at lower densities when they occur together than when they occur separately. Although this pattern would have the outward appearance of interspecific competition, since the prey are less abundant in sympatry than in allopatry, the prey do not compete, directly or indirectly. Lower abundance in sympatry is caused entirely by the greater abundance of predators supported by both prey populations together than by either prey population alone. This idea is interesting, but does it happen? There are very few studies designed specifically to test for apparent competition. Two studies suggest that apparent competition can in fact occur.

Russell Schmitt (1987) described a likely case of apparent competition involving marine invertebrates dwelling in rocky reefs off the coast of southern California. At first glance, the pattern in this system appears consistent with habitat segregation resulting from competition between two groups of prey, sessile bivalves (*Chama* and *Mytilus*) and mobile gastropods (*Tegula* and *Astraea*). The bivalves occur mostly in areas described as high-relief reefs, where they find some shelter from predators among crevices in the rocks. The gastropods are more abundant in low-relief reefs composed of rocky cobbles, and they usually do not seek shelter in crevices. However, competition between the two bivalves and gastropods seems very unlikely. The bivalves and gastropods consume different kinds of food. The bivalves filter particles from the water and the gastropods scrape algae from the rocks. Competition for space is also unlikely, because bivalves and gastropods favor different substrates. Gastropods forage on the surface of the rocks, whereas bivalves occupy crevices. A diverse array of invertebrate predators, including lobsters, octopods, and whelks, prey on both bivalves and gastropods. The gastropods appear much more vulnerable to predators, and both predators and bivalves appear to be more abundant on high-relief reefs.

To test whether predators become more abundant when prey from both habitats are available, Schmitt transferred the bivalves *Chama* and *Mytilus* to the gastropod-dominated rocky cobble reefs and observed the impact of this transfer on gastropod mortality and predator abundance. Additional bivalves were added over time to offset losses to predators and maintain a high density of prey. It was not possible to perform the reciprocal transfer of gastropods to the high-relief reefs where *Chama* and *Mytilus* usually occurred; therefore, to measure possible interactions between the bivalves and gastropods at low predator densities, Schmitt transplanted bivalves to areas with high and low natural densities of gastropods. As expected, transplants of bivalves to cobble reefs increased predator abundance. Relative to control areas that did not receive bivalves, gastropod densities declined significantly over the 65-day duration of the experiment (Figure 8.3). Gastropods had a similar indirect negative effect on bivalves, with more bivalves being consumed by predators in areas of high gastropod density (45.1 snails/m^2) than in areas of low snail density (4.7 snails/m^2). The results are consistent with a somewhat asymmetric indirect negative effect of bivalves on snails, mediated by the rapid aggregation of predators in areas with high densities of their preferred prey, the bivalves.

Sharon Lawler (1993a) also found evidence for apparent competition between two prey species in a laboratory study of interactions among protists. Lawler examined interactions between two prey species, the flagellate *Chilomonas* and the ciliate *Tetrahymena*, and their shared predator, the ciliate *Euplotes*. *Chilomonas* and *Tetrahymena* coexist in laboratory microcosms, which suggests that competition is not sufficiently intense to drive either species extinct under these conditions. Each species, when occurring in the absence of the other prey, also managed to coexist with predators for long periods of time. *Euplotes* attained much higher densities when it fed on

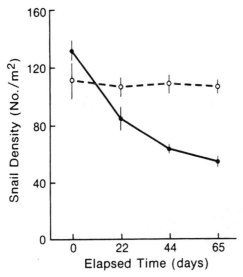

FIGURE 8.3. *Decrease in gastropod density (solid circles) at sites receiving alternate prey (bivalves) contrasted to unchanged gastropod densities (open circles) at control sites without alternate prey. The decrease is attributed to apparent competition. (Reprinted from Schmitt, 1987, with permission of the Ecological Society of America.)*

Tetrahymena, which suggests that *Tetrahymena* is a better food source than *Chilomonas*. However, when both prey species occurred together with *Euplotes, Chilomonas* was rapidly eliminated, apparently because the presence of *Tetrahymena* led to a predator density that was sufficient to drive *Chilomonas* extinct.

Holt and Lawton (1994) have recently reviewed evidence for indirect effects resulting from shared predators. They point out that although the potential consequences of natural enemies for shared prey have long been recognized, there have been surprisingly few quantitative studies of these interactions. Much of the evidence for apparent competition that they review is anecdotal, consisting of observations of high mortality among certain focal prey when alternate prey are also more abundant. There is a great need for experimental studies of apparent competition. Apparent competition may have important practical implications in managed agricultural communities. For example, attempts at biological control might be made more effective by creating situations in which introduced predators remain at higher densities by virtue of being able to feed on multiple prey species. This is, in fact, counter to the usual strategy, in which a very specific predator of a particular prey is sought.

Karban et al. (1994) have found that under some circumstances, predators are more effective at controlling a certain target prey when they are released together with an alternate prey than when released alone. They studied the effectiveness of a predatory mite, *Metaselius occidentalis*, used to

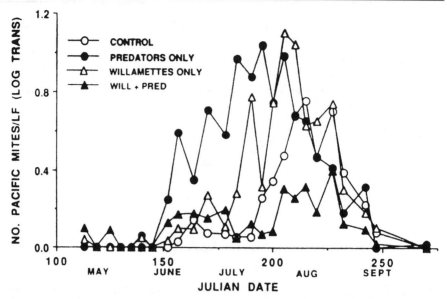

FIGURE 8.4. *Enhanced reduction of target prey (Pacific mites, which feed on grapevines) by predators when predators are introduced on grapevines together with an alternate prey (Willamette mites). The enhanced reduction may be a consequence of apparent competition mediated by the presence of an alternate prey species. (Reprinted from Karban et al., 1994, with permission. © Springer-Verlag.)*

control infestations of herbivorous Pacific mites (*Tetranychus pacificus*) feeding on grapevines. They found that introduction of the predatory mite together with an alternate food source, the Willamette mite (*Eotetranychus willamettei*) resulted in a much greater reduction in the abundance of Pacific mites than when the predator is introduced in the absence of the alternate prey (Figure 8.4). Karban et al. emphasize that a result consistent with apparent competition occurred only once in the course of several such introductions. They suggest that the low frequency of this kind of apparent competition in their field trials may reflect the impact of poorly understood aspects of environmental variation on these indirect effects.

INDIRECT MUTUALISM AND INDIRECT COMMENSALISM

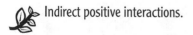 Indirect positive interactions.

Although some indirect effects have net negative consequences for the species on the receiving end of the interaction chain, other kinds of positive effects, such as indirect mutualism (see Figure 8.2), are possible. Such indirect effects were suggested by a series of observations of alternate patterns of

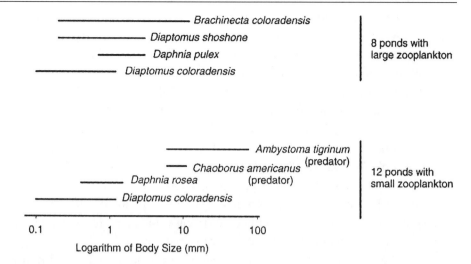

FIGURE 8.5. *Alternate communities found in alpine ponds with vertebrate predators (the salamander* Ambystoma tigrinum*) or without vertebrate predators (Dodson 1970). Ponds with* Ambystoma *have zooplankton assemblages dominated by small species, whereas ponds without* Ambystoma *contain mostly large zooplankton species. The planktivorous midge* Chaoborus *can feed only on smaller zooplankton species, such as* Daphnia rosea.

community composition made by Stanley Dodson (1970). Dodson noted that communities found in small alpine ponds in Colorado tended to fall into two groups (Figure 8.5). One series of ponds contained two predators, *Ambystoma* and *Chaoborus*, known to feed primarily on zooplankton with very different body sizes. Larval salamanders (*Ambystoma tigrinum*) feed primarily on larger zooplankton, including some large predatory copepods that can greatly reduce the abundance of smaller zooplankton species. Larvae of the phantom midge, *Chaoborus*, are restricted to feeding on smaller zooplankton species that usually do not coexist with larger species. In natural ponds, *Ambystoma* and *Chaoborus* are almost always found together, and *Chaoborus* usually does not occur in ponds without *Ambystoma*. Dodson explained this pattern as a consequence of *Ambystoma* maintaining the feeding niche provided by small-bodied prey consumed by *Chaoborus*. Presumably, the large-bodied zooplankton that predominate in ponds without *Ambystoma* are inappropriate prey for *Chaoborus*. An alternate hypothesis is that ponds without *Ambystoma* tend to freeze solid during the winter, and the same freezing that excludes *Ambystoma* may also eliminate *Chaoborus*.

Although Dodson did not experimentally test the indirect mutualism hypothesis, Giguere (1979), working in a similar system, did find evidence that *Chaoborus* performs better in the presence of a different *Ambystoma* species. Although his experiment was unreplicated, Giguere was able to show that removal of *Ambystoma* from one pond shifted the body size distribution of zooplankton toward species of large size. The shift to larger zooplankton coincided with a decrease in the abundance of *Chaoborus*. Vandermeer (1980) subsequently explored a simple model of the indirect mutualism suggested by Dodson's observation. The model considers two predator species preying on two prey species. The kind of positive indirect effect suggested by Dodson is expected to occur when each predator is highly dependent on a different prey species and when the prey compete so strongly that one is likely to exclude the other in the absence of exploitation by its predator.

The kind of interaction outlined above really has more in common with a commensalism than a mutualism, since one predator indirectly facilitates the other, whereas no reciprocal interaction seems to occur. Dethier and Duggins (1984) describe another example of an indirect commensalism and also provide a conceptual framework to predict the conditions that influence whether interaction will lead to indirect mutualism, indirect commensalism, or simple consumptive competition. In their rocky intertidal system, *Katharina*, a chiton that consumes larger competitively dominant algae, positively affects the abundance of limpets, which graze on small diatoms that are competitively excluded by larger algae. The limpets have no reciprocal indirect effect on *Katharina*, which is what makes this interaction an indirect commensalism rather than a mutualism.

Dethier and Duggins (1984) suggest that the conditions favoring indirect mutualism or commensalism versus consumptive competition are predictable from the degree of resource specialization of the consumers (Figure 8.6). If the consumers are generalists that feed on both kinds of resources, the consumers will simply compete. If the consumers are sufficiently specialized so that each requires different sets of competing resources, then a positive indirect effect will result. Whether the effect is reciprocal (a mutualism) or asymmetric (a commensalism) depends on the extent to which the resource species compete in a hierarchical or asymmetric fashion. Asymmetric competition among the resource species should favor an indirect commensalism, whereas symmetric competition among the resources should lead to a more mutualistic interaction among highly specialized consumers.

TROPHIC CASCADES, TRI-TROPHIC INTERACTIONS, AND BOTTOM-UP EFFECTS

Robert Paine (1980) coined the term **trophic cascade** to describe how the top-down effects of predators could influence the abundances of species in

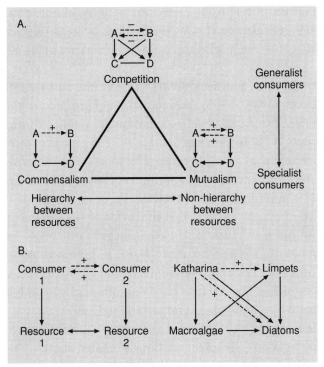

FIGURE 8.6. *(A) Conceptual model relating the extent of consumer specialization and the asymmetry (hierarchy) of competition among resources to the kinds of indirect interactions expected among consumer species (Dethier and Duggins 1984). (B) The kinds of interactions that result in an indirect commensalism between* Katharina *and limpets. (Adapted from Dethier and Duggins, 1984, with permission of the University of Chicago Press.)*

🌿 Indirect effects involving three trophic levels.

lower trophic levels. Others have focused on indirect effects that propagate from the bottom up through multiple trophic levels, called **tri-trophic effects** when the interaction involves three trophic levels (Price et al. 1980). Regardless of the direction of transmission, once an effect proceeds beyond the adjacent trophic level, it becomes indirect. Hairston et al. (1960) and Fretwell (1977) clearly invoke the trophic cascade phenomenon in their writings about population regulation, although they did not call the process by this particular name. Evidence for top-down trophic cascades is surprisingly scarce, and comes primarily from aquatic systems (Strong 1992). There is at least one convincing terrestrial example (Marquis and Whelan 1994). It has been suggested that the scarcity of trophic cascades in terrestrial systems represents a real difference in the structure of terrestrial and aquatic food webs. Strong (1992) suggests that aquatic food webs may tend to be more linear than terrestrial ones. Trophic cascades might be likely to develop in linear food chains, in which effects of one trophic level are readily passed on to other levels. In contrast, in reticulate food webs, distinctions between trophic levels are blurred and effects of one species are likely to diffuse over many adjacent species.

Studies of stream communities provide some of the best examples of trophic cascades. Power et al. (1985) showed that a top predator, in this case largemouth bass, had strong indirect effects that cascaded down through the food web to influence the abundance of benthic algae in prairie streams. The system is best caricatured as a simple three-level food chain, running from algae (mostly *Spirogyra*) to herbivorous minnows (*Campostoma* sp.) to bass (*Micropterus* sp.). The prairie streams typically experience periods of low water flow, during which the streams become series of isolated pools connected by shallow riffles. At such times, two categories of pools become obvious: bass pools with bass and luxuriant algae, and minnow pools with abundant herbivorous minnows but without bass or much algae. The pattern suggests a cascading effect of bass, which prey on minnows and could thereby promote algal growth by eliminating an important herbivore.

To test this idea, Power et al. selected three pools for observation and experimental manipulations. The manipulations consisted of the addition of bass to a minnow pool and the removal of bass from a bass pool, which was then divided in half. One half of the pool received minnows; the other half remained minnow free. A third minnow pool remained unmanipulated as a control. The response of interest was the height of filamentous algae in the pools over time. After bass removal, *Campostoma* greatly reduced algal abundance to low, heavily grazed levels similar to those observed in a natural control pool with abundant minnows (Figure 8.7). Addition of bass to a minnow pool resulted in a rapid increase in algal abundance, whereas algae remained scarce in the control pool without bass. These results are consistent with a cascading indirect effect of bass transmitted through minnows to the algae. The actual mechanism involved appears to be largely a behavioral avoidance of bass by minnows. Minnows leave pools with bass, and, when confined with bass, limit their foraging to shallow water where the risk of bass predation is least.

Similar kinds of trophic cascades may occur in lakes (Carpenter et al. 1985; McQueen et al. 1989) and have been proposed as a possible way to control nuisance blooms of algae in eutrophic waters. Trophic cascades are less dramatic in lakes than in prairie streams, and the influence of top predators generally fails to propagate all the way down to the algae. In lakes, the basic food chain (ignoring the microbial loop) runs from algae to zooplankton to planktivorous fish to piscivorous fish. Where strong trophic cascades occur, lakes with abundant piscivorous fish should have less algae than lakes in which planktivorous fish form the top trophic level, since zooplankton

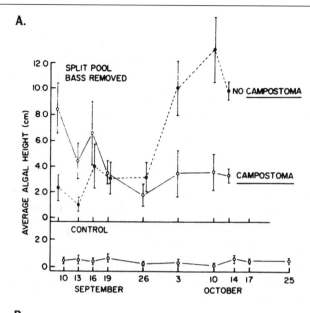

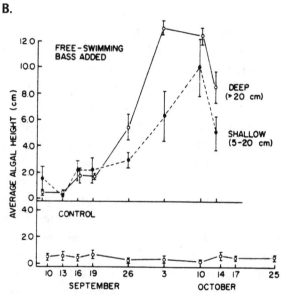

FIGURE 8.7. *(A) Direct negative effects of herbivorous minnows,* Campostoma, *on algal abundance in a stream pool. Removal of minnows leads to an increase in algal height, whereas algal height stays unchanged in a control pool with minnows and without bass. (B) Indirect positive effect of bass on algal abundance, a trophic cascade, contrasted with the same control pool shown in (A). (Reprinted from Power et al., 1985, with permission of the Ecological Society of America.)*

should be more abundant and should reduce phytoplankton to lower levels. However, the predicted cascading effects seldom appear as decreased phytoplankton abundance (Carpenter et al. 1987). One reason is that the phytoplankton consists of an array of species that differ in their vulnerability to grazing by zooplankton, and differences in zooplankton grazing pressure simply select for complementary communities of algae that differ in grazer resistance. This situation has been modeled by Mathew Leibold (1989). When zooplankton are abundant, the phytoplankton is dominated by grazer-resistant species. When zooplankton are less abundant, the phytoplankton is dominated by competitively superior species that are vulnerable to grazing. Phytoplankton remains abundant, but is dominated by different sets of species. Consequently, the prospects for manipulating fish populations to control the abundance of nuisance algae seem limited.

The best example of a terrestrial trophic cascade comes from a study by Robert Marquis and Christopher Whelan (1994). They found strong effects of insectivorous birds that were transmitted through herbivorous insects to white oak trees. Birds were excluded from some trees by netting (cage treatment), while other trees remained available to the birds (control treatment). Birds significantly reduced the abundance of herbivorous insects on the oaks. In turn, oaks with birds and reduced herbivorous insects had less leaf damage from insects and subsequently attained a higher biomass (Figure 8.8). The effect of birds on insects was further corroborated by including an insect removal treatment (spray treatment) consisting of applications of a spray insecticide combined with the hand removal of remaining insects.

Strong experimental evidence for **bottom-up indirect effects** also comes from manipulations of stream communities. Wootton and Power (1993) manipulated the amount of light available to algae by differentially shading small portions of a natural stream. They then measured how these manipulations affected the abundance of algae, herbivores, and carnivores. Increases in light created increases in algal abundance and increases in carnivore abundance, while herbivore abundance remained unchanged (Figure 8.9). These results are generally consistent with the models of Rosenzweig (1973) and Oksanen et al. (1981), which suggest that an increase in productivity in a three-level food chain should create an increase in the abundance of plants and carnivores, whereas the abundance of herbivores should not show much of an increase.

For example, consider the following simple model of a three-level food chain:

$$dn_1/dt = rn_1 - an_1n_2 - \left(r(n_1)^2\right)\big/k \tag{8.1}$$

$$dn_2/dt = ean_1n_2 - xn_2 - cn_2n_3 \tag{8.2}$$

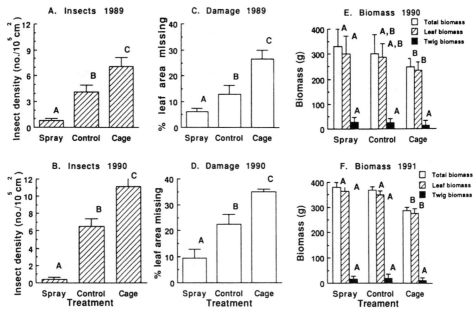

FIGURE 8.8. *Indirect effects of bird exclusion on leaf damage and biomass of small oak trees in two consecutive years. The indirect effect of birds is mediated through the reduction of herbivorous insects on trees where birds forage. Treatments labeled "Cage" excluded insectivorous birds. Spray treatments used insecticide to remove insects. Controls were uncaged and contained natural densities of herbivorous insects maintained at low levels by birds. Bars with identical letters did not differ significantly. (Reprinted from Marquis and Whelan, 1994, with permission of the Ecological Society of America.)*

$$dn_3/dt = gcn_2n_3 - yn_3 \qquad (8.3)$$

Here, n_1, n_2, and n_3 refer to the abundances of species on trophic levels 1, 2, and 3. Trophic level 1 has a carrying capacity of k and a rate of increase r. Trophic level 2 consumes trophic level 1 at a rate given by an_1n_2, the simple linear functional response of the Lotka-Volterra predator-prey equations. Consumption of trophic level 1 is transformed into new individuals of trophic level 2 at some efficiency, e, yielding ean_1n_2 as the rate of birth of trophic level 2. Individuals on trophic level 2 die at a constant rate x in the absence of trophic level 3, and at a rate of cn_2n_3 when consumed by trophic level 3. Individuals on the third trophic level are born at a rate of gcn_2n_3, assuming that consumed individuals of trophic level 2 are converted into new predators on trophic level 3 with some efficiency g. Predators on the top trophic level also die at some rate, y. This model is simpler than the one used by Oksanen et al. with respect to details of the functional responses of the predators to prey, but it makes roughly comparable predictions about how the

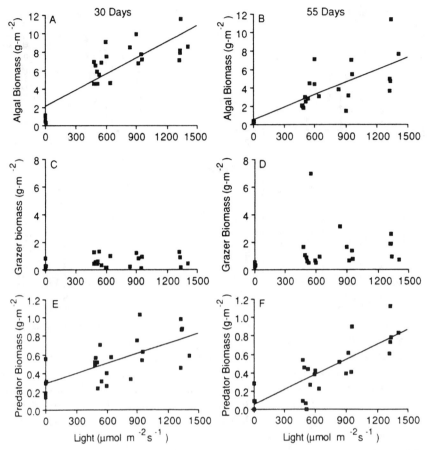

FIGURE 8.9. *Effects of increasing light levels on the biomass of primary producers (algae), herbivores (grazers), and predators in streams. In these three-level food chains, species on odd-numbered levels increase with increasing productivity, whereas species on the even-numbered level remain constant, as suggested by Fretwell (1977) and Oksanen et al. (1981). (Reprinted from Wootton and Power, 1993. © 1993 National Academy of Sciences, U.S.A.)*

abundance of each trophic level at equilibrium will change as *K*, a measure of productivity, changes (Figure 8.10).

Peter Abrams (1993) has used models to explore how different arrangements of interacting species in simple three-level food chains would affect the likelihood of bottom-up effects like those observed by Wootton and Power (1993). His theoretical results suggest that bottom-up effects will depend strongly on the amount of heterogeneity among species in their responses to species on other trophic levels. Three situations involving rather minor departures from a simple linear three-level food chain can prevent bottom-up effects (Figure 8.11). Competition among multiple species on the top trophic level may create situations in which the decline in one competitor is not offset

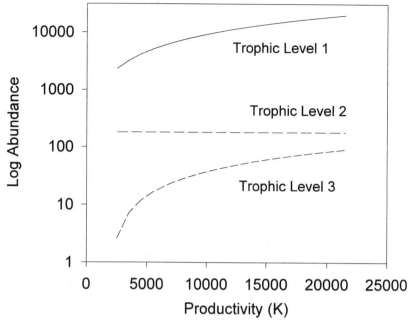

FIGURE 8.10. *Relations between abundance at equilibrium and primary productivity (k), using the model described in Equations 8.1 to 8.3. Parameter values are as follows: r = 2.5, a = 0.001, e = 0.05, x = 0.09, c = 0.01, g = 0.5, and k varies between 2500 and 21,500.*

by the increase in another. In other situations, the presence of an invulnerable species on the intermediate trophic level may divert increases in productivity from reaching the top trophic level. Alternately, if species on the intermediate trophic level share the same prey and predators, then increases in productivity simply cause a shift toward more predator-resistant species, with no net effect on the abundance of the top trophic level.

INTERACTION MODIFICATIONS: HIGHER-ORDER INTERACTIONS AND NONADDITIVE EFFECTS

Interaction modifications have attracted the attention of ecologists for many years because their existence gravely complicates the prediction of interactions in complex communities from knowledge of pairwise interactions in simple communities. For example, the multispecies formulation for the dynamics of a set of Lotka-Volterra competitors depends on the assumption that per capita competitive effects are immutable properties of pairs of species (the competition coefficients) that do not

> Some indirect effects involve changes in the ways that species interact.

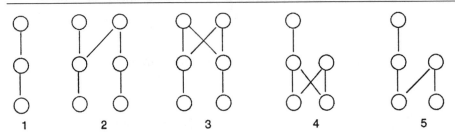

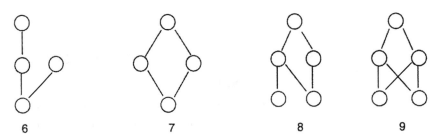

FIGURE 8.11. *Departures from simple linear three-level food chains (web 1) that can prevent bottom-up (trickle-up) positive effects of enhanced productivity on top predators (Abrams 1993). Webs 2 and 3 incorporate competition among top predators. Webs 4, 5, and 6 have an invulnerable prey on the intermediate trophic level. Webs 7, 8, and 9 have intermediate species that share predators and prey.*

change as other species in the community come or go. If the per capita effects of competitors do change with species composition, then phenomena often referred to as **higher-order interactions**, or **nonadditive interactions**, become a real concern. Higher-order interactions are a subset of the kind of indirect effects called interaction modifications. Unfortunately, both of these terms emphasize the ways in which observed interactions depart from statistical or analytical models of interactions rather than emphasizing the biological basis of indirect effects that involve interaction modifications (Abrams 1987). Typically we abstract the way in which a pair of species will interact as a coefficient, such as a competition coefficient or as an attack rate in a functional response term. The usual assumption is that those coefficients are properties of the particular pair of interacting species. However, those interactions, and the coefficients that describe them, may depend on the mix of

TABLE 8-1. Competition coefficients (α_{ij}) estimated for crustaceans interacting in combinations of two or three species. Changes in the coefficients measured for pairs or trios of species mean that competitive interactions among pairs of species (shown to the left of the slash) are changed by the presence of additional species (shown to the right of the slash), a kind of interaction modification.

Species i	Species j			
	Alonella	Ceriodaphnia	Simocephalus	Hyalella
Alonella	1	1.08/0.79	0.06/0.91	9.16/11.08
Ceriodaphnia	0/0	1	0.37/0.84	3.38/5.40
Simocephalus	0/0	0.13/0.12	1	4.30/4.40
Hyalella	0/0	0/0	0/0	1

Reprinted from Neill (1974), with permission of the University of Chicago Press.

other species present when the interactions are measured because additional species can modify the ways in which the focal pair of species interact. Additional competitors can change the ways that species compete. Additional predators or prey can alter the way a given predator interacts with a particular prey. If these interaction modifications are important, it is difficult to predict the outcome of interactions among large sets of species from information about interactions between isolated pairs of species. In a sense, every interaction becomes a special case, the outcome of which depends on the particular features of the biotic and abiotic environment where the interaction occurs.

The kind of information needed to demonstrate the existence of interaction modifications is difficult to obtain (Billick and Case 1994; Wootton 1994a; Adler and Morris 1994). There are a few convincing cases, though. One of the first efforts was by William Neill (1974). Neill showed that the competition coefficients measured for interactions among four species of small crustaceans in aquatic microcosms changed with the number of species present in the system (Table 8-1). This result means that if one wanted to predict the outcome of competition among three or four species by summing up all the pairwise competitive interactions among species, errors would arise. Neill's study has been criticized on methodological grounds because it is very difficult to measure sets of competition coefficients accurately from experimental data (Pomerantz 1981). Nonetheless, the study has few equals, and it raises important questions about the validity of assumptions used to create multispecies Lotka-Volterra models of competing species.

Case and Bender (1981) have been critical of much of the evidence presented for interaction modifications. They outline an approach to the study of interaction modifications that circumvents the problem of estimating interaction coefficients by focusing on how initial population growth rates change in different combinations of one, two, or three species. They have shown that in simple laboratory systems of one to three species of *Hydra*, population growth rates in two-species communities are significantly greater than expected from growth rates observed in single-species and in three-species communities. They suggest that some sort of mutualistic interaction occurs at low densities in two-species communities that does not materialize in three-species communities.

Other evidence for interaction modifications comes from Tim Wootton's (1992, 1993, 1994b) studies of interactions among predatory birds, limpets, mussels, and gooseneck barnacles in the rocky intertidal zone. Important direct and indirect interactions in this system are outlined in Figure 8.12. Predatory birds change the abundance of two sessile species, *Mytilus californianus* and *Pollicipes polymerus*, which in turn are the preferred substrates of two different limpet species. *Lottia digitalis* is light in color and is cryptic on light-colored *Pollicipes*. *Lottia pelta* is dark in color and is cryptic on *Mytilus*. Birds selectively reduce the abundance of *Pollicipes*, favoring *Mytilus*. In turn, *Lottia pelta* becomes more abundant, since the presence of *Mytilus* makes it more difficult for visually foraging birds to locate the limpet that is cryptic on *Mytilus*.

Other kinds of interaction modifications materialize in terrestrial communities. In some cases, species can indirectly affect others either after death (Bergelson 1990) or through the effects of nonliving material such as dead leaves and other forms of plant litter (Facelli 1994). Jose Facelli found that leaf litter produced by forest trees fundamentally changed the interactions between tree seedlings and herbaceous competitors in open fields. Where litter is abundant, seedlings of herbaceous plants have a reduced competitive impact on seedlings of the tree *Ailanthus altissima*. Although litter creates a more favorable microclimate by increasing soil moisture, it also tends to shade out herb seedlings, making them weaker per capita competitors. The effects of litter on herbivores are even more complex. Although seedlings benefit from reduced competition from herbs, they suffer increased damage from herbivorous insects. The favorable microclimate provided by plant litter leads to increased insect abundance, which can cause increased damage to a variety of plant seedlings. The net result is that plant litter has multiple indirect effects on *Ailanthus* seedlings, some positive and some negative (Figure 8.13). Facelli was able to dissect these effects by factorial manipulations of herbaceous competitors, litter, and insecticide in a field experiment conducted under natural conditions.

A.

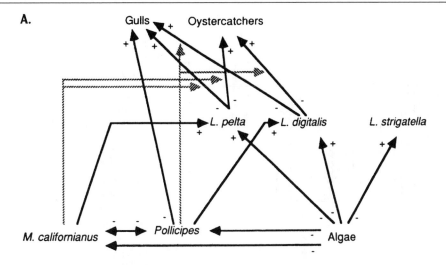

B.

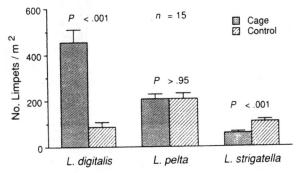

FIGURE 8.12. *(A) Direct and indirect interactions affecting the abundance of three limpet species. Solid arrows indicate direct effects; dashed arrows indicate indirect effects resulting from the modification of direct interactions between pairs of species. (B) Results of bird exclusion (cage) on the abundance of three limpet species.* L. digitalis *increases because its favored substrate,* Pollicipes, *increases when birds are excluded. (Reprinted from Wootton, 1992, with permission of the Ecological Society of America.)*

INDIRECT EFFECTS AND THE INTERPRETATION OF MANIPULATIVE COMMUNITY STUDIES

Indirect effects can complicate the interpretation of ecological experiments. Bender et al. (1984) have pointed out that, depending on the kind of experimental manipulation that is performed, responses can include a mixture of indirect and direct effects or just direct effects. Bender et al. recognize two kinds

> Indirect effects complicate the interpretation of ecological experiments.

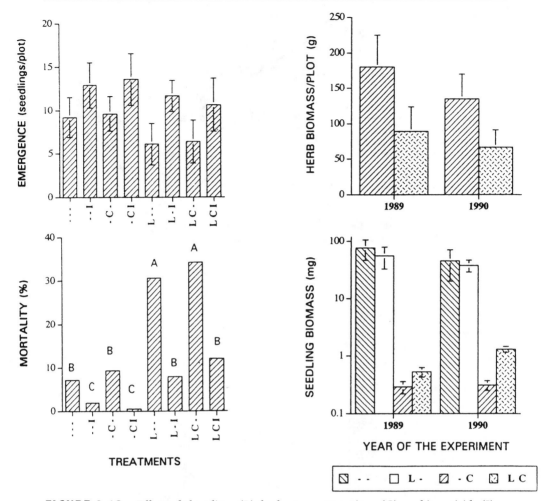

FIGURE 8.13. *Effects of plant litter (L), herbaceous competitors (C), and insecticide (I) on the emergence, mortality, and biomass of seedlings of the tree* Ailanthus altissima. *Litter indirectly increases seedling mortality by increasing the abundance of herbivorous arthropods. However, litter indirectly enhances the biomass of surviving seedlings by reducing the biomass of competing herbs. (Reprinted from Facelli, 1994, with permission of the Ecological Society of America.)*

of experimental manipulations: press experiments, in which the density of a species is permanently changed, and pulse experiments, in which the density of a species is altered and then allowed to return to its previous state (Figure 8.14). Additions or removals of species correspond to press experiments, the most common kinds of manipulations done by ecologists. Pulse experiments would correspond to a one-time increase or decrease in density of a species already present in the community, without adding or removing a species from the community or maintaining the altered density at a particular level. Bender et al. argue on theoretical grounds that responses to

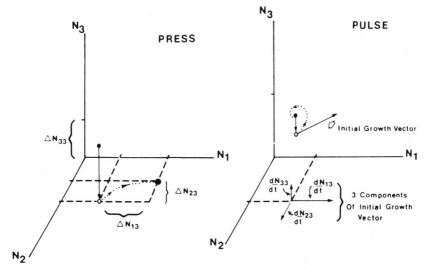

FIGURE 8.14. *Phase space representation of the difference between press and pulse experiments. The axes N_1, N_2, and N_3 refer to the abundances of species 1, 2, and 3. In a press experiment, species 3 is eliminated, and the responses of species 1 and 2 are shown by the dashed line on the N_1–N_2 plane. In the corresponding pulse experiment, species 3 is reduced in abundance but not eliminated, and all three species can change in abundance over time. The initial growth vector can be decomposed into the responses of each species to a reduction in the density of species 3. (Reprinted from Bender et al., 1984, with permission of the Ecological Society of America.)*

press experiments include direct and indirect effects, which makes them difficult to interpret. In contrast, pulse experiments should highlight only direct effects.

The theoretical arguments used to make this distinction rely on some assumptions that may be difficult to justify in natural systems, such as the existence of a stable equilibrium to which the community tends to return following a pulse perturbation. Also, the approach is limited in utility to those species that reproduce with sufficient rapidity so that responses are likely to be seen in a reasonable amount of time, say a few years. There is also the very real practical problem of engineering pulse perturbations that will be strong enough to elicit a detectable set of responses. Most press perturbations arise from the consequences of being able to either add or delete species from communities, either by additions or deletions of free-ranging organisms or by selective barriers such as cages. Species additions or deletions are often easy to engineer and yield detectable responses, important considerations for experimental ecologists. The point that press experiments must be interpreted carefully is well taken, and subsequent studies have shown that long-term press experiments often show changing patterns that can be attributed to the influence of indirect effects.

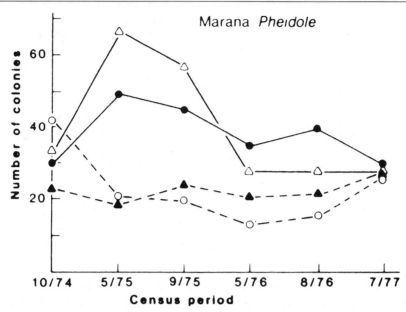

FIGURE 8.15. *Initial and long-term responses of granivorous ants,* Pheidole *sp., to removals of granivorous rodents (solid lines). Controls with rodents are shown by the dashed lines. Initial increases in ant density gradually return to control levels, as small-seeded plants that are preferred by ants are replaced by competitively superior large-seeded plants. (Reprinted from Davidson et al., 1984, with permission of the Ecological Society of America.)*

One field experiment showing the influence of unanticipated indirect effects involves a study of interactions between granivorous rodents, ants, and the plants that produce the seeds that these granivores eat. Davidson et al. (1984) initially found strong negative competitive effects of rodents on ants in an array of large field exclosures where seed-eating rodents were present at natural densities or excluded. Positive effects of rodent removals on ants were strong early in the experiment but then gradually disappeared over time, despite the fact that rodent removals continued (Figure 8.15). The gradual decline in ant abundance was due to an indirect effect of rodent removal on the small-seeded plants that are selectively consumed by ants. Rodents prefer to eat the seeds of large-seeded plants, which tend to competitively displace small-seeded plants when rodent predation fails to keep the large-seeded plants in check. Although these indirect effects eventually led to a rather different pattern than would be expected from the initial responses to rodent removal, their cause is easily interpreted. The indirect effects also took several years to become pronounced. This suggests that although caution is called for in interpreting any long-term press experiment, initial strong responses probably reflect direct effects even when indirect effects may eventually become important.

Other studies (e.g., Wootton 1994b) have shown that reliable statistical tools, such as path analysis, can be used to tease apart and identify indirect effects. In any complex community there may be many possible causal pathways of interactions among species, which collectively form an interaction web. Wootton (1994b) has used path analysis to identify some of the more likely causal relations in complex communities. Path analysis can be used to make predictions about how changes in the abundance of key species will affect the abundance of other species in the community. To the extent that these predictions differ among proposed chains of interactions, it is possible to test which scenario, or interaction web, is most likely. Path analysis has limitations. It is no more than a descriptive technique that can summarize the ways in which temporal or spatial variation in abundance is correlated among species. Nonetheless, it can be used to generate testable hypotheses.

Wootton noted four important changes in the abundance of different intertidal organisms in response to the experimental exclusion of birds (Figure 8.16). At least three different interaction chains could have produced these differences. Fortunately, the different interaction chains made different predictions about how particular species would respond to additional experimental manipulations (Table 8-2). When those manipulations were done, the results were consistent with the simplest, shortest, interaction chain, which was also the scenario favored by the path analysis. The power of this approach is that it allows the generation of alternate hypotheses, which can then be tested by field experiments. Path analysis is a descriptive technique that by itself cannot determine whether a particular interaction chain is responsible for a particular pattern. It can, however, indicate whether certain chains are more plausible than others.

CONCLUSIONS: FACTORS CONTRIBUTING TO THE IMPORTANCE OF INDIRECT EFFECTS

We still know very little about why indirect effects occur in some settings and not in others. A few tentative generalizations seem appropriate, if only to encourage further research. Strong direct effects are probably required to produce noticeable indirect effects; weakly interacting species are not likely to generate sufficient changes in abundance or behavior of transmitter species for those effects to appear in receivers. The impact of simple versus complex food chain structure on the transmission of indirect effects requires much further study. Strong indirect effects clearly materialize in systems with complex reticulate chains of interacting species, such as rocky intertidal communities (Wootton 1994b), as well as in systems with relatively linear food chains, such as the stream communities studied by Power et al. (1985). Where cascading effects fail to materialize, as in some large lakes, the failure may be

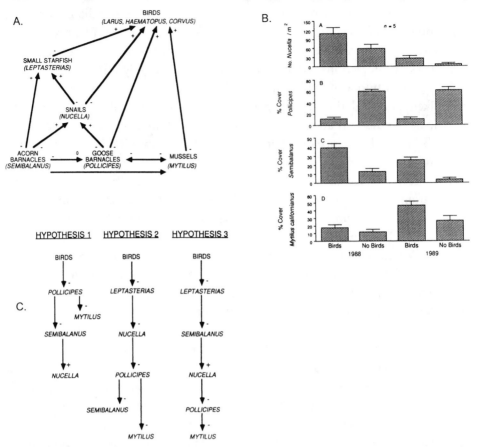

FIGURE 8.16. *(A) Pathways of potential interactions among a group of species found in the rocky intertidal zone of Washington. Horizontal arrows indicate competition; other arrows indicate predator-prey links. (B) Responses of four species to bird removal. (C) Three alternate sets of causal links (pathways) that could account for the responses of species to bird removal. The shortest path, hypothesis 1, is most consistent with the results of experiments designed to distinguish among these alternate pathways. (Reprinted from Wootton, 1994b, with permission of the Ecological Society of America.)*

due to compensatory changes in the abundances of species within trophic levels that differ in their resistance to consumers, as suggested by Leibold (1989) and Abrams (1993). Although most of the examples described here involve indirect effects among chains of three or four species, it is unclear whether this represents a limit to the transmission of indirect effects in natural communities or whether longer chains of interactions occur. McQueen et al. (1989) suggest that both top-down and bottom-up indirect effects are transmitted for very limited distances before they become undetectable. Thus, bottom-up effects would be most likely to be seen low in food chains, whereas top-down effects would be stronger for species high in food chains.

TABLE 8-2.   Predicted responses to experimental manipulations of intertidal organisms that could be used to distinguish among the three alternate interaction pathways that might explain the indirect effects of birds (see Figure 8.16).

Manipulation and Target Species	Predictions		
	Hypothesis 1	Hypothesis 2	Hypothesis 3
Reduce *Nucella*			
1. *Pollicipes*	0	+	+
2. *Semibalanus*	0	−	0
3. *Mytilus*	0	−	−
Reduce *Semibalanus* independently of birds, *Pollicipes*, and *Mytilus*			
4. *Nucella*	−	0	−
Reduce *Pollicipes* independently of birds			
5. *Semibalanus*	+	+	0
6. *Mytilus*	+	+	+
7. *Nucella*	+	0	0
Reduce birds independently of *Pollicipes*			
8. *Semibalanus*	0	0	−
9. *Mytilus*	0	0	0
10. *Nucella*	0	−	−
11. *Leptasterias*	None	+	+

Reprinted from Wootton (1994b), with permission of the Ecological Society of America.

Ultimately, we need to know the extent to which the dynamics of species in complex webs are tightly or loosely connected in order to understand whether indirect effects will propagate widely through the community or rapidly peter out. A useful analogy is to consider a web of interactions among species, in which the functional connections between species can be thought of as mechanical connections, like rigid rods or very flexible springs. Where species are linked by rigid rods, the case for tight connections among species, a force applied to any one species will be transmitted to many others. Where species are loosely connected by flexible springs, a force applied to a species may leave most others unaffected. The question is, in real communities, how many connections are rigid and how many are flexible?

PART II

FACTORS INFLUENCING INTERACTIONS AMONG SPECIES

CHAPTER 9

Temporal Patterns: Seasonal Dynamics, Priority Effects, and Assembly Rules

OVERVIEW

This chapter focuses on the changes in communities that occur over relatively short timescales. Short-timescale phenomena include seasonal patterns, which become apparent over periods of days or months, and more prolonged sequences of arrival and establishment of species during the early phases of assembly in developing communities. Long-term successional changes may require decades to centuries to occur and are the subject of a subsequent chapter. This chapter considers how short-term variation in the timing of species' invasions, arrivals, or activities influences the outcome of interspecific interactions and the establishment of species in communities. Simple differences in the order of arrival of species in developing communities can sometimes have lasting consequences for patterns of community composition. Case studies illustrate the underlying causes and consequences of temporal variation in community composition. The chapter concludes with a discussion of assembly rules, which can describe the nonrandom ways in which species succeed or fail to become established in developing communities.

THE IMPORTANCE OF HISTORY

This chapter focuses on patterns and processes related to ways in which short-term historical differences influence species interactions in developing communities. These complications arise because species differ in their time of arrival in developing communities, either because of predictable seasonal variation in abundance or by chance. These situations occur during the assembly of communities in newly available habitats as well as during cyclic patterns of seasonal community development that recur in the same locations year after year. In all these cases, we would like to know whether recent historical events strongly influence community composition (Ricklefs 1989; Ricklefs and Schluter 1993). History shapes every community to some extent, either by influencing which species can colonize developing communities or by setting the sequence of species arrival as species accumulate and interact.

One interesting question concerns whether initial differences in species composition will persist or propagate as different trajectories of community development. Alternately, communities assembled from a similar species pool may tend to converge in appearance over time as the full cast of potential colonists has the opportunity to become established. The tendency for initial differences to either persist or disappear is important because persistent differences among communities caused by chance events early in community development could be responsible for variation among comparable habitats in species composition, which is sometimes called beta diversity. Regional variation in the composition of a set of otherwise similar habitats, such as ponds or abandoned fields, might depend on when sites first become available for colonization and the order in which colonists arrive.

Simple models of competitive interactions predict that small historical differences, such as differences in the initial abundances of two competing species, can produce very different communities. Lotka-Volterra models of interspecific competition show that under some conditions history plays no role in generating community patterns. When conditions for a stable two-species equilibrium occur, communities will reach the same final equilibrium composition regardless of historical differences in the initial (nonzero) abundances of species (Figure 9.1A). Under such conditions, the model's predictions are history free, in the sense that the same eventual outcome occurs regardless of the initial abundance of each competing species. However, when different parameters create an unstable equilibrium in the same basic model, historical effects become very important because initial differences in the abundance of species can determine which species will competitively exclude the other (see Figure 9.1B). More complex models can also predict alternate outcomes corresponding to alternate community structures that depend on initial conditions (e.g., Holt et al. 1994). It is unclear what percentage of

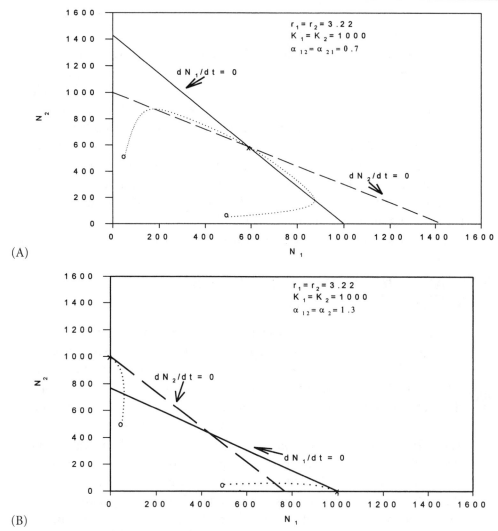

FIGURE 9.1. *Lotka-Volterra models of two competitors,* N_1 *and* N_2, *in which the outcome of competition is independent of initial conditions (A), or depends on initial conditions (B).*

natural communities is mostly shaped by history-free processes and what percentage is strongly influenced by historical processes. The examples presented here survey the kinds of historical effects that can influence interactions within communities and also point to some processes that can limit the impact of history on community composition.

The historical events that influence communities can occur on vastly different timescales. Differences of a few days in the arrival times of mycophagous *Drosophila* in the communities that develop in decaying mushrooms can alter the outcome of competition (Shorrocks and Bingley 1994). Other successional patterns involving slow-growing, long-lived organisms

Historical events that influence community patterns occur over a range of timescales.

may require many years to play out (Clements 1916; Gleason 1917; Keever 1950). Because of the practical constraints that limit studies of processes operating on very long timescales, we know much more about historical effects in systems composed of species with short generation times and rapid dynamics. Some of these historical effects have been called **priority effects**, because species present at some early phase of community development influence other species that arrive at some later time. Much of what we know about priority effects comes from simple experimental manipulations of the order of arrival of species in developing communities. These studies are often inspired by curiosity about what might happen if some typical seasonal pattern of abundance were altered.

INTERACTIONS AMONG TEMPORALLY SEGREGATED SPECIES

Naturalists have always been fascinated by temporal changes in the abundance of species. Some seasonal patterns, such as annual sequences of flowering by different plants, the flight seasons of insects, the breeding seasons of pond-dwelling amphibians, and the springtime return of migratory birds, repeat from year to year with great regularity. The term **phenology** refers specifically to these well-known seasonal patterns of abundance or activity. Such patterns have been described for many different organisms (Figure 9.2). The speculation about the causes of phenological differences among organisms has often inspired debate and considerable field experimentation, much of it addressing the potential consequences of phenological variation for interspecific interactions. Other temporal changes in species abundance occur over longer time frames and differ from seasonal patterns by being noncyclic. These long-term patterns occur over timescales ranging from a few to many thousands of years, as species wax and wane in abundance during succession or in response to climate change. Paleoecologists have described long-term changes in the abundance of species in response to climate change or natural disturbances. The examples in Figure 9.2 show that species peak in abundance at different times. Such differences may be important if earlier species either inhibit or facilitate the species that follow.

 Many species display seasonal variation in abundance or activity.

Given that species often differ in their timing of first arrival or maximum activity in communities, phenological differences create situations in which the outcomes of interspecific interactions may depend on temporal patterns

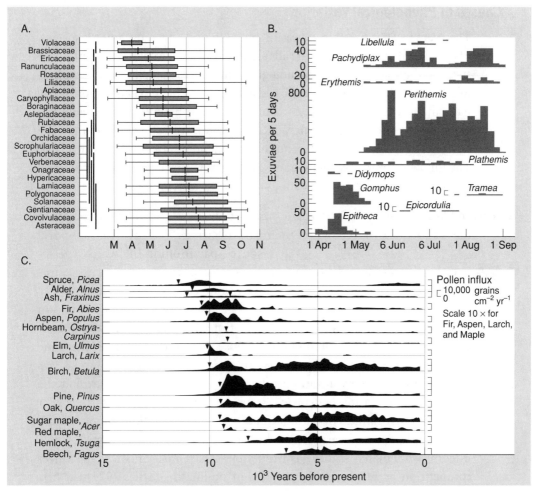

FIGURE 9.2. *Patterns of temporal variation in different species assemblages. (A) Flowering seasons of different families of plants. (Reprinted from Kochmer and Handel, 1986, with permission of the Ecological Society of America.) (B) Emergence times of larval odonates. (Adapted from Morin, 1984b, with permission of the Ecological Society of America.) (C) Abundance of trees over geological time reconstructed from pollen deposits. (Adapted from Davis, 1981, with permission. © Springer-Verlag.)*

of species. In the simple case of two species with different phenologies, three basic interactions are possible: The earlier species can facilitate, inhibit, or have no effect on the later species (Connell and Slatyer 1977). Although phenological differences make such interactions possible, species can differ in phenology for reasons that are unrelated to interactions with other species. It is important to keep in mind that the causes of phenological differences may be quite separate from their community-level consequences.

Causes of Phenological Variation

There are several possible causes of phenological variation among species. Phenological differences may be adaptive responses to interactions with other species. Alternatively, some phenological differences may simply reflect physiological constraints or stochastic events. Because phenological patterns may have many causes, it is risky to interpret any phenology as an adaptive response to interspecific interactions without direct evidence that the outcome of the interaction depends on the phenology of the species involved. The following examples include some of the potential causes of phenological variation, along with evidence supporting those interpretations.

Temporal Resource Partitioning. It is tempting to interpret seasonal differences in the abundance of ecologically similar species as evidence for **temporal resource partitioning**, in which species manage to coexist by using the same limiting resources at different times of the year. The assumption is that competition among species that are active at different times would be less intense than if the same set of species all attempted to use the same resource at the same time. This scenario is most plausible for situations in which resource levels rapidly recover from utilization; otherwise, the first species to exploit the resource would effectively deplete resource availability for species active later in the season, regardless of the amount of temporal separation among species. For example, seasonal variation in the flowering phenology of tropical hummingbird-pollinated plants has been viewed as a mechanism that reduces competition by plants for the services of pollinators (Stiles 1975). At first glance, the distribution of flowering times appears temporally staggered, such that different species may avoid extensive temporal overlap for pollinators (Figure 9.3). Such interpretations, while plausible, are plagued by the difficulty of demonstrating that the process driving the pattern is the reduction of interspecific competition. An experimental test of this hypothesis would need to show that species with similar seasons of flowering activity compete more strongly for pollinators than species with displaced periods of activity. The observation that species differ in flowering times is not sufficient evidence in itself for temporal resource partitioning, since species might be expected to differ in phenology to some extent simply by chance (Poole and Rathcke 1979). Indeed, other ecologists examining the same data have concluded that the temporal separation of flowering times is no greater than would be expected for a random pattern (Poole and Rathcke 1979).

Norma Fowler (1981) found some evidence to suggest that temporally segregated plant species compete less strongly than plants that are actively growing at the same time of year. Fowler studied competitive interactions

Are phenological differences random or not?

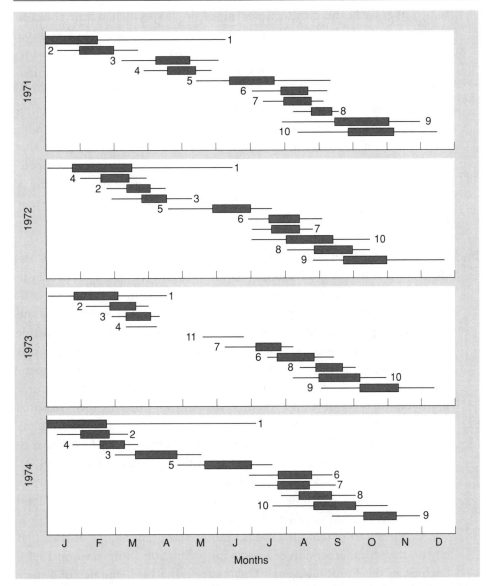

FIGURE 9.3. *Flowering times of different hummingbird-pollinated plants in Costa Rica. Different numbers refer to different species. (Adapted with permission from Stiles, 1977. © 1977 American Association for the Advancement of Science.)*

among plants growing in a mowed field at Duke University in North Carolina. The plants growing at this site mostly fell into two groups distinguished by their different growing seasons: cool-season plants growing primarily during the winter months, and warm-season plants growing in the spring and summer. By selectively removing different species of plants from experimental plots and assessing the growth responses of remaining species, Fowler showed that competitive interactions within the entire system tended to be rather weak and diffuse. Removals of single species had little effect on the remaining plants, whereas responses to removals of groups of multiple species tended to be stronger. The interesting pattern was that when plants could be shown to compete, via positive responses to plant removals, competition occurred primarily within groups of species sharing the same growing season. Plants growing at different times of the year competed relatively infrequently, in only 3 of 12 cases where comparisons were possible. Fowler's interpretation of these results, which is tempered by the observation that competitive interactions were generally weak in this system, is that competition occurred primarily within guilds of species growing at the same time of year. This pattern is consistent with the idea that temporal separation may reduce competition among species.

Tracking of Seasonal Resources. Another possible reason for temporal displacement is that different species simply require different resources, which vary in abundance over time. This situation is different from temporal resource partitioning because different species may be tracking different resources and therefore might not compete even if those different resources were available at the same time. Thus, phenological differences among the species using those resources may not be related to the reduction of interspecific competition. Instead, one can imagine sets of species with highly specialized resource needs that depend on other resource species that differ in their seasonal phenology for some reason. For example, this situation appears to apply to oligolectic bees, which often depend on only one or a very few species of plants for the pollen and nectar used to rear their offspring (Cruden 1972). Flight seasons of the bees coincide closely with the flowering seasons of the particular plants on which they rely, and species using different plants as food sources differ in phenology accordingly.

A similar sort of temporal specialization may occur among orchids that rely on the sexual deception of insects for pollination (Borg-Karlson 1990). Orchids in the European genus *Ophrys* are remarkable mimics of female wasps, to the extent that they induce males wasps to attempt to copulate with the flowers, which mimic the pheromones and approximate appearance of female wasps. Pseudocopulation with several flowers results in pollen transfer among plants. In this case, it is probably advantageous for one set of species, the flowering *Ophrys*, to bloom at times that coincide with the breeding season of the particular wasp or bee, which may be temporally segregated for other reasons.

Lack (1966) suggested that the breeding phenology of some birds is driven primarily by seasonal variation in the peak availability of food needed by rapidly growing offspring. Wolda (1987) reviews a variety of other seasonal patterns that may be driven by differences in resource availability. What is striking about most of these cases is how seldom seasonal resource tracking is clearly demonstrated.

Predator Avoidance. Predators and other natural enemies can vary in abundance over time and, in turn, affect the seasonal abundance of their prey. Avoidance of seasonally intense predation can account for some aspects of prey phenology. Despite Lack's suggestion that birds time their breeding to coincide with peak periods of prey abundance, some birds appear to avoid breeding when prey are most abundant if their enemies are also abundant at the same time (Nisbet and Welton 1984). The consequences of phenological variation for growing predators or prey can be quite complex. Very small differences in phenology can lead to major differences in prey survival when the outcome of predator-prey interactions depends on the relative sizes of prey and predators (Thompson 1975).

 Phenology can influence predator-prey interactions.

Ross Alford (1989) experimentally simulated how variation in predator phenology influenced the survival and species composition of frog tadpoles in experimental ponds. Initially, hatchling tadpoles are much smaller than their predators, but they can rapidly grow to a size that renders them invulnerable to attack. The main predator in this system, the adult newt (*Notophthalmus*), is most abundant in ponds earlier in the year, tends to leave ponds for terrestrial sites later in the season, and grows little during its time in the pond, relative to its rapidly growing prey. Newts may leave ponds at different times in different years, and anurans can also breed at different times due to variation in weather favorable for breeding. In years when frogs breed early and predators delay their departure from ponds, predators and prey may interact for prolonged periods of time. In other years, when anuran breeding is delayed or when predators leave the ponds relatively early, tadpoles and predators overlap and interact for shorter periods of time. As a result, predators and prey overlap for different amounts of time and interact at different relative sizes from one year to the next.

Alford compared patterns obtained in communities where newts remained in ponds and fed on tadpoles for different amounts of time. In three different treatments, predators fed on tadpoles for the first 9 days, the first 51 days, or the entire duration of community development. A fourth treatment without predators served as a control. A second complication is that the four prey species tend to breed at different times. Species that breed early in the season can have prolonged interactions with predators, whereas later breeders may miss the predators entirely except in treatments where predators remained present throughout the entire experiment.

Communities without predators or with predators for a short period of time were more similar than communities with predators for longer periods of time (Figure 9.4). Community composition depended in a complex way on how long predators and prey interacted within these developing communities. Brief exposure to predators actually enhanced the survival of some species by reducing the survival of others.

In other cases, differences in the timing of the arrival of interacting species may determine whether those species interact as competitors or as predators and prey. Stenhouse et al. (1983) studied interactions between two species of larval salamanders, *Ambystoma opacum* and *Ambystoma maculatum*. Both species prey on a variety of aquatic prey, but a sufficient disparity in size will allow large larvae of one species to consume small larvae of the other. In most years, *A. opacum* hatches months before *A. maculatum*. Consequently, *A. opacum* larvae are much larger than *A. maculatum* when they

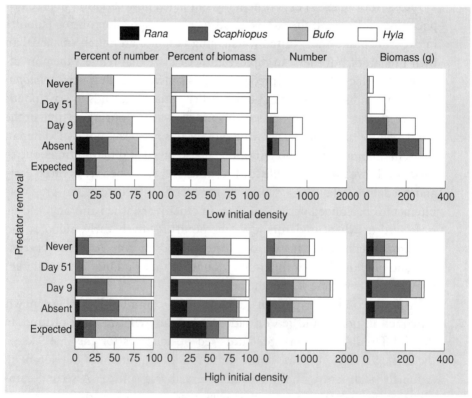

FIGURE 9.4. *Effects of removal of predators after 9 days, 51 days, or never on several measures of anuran assemblage composition in experimental ponds. Treatments labeled "Absent" did not contain predators. "Expected" shows the patterns expected from initial relative numbers of anurans added to the communities. (Adapted from Alford, 1989, with permission of the Ecological Society of America.)*

begin to interact. This size disparity favors predation by *A. opacum* on *A. maculatum*, which greatly reduces *A. maculatum* survival. However, in some years *A. opacum* hatching is delayed, and the two species begin to interact at fairly similar body sizes. When similar in size, the two salamanders interact primarily as competitors, and *A. maculatum* survives much better than when *A. opacum* enjoys a substantial size advantage.

Facilitation. Sometimes a species that is already present in a community will facilitate the establishment of a new arrival. Such facilitative interactions figured prominently in early ideas of mechanisms of succession in plant communities (Clements 1916), in which some species were thought to pave the way for others, primarily by making the habitat more favorable for later arrivals. We now know that many species that arrive early in community development simply hinder rather than hasten the establishment of others (Connell and Slatyer 1977; Sousa 1979a; Dean and Hurd 1980). However, there are some clear examples of facilitation in a variety of communities. Some examples of facilitation were described in Chapter 7 in the context of mutualisms and other positive interspecific interactions (Bertness and Callaway 1994). Other examples exist, although the underlying mechanisms for these facilitative interactions are not well understood. In some cases, facilitating species change the habitat in ways that promote the successful colonization by other species. In other cases, the facilitating species is a resource used by later colonists.

Some phenological patterns may reflect positive interactions among species.

Dean and Hurd (1980) studied patterns of colonization by sessile marine organisms on small tiles. They found examples of all three kinds of temporal interactions: negative, neutral, and positive. Positive effects were relatively infrequent. However, they did notice that plates previously colonized by two species, *Mogula* (a tunicate) and *Tubularia* (a hydroid), tended to be colonized much more rapidly by *Mytilus*, a bivalve, than were recently immersed plates that were devoid of these species. The mechanism for the apparent facilitation remains mysterious. In contrast, once established, *Mytilus* tends to inhibit the settlement of other species. The facilitation of *Mytilus* by other species is facultative, rather than obligate, since it eventually becomes established in sites without *Mogula* and *Tubularia*, although at a slower rate.

Other kinds of facilitative interactions, although experimentally undocumented, seem logically inescapable during the primary succession that occurs in newly available habitats. Thus, primary producers must precede and thereby facilitate herbivores, and herbivores must precede the arrival of higher predators, if species higher in the food chain are likely to invade a new community successfully. Such temporal patterns are one, perhaps trivial, example of what have been called assembly rules.

Physiological Constraints. Some phenological patterns are driven by interactions between physiological constraints and seasonal variation in the physical environment, such as seasonal variation in temperature, photoperiod, or precipitation. The physiological constraints influencing the seasonal activity of organisms may be historical artifacts of evolution and may not reflect optimal adaptations to a particular habitat. One example of such a pattern is the breeding phenology of frogs at temporary ponds in eastern North America. A few species can breed in late winter or early spring at nearly freezing temperatures, whereas other species tend to breed progressively later in the year as air and water temperatures increase. Although it is tempting to view this temporally staggered breeding phenology as a possible example of temporal resource partitioning of breeding ponds, this interpretation appears unlikely. In general, species breeding later in the year are at a distinct disadvantage because late breeders suffer from resource depletion by early breeders (Morin 1987; Morin et al. 1990; Lawler and Morin 1993a), the accumulation of predatory invertebrates (Morin et al. 1990), and an increasing risk of mortality from rapid pond drying in the heat of summer (Wilbur 1987). These seasonally increasing risks of mortality suggest that all frogs should breed as early in the season as possible. Some evidence suggests that frogs do breed as early as possible, but physiological differences among species constrain the timing of earliest breeding to fall at different times for different species.

> Some phenological patterns may reflect differences in the physiological tolerances of species.

The staggered breeding phenology of frogs in temporary ponds appears to be a consequence of physiological constraints that affect the temperature dependence of locomotion by adult frogs. John-Alder et al. (1988) have examined the temperature dependence of locomotion in frogs, and found that species differ in their ability to jump at low temperatures. Early-breeding species, such as the spring peeper (*Pseudacris crucifer*), can jump at nearly maximum levels of performance at low, near-freezing temperatures. Later-breeding species only begin to approach their maximum jumping ability at higher temperatures, generally above 15°C. Each species appears to breed as early in the season as possible, subject to different constraints on locomotion imposed by their different physiological tolerances for locomotion at low temperatures. These constraints may reflect the different evolutionary histories of the species, as well as the apparent difficulty of evolving low-temperature tolerance. For example, in North America, the early-breeding species within the tree frog family Hylidae belong to a single genus,

Pseudacris. The ability to move overland to breeding ponds at low temperatures is apparently something that evolved once within this group of related species (*Pseudacris*), and not in other genera (*Hyla, Acris, Limnaeodus*), despite the advantage that early breeding would confer.

Analysis of other phenological patterns also suggests an important role of evolutionary history and phylogenetic constraints in determining the timing of flowering in plants. Kochmer and Handel (1986) analyzed the flowering times of animal-pollinated plants in two widely separated locations, Japan and the Carolinas in the United States, as documented in published flora for both locations. They found that much of the seasonal variation in the timing of flowering was determined by taxonomy. In general, flowering times were more similar within families of plants than between families. This pattern suggested that flowering times were set by taxonomic constraints rather than by displacement of individual species within families to avoid competition for pollinators.

Chance. As noted previously, the distribution of biological activity over time may appear nonrandom, but statistical tests of such patterns may lead to different conclusions. For example, the pattern of

 Some temporal patterns may be indistinguishable from random.

flowering described by Stiles (1977) seems evenly spaced, but Poole and Rathcke (1979) applied a statistical test to the data that suggested that the pattern was no more regularly spaced than might be expected by chance. This approach, a variation on the null model analyses described previously for other observational studies of competitive interactions, has its own limitations. Conclusions about the regularity of spacing over time depend critically on whether the analysis applies to the entire year or only to those periods of time when it is physiologically possible for species to be active.

CONSEQUENCES OF PHENOLOGICAL VARIATION: CASE STUDIES OF PRIORITY EFFECTS

Priority effects occur when a species that is already present in a community either inhibits or facilitates other species that arrive in the community at

 Some case studies.

some later time. There are numerous examples of inhibitory priority effects. Slight differences in the timing of seed germination are often sufficient to dramatically alter the yield of two competing species. For example, Harper (1961) planted seeds of two grasses, *Bromus rigidus* and *Bromus madritensis*, either simultaneously or displaced by three weeks, with *B. rigidus* sown after

B. madritensis. When planted simultaneously, *B. rigidus* grew to account for 75% of the total biomass of the two species. When delayed, *B. rigidus* only accounted for 10% of the total biomass attained by both species. The three-week delay was sufficient to shift *B. rigidus* from dominant to subordinate status in this simple two-species system.

Animals in seasonal communities are also influenced by priority effects. Adult dragonflies (Odonata: Anisoptera) are conspicuous insects found flying near ponds and lakes in the warm summer months. The adults are often strongly seasonal in abundance, with predictable, well-defined seasons of emergence from the aquatic larval stage (Figure 9.5A). In eastern North America, dragonflies fall into two species groups, which are distinguished by the timing and relative synchrony of emergence from the aquatic larval stage. Early species emerge synchronously over a period of a few days in early spring and tend to complete their breeding before late species begin to emerge. Late species emerge nonsynchronously throughout most of the remaining summer months. They typically breed and oviposit after the eggs and hatch-ling larvae of early species are already present in the pond. Both aquatic larvae and terrestrial adults are predators. Adults and larvae can consume smaller odonates. This means that odonates may interact both as competitors for shared prey and as predators and prey. Depending on when a pond first fills with water, it could be colonized by different sets of species, early or late ones, because of predictable phenological differences. Although seasonal differences in the timing of emergence were originally interpreted as a means of temporal resource partitioning, experiments have shown that larval dragonflies interact strongly despite the temporal displacement in their seasonal arrival in ponds and lakes. These interactions show that early arrivals exert strong inhibitory priority effects on later arrivals, a result consistent with early arrivals obtain-ing an initial growth advantage that allows them to either competitively exclude or consume smaller later arrivals.

Arthur Benke (1978) used field experiments to explore the effects of early dragonfly species on late ones. Early breeders are easily manipulated by placing screened pens in natural ponds and varying when female dragonflies can deposit eggs in the pens. Egg deposition by females is reduced by covering the open tops of pens with screening during the breeding season. Early species can be reduced by covering some pens during the early flight season and then uncovering pens so that late species can oviposit. Other pens without screen lids collect eggs and larvae of both early and late species. The early species, by virtue of their head start, are larger than the late species when they begin to interact. Benke showed that early species significantly depress the abundance of late species, relative to experimental cages where early species were excluded (see Figure 9.5B). Benke interpreted this as a competitive interac-

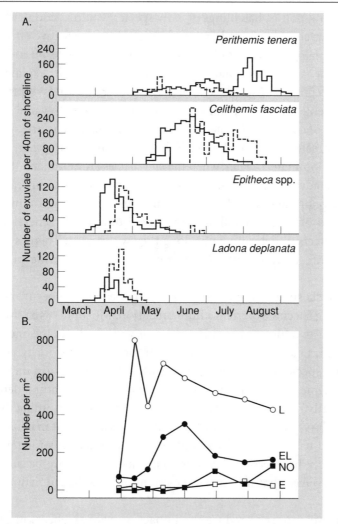

FIGURE 9.5. *(A) Emergence phenology of* Perithemis tenera *and* Celithemis fasciata, *both late, or summer, species, and two early, or spring, species,* Epitheca *sp. and* Ladona deplanata. *Solid and dashed lines show phenologies for two successive years. (Adapted from Benke and Benke, 1975, with permission of the Ecological Society of America.) (B) Effects of early species on late-species abundance. Treatments labeled EL contain early and late species; treatments labeled L contain reduced numbers of early species. (Adapted from Benke, 1978, with permission of Blackwell Science Ltd.)*

tion, but given the predilection for large dragonflies to consume small ones, predation could not be ruled out.

The strong impact of early species on late ones in Benke's study occurred under somewhat atypical conditions. In both cases, the early dragonflies colonized pens without an existing dragonfly fauna and without other predators that might greatly reduce the abundance of interacting dragonfly larvae. In established natural ponds, hatchlings of early species would encounter large overwintering larvae of late breeders, which might have a negative effect on early breeders analogous to the effect of early species on late ones documented in the pens. This seasonal shift in vulnerability might explain the ability of the two groups of species to coexist. However, the kinds of priority effects seen in Benke's experimental system could contribute to variation in the composition of communities that form at different times. The first species to reach newly created ponds will differ if the ponds fill in late spring, when early species predominate, or later, when the late species are abundant. If early arrivals enjoy an advantage over later colonists, priority effects acting early in community development could create lasting differences in community composition. No one has yet followed the initial differences generated by these priority effects to determine how long they will persist. There is other evidence, however, that suggests that some factors can override priority effects.

Priority effects such as the ones described by Benke may occur mostly where interacting species experience little predation. Morin (1984b) studied priority effects in an odonate assemblage that was very similar to the one used by Benke (1978). Like Benke, Morin found strong negative priority effects exerted by early species on late ones (Figure 9.6). However, Morin included additional experimental treatments consisting of pens with a natural density of fish. Fish greatly reduced odonates, probably by directly consuming them, so that odonate larvae were an order of magnitude less abundant in pens with fish than without fish. Where fish reduced odonate abundance, strong priority effects of early breeders on late ones vanished, apparently because so few odonates survived that negative interactions among the survivors were minimized.

Predators also appear to reduce the impact of priority effects in other aquatic systems. Morin (1987) was able to show that larvae of an early-breeding frog species, *Pseudacris crucifer,* negatively affected the growth of a later-breeding species, *Hyla versicolor,* in a series of experimental ponds. The negative interaction probably resulted from nutrient depletion, because the early-breeding species strongly depressed the growth of the late breeder even though the two species had virtually nonoverlapping larval periods. In ponds with predatory salamanders (*Notophthalmus*), very few of the later-breeding species survived, and those survivors showed no ill effects on the early breeders. Priority effects were density dependent in this system, and factors that reduce the density of interacting species, such as predation, will tend to moderate the impact of early colonists on later species.

Wayne Sousa (1979a) found clear evidence for negative effects of early colonists on later arrivals in another system, the algae that colonize boulders in the rocky intertidal zone near Santa Barbara, California. Sousa found that

	Without fish		With fish		ANOVA		
	Early	Late	Early	Late	Timing	Fish	Interaction
Plathemis lydia	0.3 (0.0, 2.2)	0.0 (—)	0.0 (—)	0.0 (—)	. . .	. . .	. . .
Pachydiplax longipennis	2.9 (0.0, 26.4)	61.6 (8.5, 413.5)	0.2 (0.0, 1.1)	0.0 (—)	*	***	**
Perithemis tenera	0.4 (0.0, 1.7)	3.7 (0.2, 17.4)	5.6 (0.3, 32.6)	1.3 (0.0, 5.9)	NS	NS	. . .
Erythemis simplicicollis	8.6 (2.4, 25.7)	24.7 (5.8, 96.3)	0.0 (—)	0.7 (0.0, 7.8)	NS	***	NS
Tramea lacerata	0.0 (—)	1.3 (0.0, 12.4)	0.0 (—)	0.0 (—)	. . .	. . .	. . .

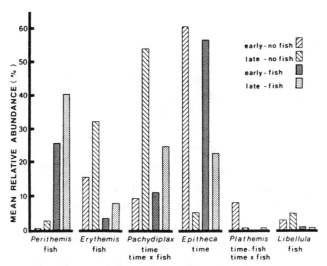

FIGURE 9.6. *Effects of predatory fish on priority effects exerted by early-breeding dragonfly species on late-breeding species. Fish reduce the abundance of odonates, reducing the intensity of priority effects and changing the relative abundance of surviving dragonflies in experimental pens. Early treatments potentially contain early and late species; late treatments have reduced numbers of early species. (Reprinted from Morin, 1984b, with permission of the Ecological Society of America.)*

Ulva, the earliest species to colonize free space on recently overturned boulders, tended to inhibit invasion by later-arriving red algae, such as *Gigartina canaliculata*, the species that eventually occupies most of the space. If *Ulva* was experimentally removed, invasion by *Gigartina* was facilitated (Figure 9.7). Natural removal of *Ulva* happens after periods of desiccation or when it is consumed by herbivores. At this site, the crab *Pachygrapsus* is an important herbivore. When caged with algae on natural boulders, the crabs reduce the percentage of cover of *Ulva* and increase the percentage of cover of *Gigartina*.

Priority effects need not be tightly linked to seasonal phenological differences among species to affect community structure. Shulman et al. (1983) studied the recruitment of marine reef fish from the planktonic larval stage to newly created artificial reefs assembled from small piles of concrete building

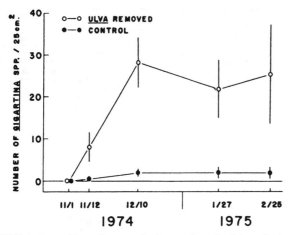

FIGURE 9.7. *Effects of removal of an early-colonizing algal species,* Ulva, *on the abundance of a later-arriving species,* Gigartina. *(Reprinted from Sousa, 1979a, with permission of the Ecological Society of America.)*

blocks. Recruitment of new species to 30 artificial reefs was inhibited by the prior residence of two kinds of fish, the beaugregory (*Eupomacentrus leucosticus*), a strongly territorial herbivorous damselfish, and predators, mostly juvenile snappers (*Lutjanus* sp.). Experimental transplants of adult beaugregories to the artificial reefs reduced settlement by surgeonfishes (*Acanthurus*) and reef butterflyfish (*Chaetodon sedentarius*). Juvenile snappers reduced the settlement of grunts and high-hats (*Equetus acuminatus*). Territories appear to open at random, when their previous owners fall prey to predators or move to greener pastures. Settlement of some species tends to occur either during times of the new moon or full moon, which means that certain fish will arrive first at a new reef depending on when that reef becomes available. Lottery models (Chesson and Warner 1981) of community composition have been proposed specifically for reef fish to account for the high diversity of coexisting forms. In these models, species that happen to have more settling larvae available when a territory opens up have a higher probability of filling that territory. To the extent that territories open at different unpredictable times, winning an open territory is a bit like winning a lottery.

Other examples of strong priority effects without a predictable phenological separation among species come from the work of Shorrocks and Bingley (1994) on interactions among larval flies that feed within decaying mushrooms. *Drosophila phalerata* and *Drosophila subobscura* lay their eggs in mushrooms on the forest floor. One species may lag several days behind the other in its arrival at a particular mushroom. These lags lead to lower survival, smaller size, and longer development times, all of which presumably reduce the fitness of late arrivals. Through clever laboratory manipulations, Shorrocks and Bingley showed that both species benefit by early arrival at the mushroom, with each enjoying a greater competitive advantage when it precedes the other by several days. When both species arrive simultaneously, *D. phalerata* invariably outcompetes *D. subobscura*. The result suggests that *D. subobscura* only manages to persist as a fugitive species under conditions in which it discovers and exploits new mushrooms before they are found by the competitively superior *D. phalerata*. Shorrocks and Bingley modeled this interaction by increasing the competitive ability, as described by a competition coefficient, of the first species arriving at a mushroom. They concluded that randomly varying priority effects, with the identity of the first-arriving species differing by chance, would not be sufficient to promote the coexistence of two unequal competitors. Instead, the inferior competitor would have to consistently arrive before the stronger one for species to coexist.

ASSEMBLY RULES

The term **assembly rule** has been applied to a variety of patterns in developing and established communities. The utility of the term, and exactly what does or does not constitute evidence for

🌿 Repeatable patterns of community development imply the existence of assembly rules.

assembly rules, continues to be debated by ecologists. Some ecologists would argue that assembly rules exist if certain sets of species that could be drawn at random from a local species pool fail to coexist at some local level (Drake 1990). In other words, any nonrandom pattern is evidence for some sort of nonrandom assembly process. At another level, any influence of early colonists on later ones suggests that community assembly depends in a potentially complex way on the identity and sequence of arrival of species as communities develop. The latter observation suggests that some deterministic pathways of community development must exist, to the extent that community assembly is not a purely random process. The problem is that for a given community and a given number of species, there may be many pathways or sequences of species invasion, establishment, and extinction, which ultimately yield particular patterns of species composition. If each pathway corresponds to an assembly rule, we may not gain much understanding if the number of pathways (rules) is very large.

Certainly, at some trivial level, some general assembly rules exist. The observation that predators cannot successfully invade a new community in the absence of prey is one example. However, it is not profound to observe that predators will soon starve in a community without suitable prey. A somewhat more interesting question concerns just how abundant prey must be before the invasion by a predator can succeed. Different models of predator-prey interactions lead to different predictions. Models in which both functional and numerical responses of predators depend on prey density predict that a threshold level of prey abundance is required for predators to become established (Oksanen et al. 1981). In contrast, ratio-dependent models of predator-prey interactions (Ginzberg and Akcakaya 1992) have no lower threshold of prey abundance, and predict that predator abundance will scale with prey abundance.

Other models suggest the possibility of other kinds of assembly rules. Law and Morton (1993) have modeled sequences of community assembly using relatively simple Lotka-Volterra models of interacting species. Their approach involves modeling the sequential invasion of producers (which compete), species that consume producers, and other species that eat the consumers of producers. For given selections of parameters that describe the net positive or negative effects of each species on the others, it is often possible to describe alternate permanent sets of species that persist indefinitely and resist invasion by other species. One such set of possible outcomes for a five-species system is shown in Figure 9.8. Law and Morton have suggested that one possible assembly rule that can be deduced from this system is that where multiple alternate permanent sets of species exist, no alternate permanent set will be a subset of any other permanent set.

Other assembly rules are based on natural history observations that suggest that certain species are found only in communities with certain properties or with certain values of species richness. Jared Diamond (1975) has described **incidence functions**, which describe the probability that a particular species will occur in a particular community, given some attribute of that community (Figure 9.9). Diamond has sketched incidence functions for a number of bird species found on islands of the Bismarck Archipelago near New Guinea. In each case, the predictive attribute of the community is bird species richness. Some species, called high-S species, occur only in speciose communities. Others, called tramp species, occur on a broad range of islands, including those of low species richness. Tramp species presumably are good colonists with very generalized requirements that are able to persist in relatively simple communities. High-S species apparently require more specialized features of communities that support a variety of other species. Diamond's incidence functions have not been applied to a variety of species, although it would be interesting to learn whether different taxa show similar kinds of patterns. The functions are not mechanistic, in the sense that they say nothing about why certain species appear more or less often than others in communities that contain a particular number of species. However, if species colonize communities independently, that is, if they do not interact, then it should be possible to predict the probability that two species will coexist based on the incidence functions for each species. That probability can be obtained from the product of the values of the incidence function for each

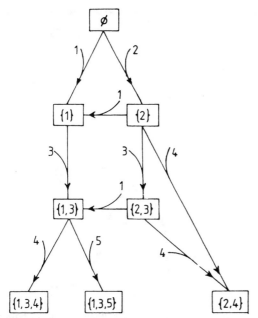

FIGURE 9.8. *Pathways of species invasions in a five-species model system that lead to three alternate permanent states. Species 1 and 2 are primary producers; species 3 consumes species 1 and 2 and is consumed by species 4 and 5. Species 4 consumes species 1, 2, and 3 and is consumed by species 5. Species 5 consumes all other species, to different degrees. The per capita rates of increase,* f_i*, for the various species are functions of the abundances,* x_i*, of the species:*

$$f_1 = 1.16 - 1.05x_1 - 0.93x_2 - 0.13x_3 - 1.30x_4 - 0.13x_5$$
$$f_2 = 1.09 - 1.18x_1 - 1.10x_2 - 0.21x_3 - 0.72x_4 - 0.01x_5$$
$$f_3 = -0.62 + 1.47x_1 + 0.70x_2 - 0.50x_4 - 0.42x_5$$
$$f_4 = -0.81 + 0.16x_1 + 1.55x_2 + 0.35x_3 - 1.20x_5$$
$$f_5 = -0.71 + 0.07x_1 + 0.44x_2 + 0.28x_3 + 0.12x_4$$

(Reprinted from Law and Morton, 1993, with permission of the Ecological Society of America.)

species in a community with a particular species richness, S. That probability provides another kind of assembly rule. To the extent that species depart from the patterns predicted by this simple rule of independent assembly, other kinds of processes and rules will need to be invoked.

The term *assembly rule* has also been applied to the observation that different sequences of species invasion in experimental communities may lead to different patterns of community composition. One example would be that a species persists in a community if it arrives first, but not if it arrives later in a sequence of species. James Drake (1991) has described how different orders of species introduction into aquatic microcosms alter the persistence and abun-

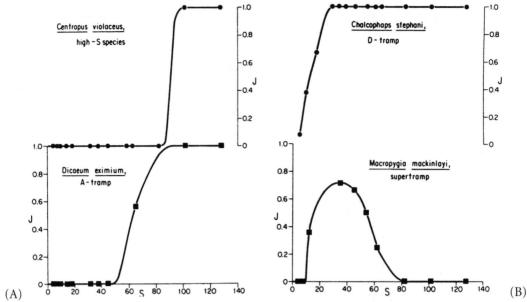

FIGURE 9.9. *Incidence functions for birds in the Bismark Archipelago. J is the probability of occurrence on an island containing S species. (A) Incidence functions for high-S species, which occur only on species-rich islands. (B) Incidence functions for tramp species that occur on islands of lower species richness. (Reprinted by permission of the publisher from* Ecology and Evolution of Communities *by M. L. Cody and J. M. Diamond (eds.), Cambridge, Mass.: Harvard University Press. Copyright © 1975 by the President and Fellows of Harvard College.)*

dance of species. In some cases, mostly those involving consumers, order of arrival is irrelevant, with species failing to become established under any circumstances. In many cases involving algae that served as primary producers, early introduction is correlated with the order of dominance of the algal species (Drake 1991; Table 9-1). In these cases, the specific mechanisms that confer dominance on early arrivals remain speculative, although a competitive advantage conferred by early arrival seems most likely.

The issue of whether assembly rules for communities can be specified in some useful fashion may seem purely academic until an ecologist is asked how to best build a community from scratch, or how to best restore a degraded community. The spotty performance of ecological restoration efforts suggests that it is not sufficient to simply return a group of species to a site and hope for the best. Such efforts often fail to establish functional self-sustaining communities. The failure is not altogether surprising, since our observations of nature suggest that natural communities develop gradually over long periods of time, probably involving many iterations of the invasion, establishment, and extinction process (e.g., Docters van Leeuwen, 1936, for plant species reinvading Krakatau; also see Thornton 1996). Instead, there is

TABLE 9-1. Different sequences of invasions by four algal species in experimental microcosms, and the order of dominance of those species that results.

Treatment/ Replicate	Sequence of Producer Invasions	Order of Relative Dominance				Proximity Invasions (days)	Proportion of Time Dominant
1-1	AK-SE-CH-SC	AK	CH	SE	SC*	120	0.98
1-2		AK	CH	SE*	SC*		
1-3		*AK*	*CH*	SE	SC*		
1-4		AK	CH	SE	SC*		
2-1	SC-SE-CH-AK	SC	CH	*SE*	*AK*	60	0.94
2-2		SC	CH	*SE*	*AK*		
2-3		SC	CH	*SE*	*AK*		
2-4		*AK*	*SC*	SE	CH		
3-1	SE-CH-AK-SC	SE	AK	SC	CH*	105	0.95
3-2		. . .	. . .	. . .	. . .		
3-3		SE	AK	*SC*	CH		
3-4		*AK*	*SE*	SC	CH		
4-1	CH-AK-SC-SE	*AK*	*CH*	SE	SC	105	0.29
4-2		AK	SC	*SE*	*CH*		
4-3		AK	SC	SE	CH		
4-4		AK	SE	SC	CH		
5-1	SC-CH-AK-SE	SC	*AK*	*CH*	SE	135	0.88
5-2		SC	*AK*	*CH*	SE		
5-3		SC	*AK*	*CH*	SE		
5-4		SC	*AK*	*CH*	SE		
6-1	SE-CH-SC-AK	SE	AK	SC	CH*	150	0.98
6-2		SE	CH	SC	AK		
6-3		SE	AK	*SC*	CH		
6-4		AK	*SC*	SE	CH*		
7-1	AK-SC-CH-SE	AK	*CH*	SC	SE	150	1.00
7-2		AK	*CH*	SC	SE		
7-3		AK	CH	SC	SE*		
7-4		AK	CH	SC	SE*		
8-1	CH-AK-SE-SC	AK	SE	SC*	CH*	105	0.40
8-2		*AK*	*SC*	*CH*	SE*		
8-3		*CH*	*SC*	AN	SE*		
8-4		*SE*	*SC*	AK	CH*		
9-1	SC-AK-CH-SE	AK	SC	SE	CH	60	0.71
9-2		*SC*	*AK*	*SE*	CH*		
9-3		*SC*	*AK*	*SE*	CH*		
9-4		AK	SC	*SE*	CH*		
10-1	AK-SE-CH-SC	SE	SC	AK	CH*	60	0.15
10-2		SE	SC	AK	CH*		
10-3		SE	SC	AK	CH		
10-4		SE	SC	AK	CH		

Note: Eventual species extinctions are indicated by an asterisk.
AK = *Ankistrodesmus falcatus*; CH = *Chlamydomonas reinhardtii*; SC = *Scenedesmus quadricauda*; SE = *Selenastrum bibrium*.
Reprinted from Drake (1991), with permission of the University of Chicago Press.

a gradual transition from simple communities to increasingly complex ones. That transition, termed **succession,** is a logical consequence of the sort of temporal interactions that we have described so far and is the subject of Chapter 13.

Another complication posed by priority effects is the enormous number of distinct sequences of species arrival in a developing community. Assuming that species arrive in a randomly determined sequence and that each species arrives only once, for a set of S different species there are $S!$ [$S!$ (factorial) = $(S(S-1)(S-2) \ldots (1)$] different sequences of species arrival. Thus, for a relatively simple assemblage of six species there are $6(5)(4)(3)(2)(1) = 720$ different sequences of species arrival. If each sequence must be examined to determine whether it yields a different community pattern, the task of specifying assembly rules based on colonization sequences becomes daunting, if not entirely hopeless. However, species may not arrive at communities in an entirely unpredictable sequence, and many sequences may yield the same outcome. In that case, there is some hope that regular rules of organization may be deduced.

CONCLUSIONS

Interspecific differences in phenology may arise from several causes, and need not be adaptive consequences of community interactions. Nonetheless, these seasonal differences in the abundance or activity of species create situations where priority effects and other kinds of short-term temporally dependent processes can affect the outcome of interactions among species. Assembly rules may be one consequence of such short-term historical events, to the extent that the order of arrival of colonists in developing communities influences ultimate patterns of community structure. Other assembly rules may not have an explicit temporal dependence, but instead describe the proclivities of species to occur in communities of differing complexity.

CHAPTER 10

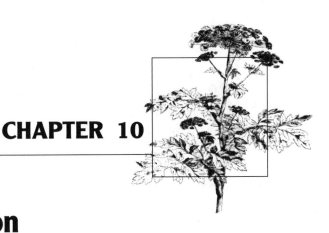

Habitat Selection

OVERVIEW

Community composition can be influenced by the behavior of potential colonists. The consequences of habitat selection can rival the impact of strong direct interactions among species in developing communities. Habitat selection is important in relatively mobile animals that easily move among habitats and actively select where they ultimately forage, reproduce, or reside. Descriptive studies suggested the importance of habitat selection by documenting associations between the abundance of animal species and other habitat attributes. Differences among habitats were often represented by the species composition or structural complexity of vegetation. Experimental studies of habitat selection show that some species selectively use habitats in ways that minimize strong negative interactions (such as predation or competition) or maximize strong positive interactions (such as prey availability). The kinds of interactions that affect habitat selection have suggested simple models that emphasize trade-offs between predation risk and prey availability. These models predict that habitat selection will depend on the benefits of foraging in a particular place discounted by the risk of mortality in that location.

ANIMALS that move freely among different habitats and exercise selectivity in their location can influence community patterns through **habitat selection**, the active choice of locations where organisms forage, grow, and ultimately reproduce. Habitat selection provides one possible explanation for the conspicuous absence of highly mobile, readily dispersing species from an apparently suitable community. As with most community patterns, chance events or exclusions caused by direct interactions with other species can produce similar results and must be ruled out as alternate hypotheses before assuming that habitat selection is at work.

Habitat selection can influence patterns of community assembly.

Habitat selection can function like a selective filter between a developing community and the species pool of potential members by sorting among species that can actively avoid or choose to colonize a particular place. Those choices often depend on the kinds of interactions that are likely to occur with other species that are already present in the community. Factors that influence habitat selection include the avoidance of physiological stress, the availability of prey or other necessary resources, and the avoidance of competitors and enemies. Animals respond to combinations of these factors in complex ways, and there is some evidence that some animals make relatively sophisticated choices by weighing foraging advantages against mortality risks in particular sites. Although much research on habitat selection has focused on higher vertebrates, especially birds, field observations and experiments show that even animals with relatively modest sensory abilities have impressive abilities to select favorable habitats.

Evidence for habitat selection comes from studies using a variety of approaches. One approach draws on natural history observations to associate the presence of particular species with biotic or abiotic features of the habitat. Correlative studies linking the abundances of organisms to habitat features can be highly quantitative, often using multivariate statistical analysis of community patterns. In these studies, plant species composition, which is easily measured, provides a measure of habitat attributes. These studies generally assume that any associations between animals and habitats are a consequence of habitat selection, since mobile organisms can move freely among habitats. A more direct approach involves experimental manipulation of the factors thought to influence habitat choice, and subsequent observation of whether organisms respond to habitat alterations.

CORRELATIONS BETWEEN ORGANISMS AND HABITATS

The observation that some mobile organisms are found in certain habitats and not others is de facto evidence for habitat selection. Early work by

MacArthur (1958) on microhabitat use by coexisting warblers is one example of this approach. Extensions of this approach attempt to describe associations between sets of species and particular attributes of the habitat, usually involving either plant species composition or variation in the gross attributes of the plant community, such as variation in foliage height, that provide different opportunities for foraging, nesting, and predator avoidance (MacArthur and MacArthur 1961). In some cases, there appear to be strong associations among sets of species and various aspects of the habitat. MacArthur and MacArthur (1961) found a strong positive relation between bird species diversity and foliage height diversity (Figure 10.1). The latter measure ignored the actual species composition of the plants responsible for the foliage, and focused instead on the relative distribution of foliage at different heights above the ground. Such structural measures of habitat variation work best in

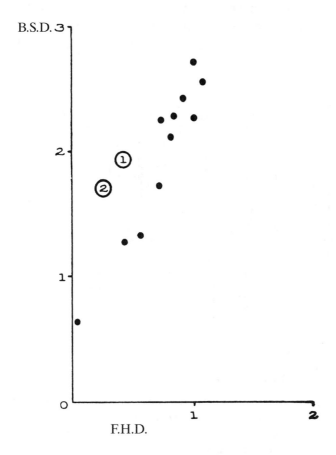

FIGURE 10.1. *Relations between bird species diversity and foliage height diversity across a range of habitats. The positive relation seen here is not universal, but its existence in this small sample of habitats suggests that more bird species occur in more spatially complex habitats. The high mobility of birds in turn suggests that this pattern is the result of birds selectively residing in more complex habitats. (Reprinted from MacArthur and MacArthur, 1961, with permission of the Ecological Society of America.)*

situations in which there is considerable variation among sites in foliage height. Others have found that foliage height diversity is only a gross correlate of bird species diversity and have suggested that more detailed descriptions of plant species composition provide a better prediction of bird community structure (e.g., Holmes et al. 1979a).

Studies of correlations between the actual species composition of forest vegetation and of birds sometimes show clear associations between particular foraging guilds and vegetation (Holmes et al. 1979a). Holmes et al. (1979a) thought that habitat variation, as measured by plant species composition, was a major determinant of bird species composition in a hardwood forest in the northern United States. Their multivariate analysis showed that different groups of bird species responded to aspects of variation in plant species composition, such as understory shrubs and different species of canopy trees (Figure 10.2). Wiens and Rotenberry (1981), working in a rather different habitat, shrub-steppe vegetation in the western United States, found little effect of habitat on bird species abundances. Unlike the northern hardwood

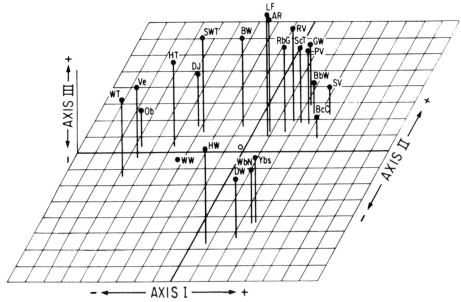

FIGURE 10.2. *Relations between habitats, primarily the species composition of forest vegetation, and the presence of different bird species (plotted points) in a northern hardwood forest in New Hampshire. Different points correspond to the average attributes of habitats used by different bird species. The habitat axes are complex combinations of many habitat variables obtained from a principal component analysis. The first axis separates ground feeders (negative values) from canopy feeders (positive values), and is related to vegetation height. The second axis separates species that forage on bark (negative values) from species that forage elsewhere. (Reprinted from Holmes et al., 1979a, with permission of the Ecological Society of America.)*

forest studied by Holmes et al. (1979a), shrub-steppe vegetation is relatively species poor and has limited vertical structure, which may account for a weak relation between habitat and species composition.

In these and similar studies, relations between animal abundance and habitat, as measured by plant species composition, are only correlations, and little can be inferred about causal mechanisms. However, experimental manipulations of biotic components of the habitat in other systems show that even species that are considerably less mobile than birds can actively select among habitats in response to particular cues. In turn, habitat selection can have important consequences for community patterns.

CUES AND CONSEQUENCES

Habitat Selection Based on Prey Availability

Kenneth Sebens (1981) described an intriguing example of habitat selection by the settling larvae of the large Pacific sea anemone, *Anthopleura xanthogrammica*. Adult anemones are nearly sessile sit-and-wait predators. They rely on waves or other disturbances to dislodge and transport large prey items, such as bivalve mollusks, to their waiting tentacles. Sebens noted that juvenile *Anthopleura* tended to occur selectively in dense patches of the bivalve mollusk *Mytilus*, which are important prey of larger adult *Anthopleura*. These prey patches tended to occur on vertical rock walls, where a slow rain of dislodged prey could support adult *Anthopleura* located below the prey patch. As the small anemones grew they gradually moved down through the *Mytilus* patch to reside in locations near the base of the rock walls where the rain of dislodged prey was likely to be greatest. Settling larvae of *Anthopleura* are presumed to selectively settle in *Mytilus* patches, even though the larvae are little more than ciliated balls of cells with very modest sensory capabilities. The fact that *Anthopleura* are capable of some degree of habitat selection, in this case involving the selection of sites with a high potential for prey availability, is quite remarkable and provides one example of a general pattern of spatial association between predators and their prey.

Habitat Selection Based on Competitor Avoidance

The example of habitat selection by juvenile *Anthopleura* described above has even more striking counterparts in other marine invertebrates, whose mobile larvae exercise considerable habitat selection before settling down to a sessile adult existence. Richard Grosberg (1981) has experimentally shown that several species of settling invertebrates will discriminate among substrates based on the

Some organisms avoid habitats that contain superior competitors.

density of potential competitors that they encounter. Grosberg studied the settling behavior of an assortment of invertebrates that form the fouling community found on solid substrates in a tidal salt pond in Massachusetts. The dominant competitor in the system is a small tunicate called *Botryllus*. It tends to overgrow and displace many other sessile species in the system. Grosberg was able to coax different densities of *Botryllus* larvae to settle on small glass plates, and then observed how other species settled in response to high or low densities of this superior competitor. The settlers fell into two groups containing roughly equal numbers of species. One group actively discriminated against plates with high densities of *Botryllus* and settled selectively on plates with no or few *Botryllus* (Figure 10.3). These are the species capable of significant habitat selection and the avoidance of strong interspecific competition; they also tend to be the species at greatest risk for overgrowth by *Botryllus*. The other group did not discriminate among substrates with different densities of the superior competitor, *Botryllus*. This nonselective group consisted mostly of species with elevated feeding structures that are not prone to overgrowth by *Botryllus*.

Habitat Selection Based on Predator Avoidance

A variety of animals, including invertebrates and vertebrates, appear to select against habitats that contain predators. Andrew Sih (1982) studied patterns of habitat use by different size classes of the predatory aquatic bug *Notonecta hoffmani*, which inhabits stream pools in California. *Notonecta* are cannibals, and large adults will attack and kill smaller juveniles, which go through five successively larger subadult size classes, or instars. Use of different portions of stream pools by large and small *Notonecta* appears to reflect compromises between selecting habitats with abundant food and avoiding cannibalistic predation by adults. Sih was able to show that the three smallest instars (1 to 3) can be attacked and killed by adults, whereas instars 4 and 5 are relatively invulnerable to attack by adults. Based on this result, instars 1 to 3 should avoid adults to minimize their risk of attack. Examination of the distribution of adults and smaller instars in stream pools shows that adults tend to preferentially forage near the center of stream pools, whereas smaller instars forage away from the center, near pool edges. When adults were experimentally removed from six pools, and left in another six pools as controls, a greater proportion of smaller instars used the central portion of pools without adults, and those smaller instars also spent more time actively moving about (Figure 10.4). This result suggests that smaller instars move less and avoid the centers of pools when adults are present.

 Other organisms select habitats that minimize the risk of predation.

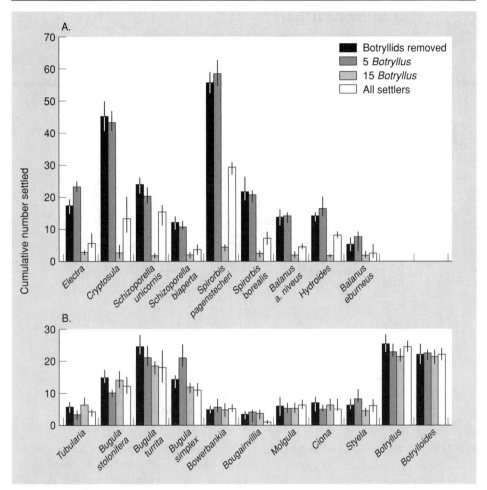

FIGURE 10.3. *Avoidance (top) or tolerance (bottom) of the spatial competitor* Botryllus *by different species of marine invertebrates that settle from the plankton onto glass plates containing different densities of* Botryllus. *Significant reductions in settlement by species on plates with high densities of* Botryllus *is evidence for habitat selection by settling larvae. (Adapted with permission from* Nature *290: 700–702, R. Grosberg. © 1981 Macmillan Magazines Limited.)*

Adult *Notonecta* presumably prefer the central portions of pools because of greater prey availability. This raises the question of whether juvenile *Notonecta* might overcome their avoidance of adults if sufficient prey were available to make those habitats particularly attractive. Sih attacked that question by creating artificial patches of abundant prey in laboratory aquaria and then observing whether notonectids of different size, and therefore different predation risk, would exploit these rich feeding patches in the presence or absence of adult *Notonecta*. Patterns of habitat utilization by small notonectids were very similar to those observed in natural pools. All size classes

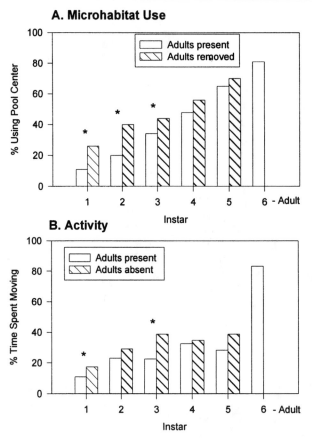

FIGURE 10.4. *Size-dependent patterns of habitat use by different instars of the aquatic bug* Notonecta hoffmani *in stream pools with or without cannibalistic adults. Small instars avoid the centers of pools when adults are present, and also move less, reducing the probability of their detection by hungry adults. (Data from Sih, 1982.)*

tended to forage in the high-density prey patches when predatory adults were absent, indicating that the juveniles were capable of selecting profitable feeding locations. However, when adult *Notonecta* were present, the smaller, more vulnerable instars once again tended to avoid the center of the tanks despite the abundance of prey. The inference is that notonectids were weighing the relative costs and benefits of foraging in patches where prey were abundant and where the risk of predation was great. This suggests that predator avoidance has a real cost that appears as reduced opportunities for foraging, which might slow larval growth and prolong the period of larval development. Without such a cost, smaller notonectids would have no incentive to switch to more profitable prey patches in the absence of predatory adults.

Other kinds of organisms appear to make similar ontogenetic shifts in habitat use that depend on the presence of predators, although the costs of predator avoidance seem slight. Morin (1986) observed that tadpoles of the spring peeper, *Pseudacris crucifer*, spent the first two weeks after hatching hidden in the bottom litter layer of artificial ponds, regardless of the presence or absence of predators. As the tadpoles grew, they tended to move up off the bottom and to forage in more conspicuous locations higher in the water column, but this transition occurred only in ponds without predators. In ponds containing predators (newts), *Pseudacris* did not gradually appear in conspicuous locations, but instead remained hidden within litter layer on the pond bottom during their entire two-month larval period (Figure 10.5). The use of alternate habitats in the artificial ponds was not associated with differences in larval growth, which suggests there is no obvious cost to the predator avoidance strategy (see Figure 10.5). Absence of a detectable cost of predator avoidance raises the question of why tadpoles should even bother to forage in exposed locations in the absence of predators, since the safe strategy would simply be to always remain inconspicuous in benthic microhabitats regardless of predator abundance. Perhaps under different conditions, such as lower food availability or higher tadpole density, a greater and more measurable cost to remaining hidden in benthic litter might materialize.

Joseph Holomuzki (1986) studied patterns of microhabitat use by larvae of the tiger salamander, *Ambystoma tigrinum*, in ephemeral ponds in Arizona. Holomuzki noticed that *Ambystoma* changed their diel patterns of microhabitat use in ponds containing an important predator, adults of the large aquatic beetle *Dytiscus*. *Dytiscus* tended to forage primarily at night in the shallow littoral areas of the ponds. After dark, when *Dytiscus* were active in the shallows, *Ambystoma* moved away from the shallows into open, deeper water where few beetles occurred, despite the greater availability of food in the shallows. During the day, when the risk of predation from *Dytiscus* was small, *Ambystoma* larvae returned to the shallows where prey were abundant. Controlled experiments in large aquaria showed that *Ambystoma* preferentially used deeper locations when beetles were present, but avoided these areas when beetles were absent (Figure 10.6).

Conflicting Demands Imposed by Multiple Causes of Habitat Selection

Other studies have shown that adults of some species can detect the presence of potential competitors and predators of their offspring and can select sites for their offspring to minimize some of those risks. Resetarits and Wilbur (1989, 1991) experimentally manipulated the abundance of potential predators and competitors that interact with larvae of the southern gray tree frog, *Hyla chrysoscelis*. Manipulations took place in an array of small artificial ponds. The frogs readily use the artificial ponds as breeding sites, with males calling

Frogs avoid breeding in ponds with certain predators and competitors.

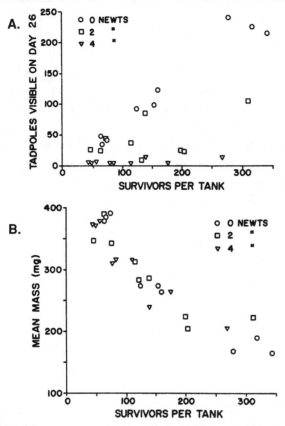

FIGURE 10.5. *(A) Differential use of exposed microhabitats by* Pseudacris *tadpoles in ponds containing different densities of predators. Habitat use is described by the relation between the minimum number of tadpoles known to be alive on that date versus the number of visible tadpoles seen foraging in the water column. Tadpoles alive but not visible were hidden in the litter in the bottom of the ponds. The majority of tadpoles in ponds with predators were not visible, indicating a shift in microhabitat use mediated by predators. (B) Relations between final tadpole density and size at metamorphosis in ponds containing different densities of predators. There is no effect of predators, or differences in microhabitat use, on the density-size relation, indicating that prey attained similar sizes at a given density despite differences in habitat use. (Reprinted from Morin, 1986, with permission of the Ecological Society of America.)*

from the pond margins, and females ovipositing in the ponds. Differences in habitat use can be measured by counting the number of calling males at ponds containing different risks (species) and determining where the females lay their conspicuous floating films of eggs. Although males select calling sites, once a small male amplexes with a much larger female the final selection of the site for egg deposition is up to the female.

Resetarits and Wilbur created seven kinds of ponds containing different species that might influence habitat choice: 1) controls without any other

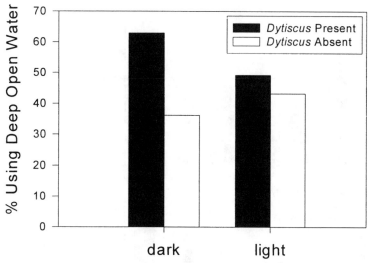

FIGURE 10.6. *Patterns of habitat use by larval* Ambystoma tigrinum *with and without its predator* Dytiscus. Ambystoma *alters its habitat use to avoid predation when Dytiscus is active in the dark, but not when it is inactive in the light. (Data from Holomuzki, 1986.)*

predators or competitors, 2) intraspecific competitors, larvae of *Hyla chrysoscelis,* 3) an interspecific competitor, larvae of the large frog *Rana catesbeiana,* and four different predators, 4) the adult salamander *Notophthalmus,* 5) the larval salamander *Ambystoma maculatum,* 6) the fish *Enneacanthus chaetodon,* and 7) the larval dragonfly *Tramea carolina.* The experiment was very ambitious, using a total of 90 small experimental ponds: 10 ponds for each of the competitor or predator treatments, and 30 ponds for the controls.

Relative to controls, males actively avoided ponds containing either conspecific larvae or the fish, *Enneacanthus* (Figure 10.7). Relative to controls, females were less active at and oviposited less frequently in ponds containing two predators, *Ambystoma* and *Enneacanthus,* as well as conspecific competitors (see Figure 10.7). Female frogs did not discriminate among the other treatments. These results suggest that both males and females are adept at avoiding intraspecific competition and predation by fish. *Hyla* usually does not breed in permanent ponds with fish, and its larvae fail to survive with fish. It is unclear why males and females differ in their ability to discriminate against ponds containing *Ambystoma,* and why some risks, but not others, are recognized. This study does make clear that the absence of some species from particular locations may be as much a consequence of habitat selection by reproducing adults as of postarrival interactions among competitors and predators.

Other studies have explored the trade-offs between opportunities for foraging and predation risk that can produce ontogenetic shifts in habitat uti-

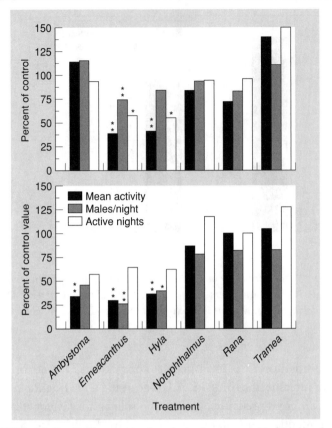

FIGURE 10.7. *Differences in the relative frequency of habitat use by male (top) and female frogs (bottom),* Hyla chrysoscelis, *using artificial ponds containing different species of potential competitors and predators. The asterisks indicate significant avoidance, that is, reduced habitat use, relative to controls. (Adapted from Resetarits and Wilbur, 1989 and 1991, with permission of the Ecological Society of America.)*

lization by growing prey. Earl Werner and colleagues (1983a,b) have studied how trade-offs between the energetic rewards of foraging where prey are abundant and the risk of predation

 Habitat selection by some fish depends on size-based risks of predation.

influence microhabitat use by the bluegill sunfish, *Lepomis macrochirus*. Small bluegills forage most efficiently on zooplankton in the open, deeper waters of ponds, but they seldom use those habitats until they become relatively large and invulnerable to their major predators, the largemouth bass, *Micropterus salmoides*. Small bluegills preferentially use nearshore vegetated habitats, despite their lower profitability for foraging, because these habitats greatly reduce the risk of predation by bass. The bluegills only move out into the open water, where bass are abundant, after they have attained a body size that

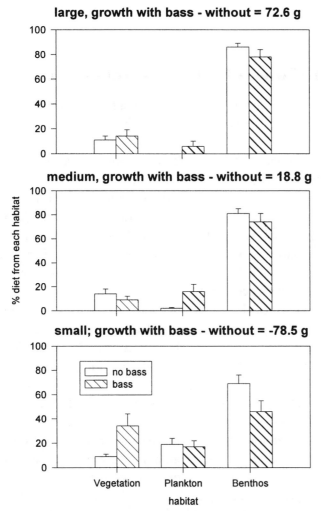

FIGURE 10.8. *Effects of the presence or absence of predators, largemouth bass, on habitat use and growth rates by three different size classes of bluegill sunfish. Differences in habitat use are indicated by differences in the relative frequency of prey items typical of open-water, benthic (bottom), and vegetated habitats. The smallest size class of bluegill forages less in open and benthic habitats and grows at a slower rate in the presence of bass. Larger, relatively invulnerable bluegills do not alter their habitat use in response to predators. (Data from Werner et al., 1983a.)*

makes predation by bass unlikely. Werner et al. have performed unreplicated experiments that strongly suggest that the presence or absence of bass shifts the patterns of habitat utilization of small bluegills. Small bluegills from a pond without bass spent more time foraging away from littoral vegetation and grew more rapidly than their counterparts in a similar pond stocked with bass (Figure 10.8). Medium and larger size classes of bluegills that are not at risk for predation by bass did not alter their foraging to avoid bass, and failed to show the growth reductions seen in small bluegills that altered their habitat use to avoid bass.

A GRAPHICAL THEORY OF HABITAT SELECTION

Earl Werner and James Gilliam (1984) developed a simple graphical theory that can be used to predict size-dependent habitat shifts such as those observed for bluegills interacting with bass. The approach has a rigorous quantitative framework derived from optimal-control theory, but it also has an elegant graphical representation that can be appreciated without much knowledge of the underlying mathematics. The basic premise is that growing organisms have size-specific rates of growth (g) and mortality (μ) in a given habitat. These rates may also differ between habitats, depending on habitat-specific differences in prey availability and predation risk. If maximizing growth were the only concern of a growing organism, it should switch habitats in such a way that its size-specific growth remains maximal over time. If that were the case, then the growing animal depicted in Figure 10.9 should switch habitats at the point at which the size-specific growth curves in the two habitats cross, in this case switching from habitat 1 to habitat 2 at size s. If growth is consistently higher in one habitat, then no switch between habitats should occur.

Habitat choice may depend on the risk of predation relative to opportunities for growth.

Of course, maximization of growth is only one problem confronting a growing organism. Another problem is minimizing mortality, so that the growing organism has a maximal opportunity to grow to reproductive size before it is killed by predators or some other agent. Werner and Gilliam show that for juvenile organisms in a stable population ($r = 0$), the habitat-switching rule that maximizes survival through a particular size involves using habitats in a way that minimizes the ratio of μ/g, where μ is a size-specific mortality rate and g is a size-specific growth rate. Depending on the form of size-specific μ/g curves in different habitats, organisms may switch habitats at sizes that are considerably different from those that might be expected based on the maximization of growth rate alone (see Figure 10.9). The μ/g criterion only holds true for prereproductive organisms in a stable population. For reproducing individuals in nonstable populations, the criterion becomes more complex and requires minimizing $(\mu + r - b/v)/g$, where r is the rate of increase, b is the size-specific natality, or birth rate, and v is the reproductive value at that particular size.

Werner and Gilliam (1984) and Werner (1986) have suggested that this approach could be used to explain a diversity of size-dependent patterns of habitat selection. These patterns range from the size-dependent use of prey-

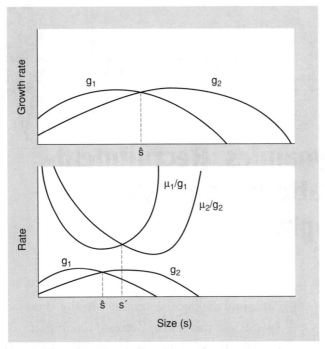

FIGURE 10.9. *Patterns of size-dependent growth,* g, *and mortality,* μ, *in two different habitats. An organism should switch from habitat 1 to habitat 2 at size* s *to maximize size-dependent growth, as long as mortality is similar in both habitats. Where size-dependent growth and mortality differ between habitats, the organism should switch habitats at size* s', *to minimize the ratio of* μ/g. *(Adapted from Werner and Gilliam, 1984, with permission, from the* Annual Review of Ecology and Systematics, *Volume 15. © 1984 by Annual Reviews.)*

rich open-water habitats by juvenile bluegills through the alternate use of aquatic and terrestrial habitats by organisms with complex life cycles.

CONCLUSIONS

Many of the examples of habitat selection described in this chapter focus on relatively small spatial scales and consider only one or a few species. Nonetheless, it seems reasonable to suppose that similar processes operate at larger spatial scales involving many animal species that are capable of exercising some degree of habitat selection. When we seek explanations for differences in the composition of communities, we need to remember that those differences may result as much from choices made by animals before they join communities as by interactions that occur after species come together in a particular place.

CHAPTER 11

Spatial Dynamics, Recruitment-Limited Patterns, and Island Biogeography

OVERVIEW

Temporal and spatial variation in the abundance of species can increase or decrease the impact of interspecific interactions on community composition. Spatial variation can occur at vastly different scales, ranging from the clumping of organisms within small patches of habitat to patterns that appear at the level of island archipelagoes or different areas within continents.

Some interactions must be considered in an explicitly spatial framework. For example, the intensity of competition experienced by sessile organisms depends on the number of competitors within an immediate spatial neighborhood. Intraspecific aggregation in spatially subdivided habitats can favor the coexistence of competitors. The distribution of organisms among subdivided habitats can also stabilize interactions that are unstable in undivided habitats. The persistence of some predator-prey interactions depends on a complex spatial framework of patchy habitats that create a shifting patchwork of temporary prey refuges. At larger spatial scales, spatial variation in recruitment influences the intensity of postrecruitment density-dependent interactions. At even larger scales, the size and isolation of island habitats influences the number of species that coexist. All these phenomena emphasize that heterogeneity in the spatial distribution of organisms can influence the composition and dynamics of communities.

THIS chapter considers how the spatial distribution of organisms, either within or among communities, alters interspecific interactions and patterns of community composition. Spatial dynamics operate at different scales and affect many processes, including interactions such as competition and predation. Species are seldom spatially distributed in a homogeneous or random pattern. Clumping, or spatial heterogeneity in abundance, can influence the persistence of interspecific interactions and the resulting community patterns in patchy or subdivided habitats. Spatial heterogeneity in the influx of individuals into local communities can also set the stage for increased or decreased intensities of density-dependent interactions. The relative importance of density-dependent interactions at a particular site may thus be determined by processes at other locations that influence the supply of colonizing organisms that reach that site.

 Spatial clumping can influence interspecific interactions.

The simple models that we have considered so far assume that species interact in systems that are closed to immigration and emigration. Consequently, the persistence of interacting species is assumed to reflect some sort of dynamic equilibrium or balance within a bounded community. In reality, communities are open to varying amounts of immigration and emigration, such that some or many of the individuals in a particular location arrive from other communities. This is particularly true in communities such as the rocky intertidal zone, where the sessile adults of many species become established at a particular location after long-distance transport during a planktonic larval stage. In these open communities, the supply of settling larvae produced by adults in distant locations may be much more important than larvae produced by local adults in determining the density and identity of species that interact on a particular stretch of rocky shore.

Most communities are open systems.

Theory suggests that the spatial distribution of species can have important implications for interactions within communities. A tendency for individuals of a species to aggregate within spatially isolated habitats can promote coexistence of competitors within a mosaic of habitat patches (Atkinson and Shorrocks 1981; Ives and May 1985). Experiments and theory both show that the subdivision of interacting populations into spatially isolated units connected by infrequent migration can promote the persistence of a predator-prey interaction that proves to be unstable in undivided habitats (Huffaker 1958; Caswell 1978; Holyoak and Lawler 1996a,b). Other models suggest that the explicit inclusion of the spatial distribution of predators and prey can promote complex dynamics, including chaos, in relatively simple systems (Comins et al. 1992).

INTERSPECIFIC INTERACTIONS IN PATCHY, SUBDIVIDED HABITATS

We have already seen one way in which the spatial arrangement of organisms can influence competitive interactions. Sessile organisms such as plants compete primarily with nearby neighbors for resources. Neighborhood models, and the experiments conducted to calibrate and test those models, make this spatial dependence explicit in competitive interactions among terrestrial plants (Pacala and Silander 1985, 1990). However, other aspects of the spatial arrangement of mobile organisms, including the tendency for species to aggregate in small discrete habitats and the effects of habitat subdivision on dispersal and aggregation, can have important effects on the dynamics of competitors and of predators and prey.

Competition

There is an extensive body of theory, largely untested in natural settings, suggesting that intraspecific aggregation, or clumping, can promote the coexistence of competitors in patchy habitats. If superior competitors tend to be clumped in a fraction of available discrete habitats, there will be empty patches of habitat, which can then be exploited successfully by inferior competitors. Atkinson and Shorrocks (1981) and Ives and May (1985) have modeled situations inspired by the breeding biology of many insects that breed in patchy habitats such as flowers, fruits, fungi, dung, or carrion. Within these patchy habitats, larval insects can attain high densities, and competition for resources can be intense. One process that might account for the coexistence of many species in such systems is that each

 Aggregation in subdivided habitats can promote the coexistence of competitors.

species may have clumped distributions, and the clumping of each species may be independent of the others. Simulation models of this process show that if the superior competitor has a highly clumped spatial distribution, the clumping effectively creates a spatial refuge for the competitively weaker fugitive species, which is able to exploit the sites left unused by the better competitor (Figure 11.1).

Simple subdivision of the environment without intraspecific aggregation is not effective in promoting coexistence. For the aggregation effect to work, the species must be independently distributed among patchy habitats; that is, their abundances should not be positively associated among patches. On average, sites with dense clumps of one species should not also contain dense clumps of other species. Independent spatial distributions of two or more species could occur if different species respond to different cues when selecting habitats for their offspring. Some evidence from natural systems suggests that the distributions of diptera and other arthropods living in forest mushrooms are in fact highly positively correlated, which suggests that the aggregative mechanism may not provide a convenient explanation for coexistence in this particular example (Worthen and McGuire 1988). Other studies show that the diptera breeding in small mammal carcasses are in fact highly aggregated intraspecifically and either negatively associated or unassociated interspecifically (Ives 1991), precisely the kind of situation required for aggregation to favor coexistence.

Tony Ives (1991) measured patterns of aggregation and the intensity of competition observed among five species of larval diptera that live in the decomposing carcasses of small mammals. The two most common species, *Phaenicia coeruliverdis* and *Sarcophaga bullata*, respond to increased intraspecific aggregation within rodent carcasses primarily by producing a

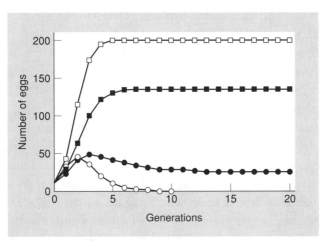

FIGURE 11.1. *Intraspecific aggregation within patches of a subdivided habitat promotes the coexistence of two competing species, as indicated by a model described by Atkinson and Shorrocks (1981). Population trajectories for two species are shown for undivided (open symbols; 1 site) and divided (closed symbols; 20 sites) habitats. (Adapted from Atkinson and Shorrocks, 1981, with permission of Blackwell Science Ltd.)*

smaller clutch size when they mature than would be the case if they grew under less crowded conditions. When both species occur in the same carcass, *Phaenicia coeruliverdis* also reduces one measure of *Sarcophaga bullata* reproductive success, which is defined by the product of adult fly abundance and clutch size produced in a single carcass. Ives estimates that intraspecific aggregation by *P. coeruliverdis* caused a 26% decrease in its own recruitment while resulting in a 74% increase in the recruitment of *S. bullata*, relative to what might be expected if *P. coeruliverdis* were randomly distributed among carcasses instead of clumped. The inference is that the aggregated distribution of fly larvae among carcasses reduces the intensity of interspecific competition experienced by *S. bullata*. The applicability of this phenomenon to other kinds of organisms in patchy habitats requires much additional study.

Predator-Prey Interactions

The persistence of predator-prey interactions has often been attributed to the existence of spatial refuges that give prey a temporary respite from predators. Gause (1934) found that the inclusion of a spatial refuge in simple laboratory cultures of the ciliated protist *Paramecium* and its predator *Didinium* could prevent the extinction of prey. However, even in this simple setting, predators usually starved to death when they were unable to exploit the few prey remaining in the refuge. For predators and prey to persist, a system with multiple sites for predator and prey interaction, connected by migration, would be necessary; such a system creates opportunities for spatially shifting refuges from predation. This system has been studied in a laboratory setting only recently (Holyoak and Lawler 1996a,b), but analogous systems using different organisms show how important spatial dynamics can be to the persistence of predator-prey interactions.

Huffaker (1958) used a predator-prey interaction between two mite species to show that subdivided habitats can promote the persistence of predators and prey. In this simple laboratory system, a herbivorous mite, *Eotetranychus sexmaculatus*, lived and fed on the surface of oranges, which provided a convenient discrete unit of habitat that can be varied in abundance and spatial configuration. The predatory mite *Typhlodromus occidentalis* fed on *Eotetranychus*. Both mites spent their time foraging on the surface of the oranges, and dispersed primarily by crawling over surfaces, although the prey could also disperse by using strands of silk to rappel from one site to another. The available surface area per orange, the number of oranges, the distribution of oranges among other habitat units (similarly sized rubber balls), and avenues for dispersal (wires and dowels), were all subjects of experimental manipulations designed to create increasing amounts of subdivision and iso-

 Subdivided habitats can promote the persistence of predators and prey.

lation among habitat units. Populations of *Eotetranychus* tended to oscillate irregularly in the absence of predators, whether they occurred in a few large, clumped habitats or in many small, widely dispersed habitats (Figure 11.2). Addition of the predator to the system led to the rapid extinction of the prey

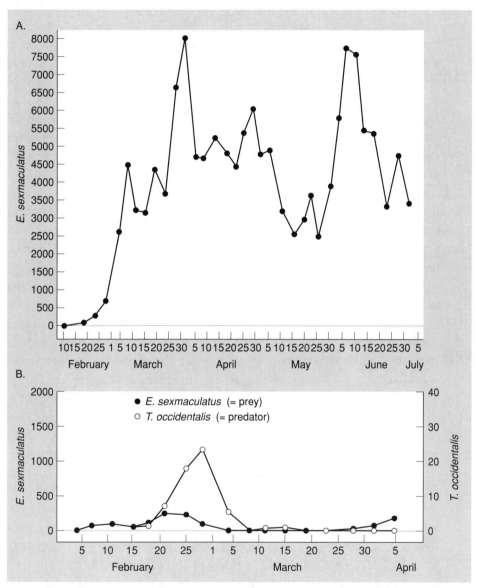

FIGURE 11.2. *Population dynamics of prey (*Eotetranychus*) alone and prey and predators (*Typhlodromus*) together in settings with relatively little spatial complexity. (A) Prey alone, with 4 oranges dispersed among 36 foodless units (rubber balls). (B) Predators and prey together, with 20 oranges distributed among 20 foodless units. (Adapted with permission from Huffaker, 1958. Hilgardia, vol. 27, p. 360. © Regents University of California.)*

mites, followed by the extinction of the predators, under a variety of combi-nations of spatially subdivided habitats (see Figure 11.2). Huffaker thought that this result was directly analogous to Gause's observation that the preda-tor *Didinium* readily overexploited its prey *Paramecium*, which led to rapid extinction of either *Paramecium* or both species.

Only when a very complex array of habitats was used did sustained prey-predator oscillations occur (Figure 11.3). The oscillations apparently resulted from various features of the spatial environment that gave a slight dispersal advantage to the prey, allowing them to temporarily increase in abundance in ephemeral predator-free refuges until those sites were eventually colonized by the predator. Vaseline barriers created a maze that limited the dispersal of both species, whereas the inclusion of small wooden posts allowed the prey, but not the predators, to disperse to other sites. Persistence results when enough prey manage to escape from predators to repopulate the system, while at least a few predators manage to persist in other sites without starving. The results are a succession of outbreaks of prey, followed by predator outbreaks that drive the prey to low levels. It would be interesting to know precisely how complex a system was necessary to promote the prolonged persistence of this predator-prey interaction, but the trial and error approach to increasing habitat subdivision makes it difficult to guess.

Other studies have explored the role of habitat subdivision under more natural circumstances. Peter Kareiva (1987) demonstrated effects of habitat subdivision on the dynamics of a different arthropod predator-prey interac-tion, which shares some of the properties of the interactions among the mites studied by Huffaker. Kareiva studied the

 Habitat subdivision effectively weak-ens the intensity of predator-prey interactions.

aphid, *Uroleucon*, which lives and feeds on goldenrod (*Solidago* sp.) and is in turn fed on by the ladybird beetle, *Coccinella*. Kareiva created continuous or patchy stands of *Solidago* by mowing plants in an old field into continuous strips or discontinuous patches (Figure 11.4). In continuous strips of host plants, where there were few barriers to dispersal, the predator *Coccinella* was able to readily disperse and aggregate on plants with high concentrations of *Uroleucon*, suppressing the frequency of prey outbreaks. In discontinuous strings of goldenrod patches, *Coccinella* dispersal was restricted, and *Uroleu-con* often attained very high densities. Although Kareiva cautioned against accepting the generalization that patchiness promotes stability in most preda-tor-prey interactions since it led to a greater frequency of outbreaks in his system, it is clear that in both his study and Huffaker's system, greater habitat subdivision, through its greater effects on predator dispersal, affected prey-predator dynamics by limiting the ability of predators to extirpate prey.

Perhaps the best demonstration that metapopulation dynamics can promote the persistence of predator-prey interactions is a recent effort that

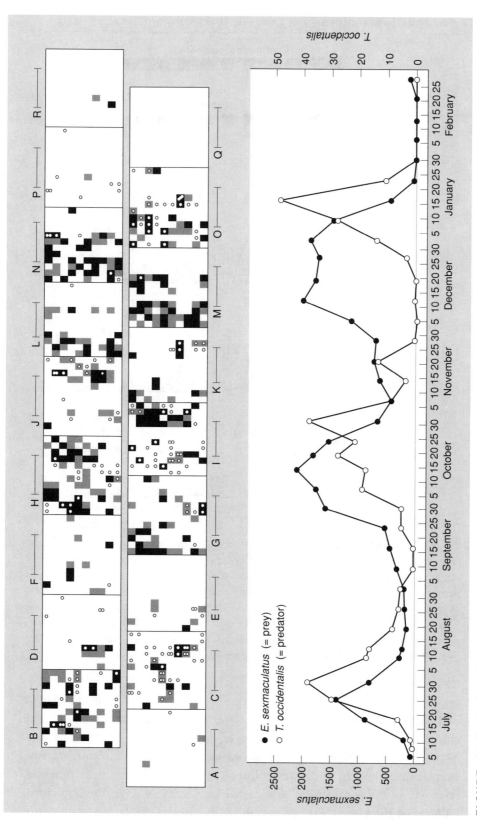

FIGURE 11.3. *Prolonged coexistence of predators and prey in a spatially complex environment, consisting of 120 habitat units partially separated by a maze of vaseline barriers, with dowels and wires added to enhance prey dispersal. Shaded squares and circles correspond to spatial variation in the densities of prey and predators, respectively. (Adapted with permission from Huffaker, 1958. Hilgardia, vol. 27, p. 370. © Regents University of California.)*

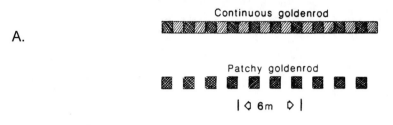

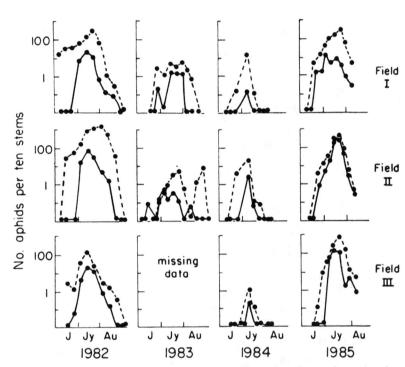

FIGURE 11.4. *Effects of habitat subdivision on the maximum abundance of prey* (Uroleu-con) *in continuous and discontinuous patches of the plant* Solidago. *(A) Experimental design showing arrangement of experimental plots, with plants mowed into separated patches or a continuous row. (B) Maximum values of* Uroleucon *abundance over time in continuous (solid lines) and discontinuous (dashed lines)* Solidago *plots. Prey reach higher densities in patchy* Solidago *because habitat subdivision interferes with the dispersal of the predator* Coccinella. *(Reprinted with permission from* Nature *326: 388–390, P. Kareiva. © 1987 Macmillan Magazines Limited.)*

builds on the pioneering work of Gause (1934) cited above. Marcel Holyoak and Sharon Lawler (1996a) used a system of isolated or interconnected culture vessels to compare the dynamics of single and subdivided populations of two ciliates, the predator *Didinium nasutum* and its prey *Colpidium striatum* (Figure 11.5). In undivided cultures, the predator-prey interaction was relatively unstable and usually persisted for about 70 days. In subdivided cultures of the same total volume but with connections between units for

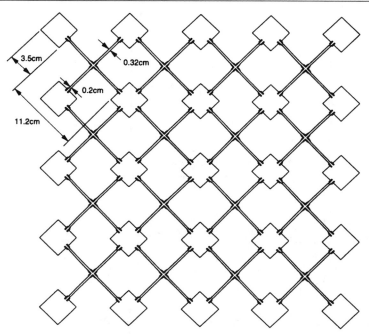

FIGURE 11.5. *Arrangement of interconnected culture vessels used to study the metapopulation dynamics of* Colpidium *and* Didinium. *Dynamics observed in arrays such as this were compared with dynamics in single vessels of comparable volume. (Reprinted from Holyoak and Lawler, 1996, with permission of the Ecological Society of America.)*

migration, the interaction persisted for at least 130 days, until the experiment was finally terminated (Figure 11.6). Observation of population dynamics within the subdivided array of cultures showed that abundances of predators and prey oscillated in an asynchronous fashion across the array, a key feature of a predator-prey system persisting as a result of metapopulation dynamics. Had the populations within the subunits simply acted as a single highly connected population, population fluctuations would tend to be synchronous across the entire array. Within individual cells of the arrays, extinctions of both predators and prey were frequent but were offset by recolonizations brought about by relatively infrequent dispersal from other subunits.

 Metapopulation dynamics can favor coexistence.

This study sets a new standard for experiments on metapopulation dynamics, since, unlike Huffaker's pathbreaking study, Holyoak and Lawler replicated their experimental arrays, measured dispersal rates, extinctions, and colonizations, and carefully described the spatial asynchrony of the population fluctuations of predators and prey. This information shows that the persistence of predators and prey was in fact directly caused by asynchronous population fluctuations in different portions of the subdivided community,

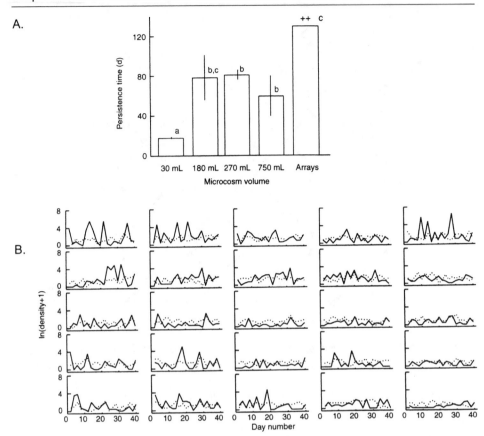

FIGURE 11.6. *(A) Habitat subdivision enhances the persistence of the predator-prey inter- action between* Colpidium *and* Didinium *in arrays of culture vessels compared with single vessels of various total volume. Arrays contained a total volume of 750 ml, and individual units in the array contained 30 ml. (B) Population dynamics of prey and predators in each culture vessel within an array, showing that population fluctuations in different parts of the array tended to be asynchronous, as would be expected if enhanced persistence was a consequence of metapopulation dynamics. (Reprinted from Holyoak and Lawler, 1996a, with permission of the Ecological Society of America.)*

together with the movement of predators and prey from areas of high abun- dance to areas where populations had crashed. Use of carefully designed control treatments shows that predator-prey persistence cannot be attributed to either a simple increase in the volume of habitat in which the interactions occurred or to an increase in the total size of the interacting populations.

The tendency for predators and prey to persist for very long periods of time within an array of habitat patches connected by migration can be modeled using a simulation approach that makes very few assumptions about the detailed biology of the interacting organisms. Hal Caswell (1978) devel- oped a model that was originally intended to show how predation can promote the coexistence of two competing species that interact in a subdivided metacommunity. Although the promotion of coexistence is interesting in itself, the model also makes the point that predators and prey can coexist in a nearly indefinite game of hide-and-seek, even when the interaction between predator and prey always leads to the extinction of both species within a single habitat patch.

Caswell's model assumes that the habitat can be divided into a series of discrete patches, called cells. The number of interconnected cells, N, can be varied to explore its effect on the model. Within a cell, two competing prey species, A and B, interact such that the better competitor, B, always excludes the inferior competitor, A, within a given period of time after the competitive interaction begins. This time span, called TC, can be varied in the model to mimic differences in the strength of the competitive interaction. Both prey species can disperse to open, unoccupied cells in the system with a probability that depends on their intrinsic dispersal ability, D_A or D_B, and the fraction of the cells in the system that currently contain either species A (N_A/N) or B (N_B/N). Thus, the probability that an empty cell will be colonized by species A in the next iteration (time unit) of the model is $D_A(N_A/N)$. The model is further constrained so that the inferior competitor can only invade empty cells, but the superior competitor can invade cells containing the inferior competitor. Simulations show that simple subdivision of the habitat is not sufficient to greatly prolong the coexistence of the two competitors within the system, although the interaction does proceed much longer in a subdivided habitat than in a single cell (Figure 11.7).

Models show that unstable interactions can persist indefinitely in subdivided habitats.

Addition of a predator, species C, to the system can greatly prolong the system-wide persistence of all three species. The predator can only invade cells already occupied by prey, and the predator eliminates either or both prey species within a cell after an interval of time denoted by TP. The cell is then open and available for recolonization. The predator disperses to open cells with a probability given by $D_C(N_C/N)$, where D_C is the predator's dispersal ability and (N_C/N) gives the fraction of the cells within the system that currently contain the predator. Caswell shows that under some circumstances, all three species—the two prey and their predator—can persist many times longer than the interactions will persist in a given single cell (Figure 11.8).

By systematically varying N, TC, TP, D_A, D_B, and D_C, it is clear that increases in the number of cells, the dispersal ability of the predator, the dispersal ability of the competitively inferior prey, or the time required for competitive exclusion to occur within a single cell all prolong the coexistence of the three species (Figure 11.9). Increases in the dispersal ability of the competitively superior prey or in the time required for predators to exclude prey

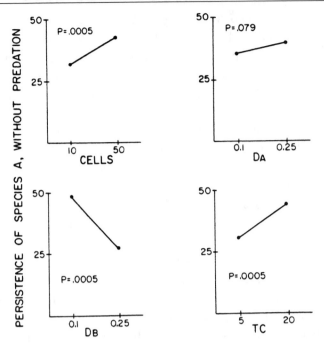

FIGURE 11.7. *Effects of habitat subdivision (N), dispersal ability (D_A, D_B), and the time required for competitive exclusion within a patch (TC) on the persistence of an interaction between an inferior (species A) and superior (species B) competitor. (Reprinted from Caswell, 1978, with permission of the University of Chicago Press.)*

from a cell shorten the duration of coexistence within the system (see Figure 11.9). Although this model contains very little natural history or detailed biological information of any sort, it captures the essence of a predator-prey interaction in a subdivided habitat and shows how predators and prey might coexist in a sufficiently complex set of cells linked by migration.

Other models have explored the conditions promoting the coexistence of predators and prey in space. Hassell et al. (1991b) have shown that for a variety of models based on the difference equation models of Nicholson and Bailey (described in Chapter 5), subdivision of a parasitoid-host interaction across a patchy environment will yield a stable equilibrium as long as the $(CV)^2$ of the parasitoid distribution among prey patches is greater than 1. What does this mean? The $(CV)^2$ is equivalent to (standard deviation/mean)^2, which can be expressed as (variance/mean^2). Parasitoids with a clumped distribution will have a variance-mean ratio greater than 1, from the properties of the Poisson distribution. This criterion thus indicates that parasitoids must be significantly clumped, and the degree of clumping (value of the variance-mean ratio) must increase as the average number of parasitoids per prey patch increases. This result holds regardless of whether the predators are

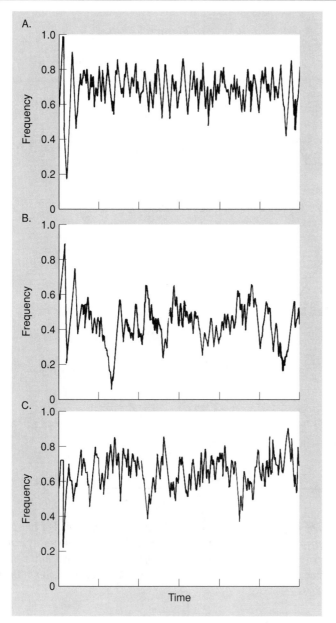

FIGURE 11.8. *Results of a simulation showing prolonged coexistence of two prey (A, B) and their predator (C) in a subdivided habitat. Values of model parameters used in this simulation were N = 50, D_A = 0.25, D_B = 0.10, D_C = 0.25, TC = 20, and TP = 20. (Adapted from Caswell, 1978, with permission of the University of Chicago Press.)*

randomly distributed among prey patches or are preferentially aggregated in areas of high prey density.

Hassell et al. (1991a) have also shown that relatively simple models of host-parasitoid dynamics in patchy environments can lead to complex

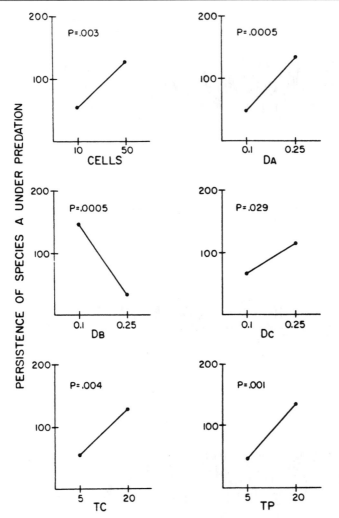

FIGURE 11.9. *Effects of varying the parameters in Caswell's (1978) model of competition and predation within a subdivided habitat. The response shown is the average persistence time of the inferior competitor in the system. Values of p indicate statistically significant changes in persistence time. (Reprinted from Caswell, 1978, with permission of the University of Chicago Press.)*

dynamics, including cyclic and chaotic fluctuations in abundance (Figure 11.10). These patterns arise whether the models used are fairly detailed or whether they correspond to cellular automata with relatively little biological content. These conclusions also may depend on the form of the models used, since Murdoch and Stewart-Oaten (1989) found that subdivided populations modeled using sets of differential equations do not appear to be stabilized by patchiness in the way that difference equations are. So far, it remains unclear whether any natural patchy systems exhibit the kinds of behavior predicted by

(a)

(b)

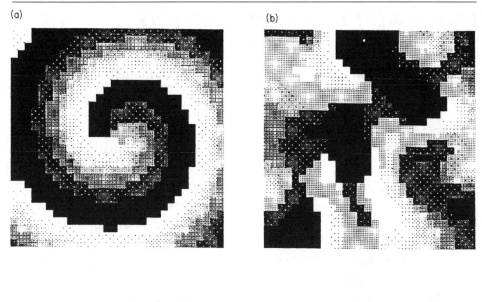

(c)

(d)

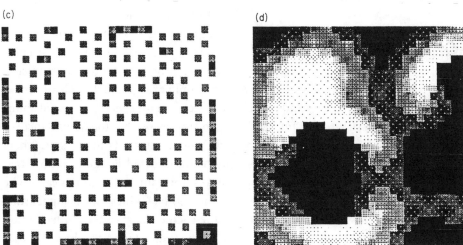

FIGURE 11.10. *Predator-prey interactions within a subdivided habitat linked by migration can produce an array of spatial patterns in the density of prey, ranging from spirals (A), to spatially chaotic patterns (B), to regular patterns of alternating high and low abundance (C). Black squares are empty habitats, darkly shaded squares contain increasing abundances of prey, and light squares correspond to patches with increasing predators and decreasing prey. (Reprinted from Comins et al., 1992, with permission of Blackwell Science Ltd.)*

the models. The ultimate resolution about which result, increased or decreased stability, follows from the subdivision of interacting predators and prey into multiple habitats must await additional empirical tests using real species interacting in appropriately patchy habitats.

RECRUITMENT-LIMITED INTERACTIONS: SUPPLY-SIDE ECOLOGY

Spatial variation in the abundance of organisms can occur over somewhat larger scales than those described above. For example, the density of sessile marine organisms that settle in the rocky intertidal zone can vary over distances ranging from several meters to many kilometers (Dayton 1971; Menge et al. 1994). Evidence for the importance of spatially variable recruitment rates in determining community patterns comes from the different outcomes of studies conducted at different places or at different times. Other studies have built on these observations and incorporated known differences in recruitment rates among sites into experimental studies of interspecific interactions.

Robert Paine's (1966) demonstration of keystone predation by *Pisaster* in the rocky intertidal zone may have been a fortuitous consequence of unusually high recruitment rates by prey in the areas where predators were removed (Dayton 1971; Underwood et al. 1983; Menge et al. 1994). Unusually high settlement would create especially intense competition among abundant settlers in sites from which predators were removed, and could exaggerate the role of predators in thinning settlers and reducing competition for space. The initial abundance of sessile rocky intertidal organisms is set by the abundance of planktonic larvae that are transported by ocean currents to the sites where settling occurs. Sometimes the supply of transported larvae is so low that space is never limiting, and competition is unimportant. This appears to be the case in some sites studied by Dayton (1971), Gaines and Roughgarden (1985), and Menge et al. (1994) along the Pacific Coast of the United States.

 Interspecific interactions exhibit density dependence . . .

All these studies failed to observe either intense competition or keystone predation at certain locations having a set of species very similar to those studied by Paine. Sites without keystone predation had relatively low settlement of potential competitors compared with Paine's sites.

. . . and densities vary widely over space and time.

Analogous differences in recruitment rates may explain the very different organization of rocky intertidal communities observed between the exposed coasts of western North America and Australia. For instance, Anthony Underwood (Underwood et al. 1983) has argued that even in the absence of predators, strong competition for space is rare in Australian intertidal communities. The settlement of Australian rocky intertidal organisms is often low and highly variable in space and time (Underwood et al. 1983). Consequently, competition for space in the Australian rocky intertidal zone is infrequent, and predators fail to enhance the abundance of competitively inferior species, partly because competition seldom happens and partly because predator recruitment is highly variable as well.

Peter Fairweather (1988) studied the importance of variation in initial prey density on interactions between the barnacle *Tesseropora rosea* and the predatory whelk, *Morula marginalba,* in Australia. Interactions were studied at two sites with different recruitment rates. In addition, variable recruitment of each species could be mimicked by experimental removals of the barnacles or whelks to simulate low recruitment by either predator or prey. Where barnacles (prey) were removed, predatory whelks emigrated to other sites where prey were more abundant (Figure 11.11). The departure of predators from sites with few prey means that low prey recruitment can create density-dependent refuges from predators, since the predators tend to overlook such sites while concentrating their foraging elsewhere. Rates of predation by

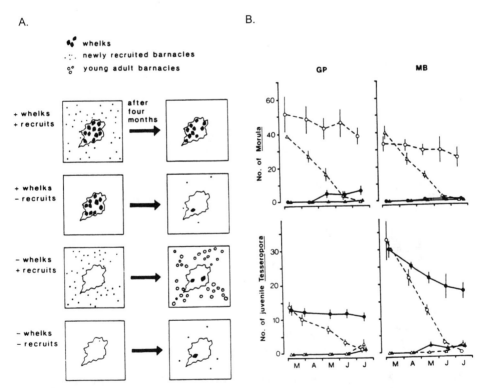

FIGURE 11.11. *Graphical summary of effects (A) and changes in densities (B) of barnacles (prey) and whelks (predators) in response to removals of barnacles or whelks. Where prey are removed, predators also become less abundant; where predators are removed, prey become established and survive at higher densities. GP and MB are different sites with different settlement rates. Open circles = + whelks, + recruits; open triangles = + whelks, − recruits; solid circles = − whelks, + recruits; solid triangles = − whelks, − recruits. (Reprinted from Fairweather, 1988, with permission from* The Biological Bulletin.*)*

whelks were highest in areas of highest initial prey density, suggesting that the predators tended to aggregate and forage in places with high prey densities.

In addition to influencing the probability that prey will compete, the initial density of settling prey can influence whether predators will be sufficiently abundant to inflict much mortality on the prey. Perhaps the best example of this influence comes from the observations of Gaines and Roughgarden (1985, 1987) regarding interactions between *Pisaster* and *Balanus* in California. The predator, *Pisaster ochraceous*, does not forage extensively in areas of low prey (*Balanus glandula*) density; consequently, low-density populations of prey experience very little predation (Figure 11.12).

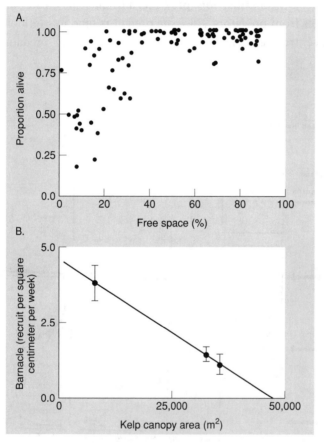

FIGURE 11.12. (A) Effects of initial density (percentage of space occupied) on survival of barnacles. Survival is lower at high-density sites because those sites also support high densities of predators. (Reprinted from Gaines and Roughgarden, 1985, with permission.) (B) Variation in offshore habitat (kelp forests) influences the settlement of barnacles. Kelp forests provide habitat for many predators, including juvenile fish, which may be responsible for decreasing the abundance of larval barnacles. (Adapted with permission from Gaines and Roughgarden, 1987. © 1987 American Association for the Advancement of Science.)

The cause of spatial variation in prey recruitment appears to be correlated with the consumption of larval prey by offshore predators, mostly fish, which have a spatially variable distribution that is correlated with the presence of offshore beds of kelp (see Figure 11.12). The pattern suggests that variation in recruitment is not purely the result of spatial variation in physical transport processes that bring settling larvae to the rocky coast. In addition, strong biotic interactions, including predation, can limit settling and thereby limit the extent of postsettling competition and predation.

All these studies of supply-side ecology emphasize the fact that many communities are open systems in which the density of interacting organisms is set by processes that are extrinsic to, or operate outside of, the site where the interactions ultimately take place. Thus, such communities must be studied at spatial or temporal scales that are sufficient to reveal variation in recruitment to fully understand why phenomena like keystone predation vary in importance among locations.

LARGE-SCALE SPATIAL PATTERNS: ISLAND BIOGEOGRAPHY

The previous sections of this chapter examined how spatial variation in abundance influences various interactions among organisms. The final section of this chapter considers a different aspect of community patterns that depends on the spatial configuration of relatively isolated communities. We turn our attention to islands and some of the properties of island communities that have been studied under the heading of

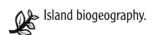 Island biogeography.

island biogeography. The islands can be the usual sort, chunks of terrestrial habitat surrounded by water, or virtual islands, fragments of one kind of habitat surrounded by another kind of habitat that is inhospitable to the organisms living within the virtual island. Comparative studies of islands have played an important role in ecology since the pioneering work of Darwin (1859) and Wallace (1878). More recent studies have focused on patterns that emerge from comparisons of the numbers of species on islands of different size and isolation (MacArthur and Wilson 1967).

Species-Area Relations

The number of species found in a particular area increases with the size of the area examined. When the areas considered are islands, the number of species found in a particular taxonomic group, such as birds, lizards, or ants, increases in a predictable way with island area (Figure 11.13). This **species-area relation** can be described by plotting the logarithm of species richness

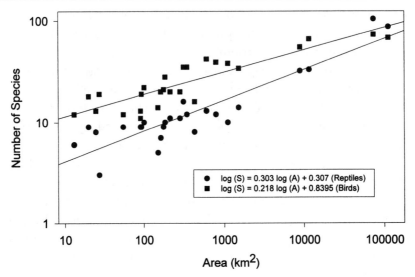

FIGURE 11.13. *Relations between island area and the number of species for birds and reptiles on islands in the Caribbean Sea. (Data from Wright, 1981).*

against the logarithm of island area. For many taxa, this approach yields a linear relation, described by the equation

$$\log(S) = z\log(A) + \log(c)$$

where S is species richness, A is island area, z is the slope of the relation, and $\log(c)$ is the y intercept of the relation. For a large number of taxa on islands of varying area, z takes on values ranging from 0.2 to 0.35. The constant c is taxon-specific and varies widely among groups.

Comparisons of species-area relations for islands and patches of habitat subsampled from larger continental areas usually differ in that islands have a steeper slope (z) and lower intercept (c) than do similar areas subsampled from a mainland. Values of z for taxa sampled from continuous areas of the mainland tend to run from about 0.12 to 0.17. Reasons for the difference remain conjectural (Rosenzweig 1995). The mainland pattern may be a statistical consequence of sampling from ever-larger areas holding more individual organisms, more habitats, or, for very large areas, more biogeographic regions. The island pattern can be explained differently, as an outcome of the interaction between rates of arrival and extinction of species on islands. This interplay is described below.

Equilibrium Island Biogeography. As stated previously, species-areas relations show that the number of species increases with island area. The equilibrium theory of island biogeography provides one possible explanation for this pattern (MacArthur and Wilson 1967). The theory assumes that the number of species on an island is a consequence of the dynamic equilibrium that results from the interplay between rates of colonization and extinction. Rates of colonization and extinction are thought to vary with the number of species on the island, as shown in Figure 11.14. As species accumulate on the island, the

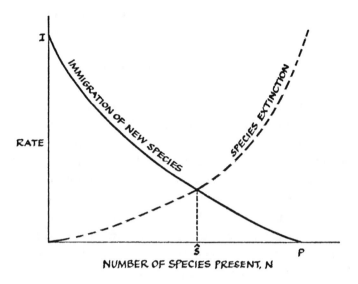

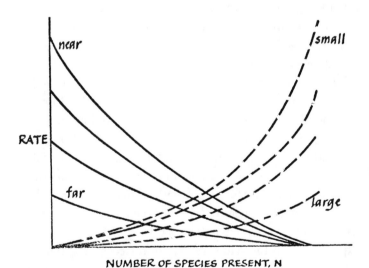

FIGURE 11.14. *Hypothetical rates of colonization (solid lines) and extinction (dashed lines) for islands of different sizes and distances from the mainland that provides the source of colonists. Where the colonization and extinction curves cross, an equilibrium number of species results. (Reprinted with permission from MacArthur, R. H., and Wilson, E. O. The Theory of Island Biogeography. © 1967 by Princeton University Press.)*

rate of colonization decreases, presumably because there are fewer species remaining in the pool of colonists to invade the island. With the increase in the number of species, extinction rates of resident species increase, perhaps because strong negative interactions with other species become more likely. Where the colonization and extinction rates are equal—the point at which the lines cross in Figure 11.14—an equilibrium number of species results. The theory predicts a particular species richness for a given location with a particular set of colonization and extinction curves, but is silent about the actual species composition because species are assumed to be constantly turning over.

We can make some speculations about why the number of species varies in an important way with island size by assuming that island size affects rates of colonization or extinction. For example, if larger islands can hold more species, extinction rates will be lower at a particular value of species richness on larger islands than on smaller ones. All else being equal, especially the colonization rates, this leads to greater equilibrium species richness on larger islands. The theory can also be modified to explain the consequences of greater isolation of an island from a source pool of colonists. More-isolated islands are assumed to have a lower colonization rate than less-isolated ones, all else being equal. For islands of a given area, this leads to a lower value of species richness for more-isolated islands.

Simberloff and Wilson (1969) performed some of the first experimental tests of island biogeography theory using the arthropod fauna on small mangrove islands near the Florida coast. Because the islands were quite small, it was possible to remove the resident arthropods with insecticide and then observe whether the empty islands returned to the level of species richness observed before the defaunation. In most cases, approximately the same number of species returned to each island as was observed before defaunation, although the identity of those species was often quite different before and after the manipulation (Figure 11.15). Simberloff and Wilson interpreted the similarity of species richness before and after defaunation as evidence for an equilibrium in species richness. They did not specifically measure either colonization or extinction rates to assess whether curves similar to those predicted by the theory would appear or whether the curves would predict the resulting species richness observed on the islands.

Other experiments in the same system addressed how species richness changed with a decrease in island size. Simberloff (1976) used a chain saw to reduce the area of mangrove islands by simply cutting away some of the trees that formed each small island. Eight islands were reduced in area, while another remained unchanged as a control. In all cases, the number of arthropods present on the islands decreased after islands became smaller (Figure 11.16). Since the islands were structurally simple, consisting mostly of a single

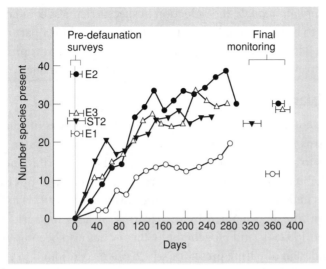

FIGURE 11.15. *Accumulation of arthropod species on small mangrove islands after experimental defaunation with insecticide. In most cases, islands returned to values similar to those measured before defaunation. (Adapted from Simberloff and Wilson, 1969, with permission of the Ecological Society of America.)*

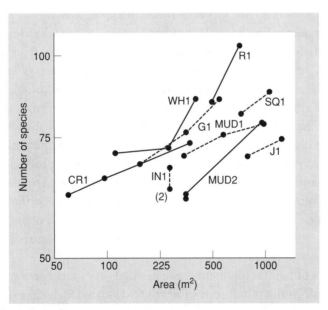

FIGURE 11.16. *Effects of reductions in the area of small mangrove islands on the number of arthropod species living on the islands. Reductions in island area led to reductions in species richness. (Adapted from Simberloff, 1976, with permission of the Ecological Society of America.)*

tree species, *Rhizophora mangle,* the reduction in species richness could not be easily attributed to a differential loss of microhabitats used by a part of the arthropod fauna. The conclusion was that reduced species richness probably resulted from increased extinction rates when islands became smaller.

Patterns in Virtual Islands

While island biogeography theory was inspired by the properties of communities of organisms found on terrestrial islands surrounded by water, there have been efforts to apply and extend these ideas to other kinds of island-like habitats. These habitats range from isolated patches of montane forest found on mountaintops in deserts of the American Southwest (Brown 1971) to fragments of tropical forests generated by the clearing of land for agriculture (Bierregaard et al. 1992). Island biogeography theory has also been used as a rationale for the design of nature preserves, which often correspond to virtual islands of preserved habitat in a sea of surrounding altered habitat (Simberloff 1988; Meffe and Carroll 1994). The only problem with this approach is that the theory often makes no specific predictions about the features of nature preserve design that it is invoked to defend (Simberloff 1988), such as the utility of many small versus a few large reserves.

Some habitat islands do not appear to function as predicted by island biogeography theory. James Brown (1971) studied patterns of species richness in the boreal mammals isolated in montane forests of peaks in the Great Basin of western North America. The patterns suggest that current values of species richness cannot be explained as an equilibrium between rates of colonization and extinction. Rather, the intervening desert habitat is so unfavorable for montane species that no colonization has occurred since these habitats were linked during the Pleistocene. In the absence of colonization, extinctions have occurred, with more extinctions occurring in smaller "islands," producing an unusually steep slope in the species-area relation for these isolated mountaintop mammals (Figure 11.17). The absence of a relation between the area-corrected species richness observed and the proximity to a source of potential colonists suggested to Brown that little or no colonization has occurred since the montane forests were isolated.

Other studies of habitat fragments have focused on experiments carried out on a large scale in the rain forests of Brazil to determine whether islands of rain forest habitat will lose species or biomass as island area declines (Bierregaard et al. 1992; Stouffer and Bierregaard 1995; Laurance et al. 1997). Replicate islands of forest habitat that were 1 hectare and 10 hectares in area were monitored before and after their isolation from nearby unfragmented forest. Despite the proximity of potential colonists in nearby forests, the number of bird species in newly isolated fragments declined dramatically during the first nine years after the fragments were isolated (Figure 11.18).

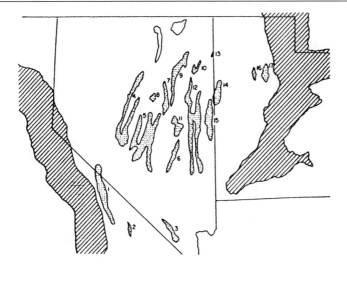

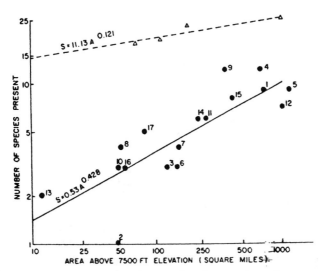

FIGURE 11.17. *Species-area relations for the number of mammal species found in montane forest habitats located in the Great Basin of the western United States. Numbers on both plots indicate different habitat islands. The dashed line shows the species-area relation within larger continuous areas of montane habitat. (Reprinted from Brown, 1971, with permission of the University of Chicago Press.)*

Some species eventually returned to fragments, but only after the surrounding habitat had regenerated.

Other work in the same system describes an unexpected collapse in above-ground biomass of forest trees (Laurance et al. 1997). Permanent study plots within 100 meters of the edge of forest fragments lost an average of approximately 14% of tree biomass within two to four years of forest

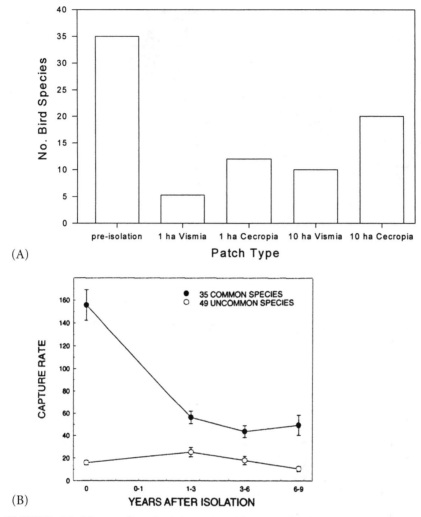

(A)

(B)

FIGURE 11.18. *Declines in (A) species richness and (B) capture rates for insectivorous birds in 1- and 10-hectare fragments of Amazonian forest. Fragments differ with respect to dominant plant species,* Vismia *or* Cecropia, *surrounding the isolated patch. (Data for (A) are from Stouffer and Bierregaard, 1995. (B) Reprinted from Stouffer and Bierregaard, 1995, with permission of the Ecological Society of America.)*

fragmentation. The losses were attributed mostly to wind throws, which differentially affect tall forest trees exposed on forest fragment edges. Comparable plots in the interior of fragments, or in undisturbed forest, did not show similar declines.

This landmark study of the consequences of tropical forest fragmentation, along with other studies, suggests that important differences exist between the biota of real islands, which tend to be colonized by highly vagile species, and habitat islands, which often hold species that disperse infrequently, if at all, through the kinds of habitats that separate fragments of interest. Patterns in habitat islands may be driven more by area-related dynamics of extinction, with little colonization occurring to offset those extinctions. However, where species do disperse readily, the size and spatial configuration of islands will interact with both colonization rates and extinction rates to influence community patterns.

CONCLUSIONS

The subdivision of communities into spatially separated subunits linked by migration (metapopulations) has important consequences for the outcome of interspecific interactions. Theory suggests that aggregation within subdivided habitats can promote the coexistence of competitors that would not persist in a single homogeneous habitat. Theory and experiments also indicate that metapopulation processes can promote the persistence of locally unstable predator-prey interactions. Spatial variation in the densities of organisms that colonize communities can also create variation in the intensity of interspecific interactions that structure communities. At even larger spatial scales, the size and spatial arrangement of island habitats can interact with colonization and extinction rates to influence patterns of species richness.

PART III

LARGE-SCALE, INTEGRATIVE COMMUNITY PHENOMENA

CHAPTER 12

Causes and Consequences of Diversity

OVERVIEW

The sources of species diversity within and among natural communities constitute one of the central unresolved problems in community ecology. Competing explanations for species diversity patterns differ in stressing mechanisms that operate when species assemblages are near or far from equilibrium. Here, stable equilibrium conditions refer to communities with a relatively constant species composition, whereas nonequilibrium situations are characterized by fluctuating species composition and extensive variation in population dynamics. The term *stable equilibrium* has a precise meaning when applied to the analysis of mathematical models, but is applied much less rigorously to describe natural communities. Simple examples based on the behavior of mathematical models illustrate stable equilibria and alternative kinds of dynamic behavior, including population fluctuations resulting from stable limit cycles and chaos.

It is uncertain whether the dynamics of species in natural communities conform to stable equilibria or to various kinds of nonequilibrium behavior. Relations between species diversity and different community attributes, such as stability or productivity, also remain controversial. Some empirical studies show that communities that are more diverse exhibit more consistent or predictable properties than simple communities, although this pattern is not sufficient to demonstrate enhanced stability. Two theoretical studies predict that individual populations in more diverse communities should be less stable, whereas aggregate attributes based on many species, such as plant biomass, may be more stable. Resolution of the diversity-stability paradox hinges on the different meanings of stability applied to empirical and theoretical patterns and on the assumptions used to develop theoretical explorations of diversity-stability relations.

Species diversity varies with latitude and productivity in ways that may have no simple, single-factor explanation. Causes of relations between productivity and diversity, and of latitudinal gradients of species diversity, are fundamental unsolved problems in community ecology.

EQUILIBRIUM AND NONEQUILIBRIUM COMMUNITIES

Most of the important ideas about the sources of diversity in natural communities stress either processes that allow many species to coexist at equilibrium (**equilibrium mechanisms**) or that

 Is diversity maintained by equilibrium or nonequilibrium mechanisms?

prevent an equilibrium from being reached and therefore prevent exclusions of species that might result as a consequence of that equilibrium (**nonequilibrium mechanisms**). The ideas involving equilibrium processes view community composition as the stable outcome of interspecific interactions, that is, as the set of species abundances reached when rates of change in population size are zero and to which the community will return if any of those populations are perturbed. In this view, communities show a tendency to return to a stable equilibrium composition after some disturbance, and the composition of the community at equilibrium corresponds to the typical composition seen in nature. The processes invoked to explain diversity are those that allow more species to coexist in a stable equilibrium of the sort used to assess the behavior of mathematical models of interacting populations. The various kinds of dynamic behavior that can occur in simple models are reviewed in Figure 12.1. These include stable equilibria and various unstable but persisting dynamics, including stable limit cycles and chaotic behavior. In contrast to equilibrium-based explanations, nonequilibrium mechanisms account for the maintenance of diversity within communities by focusing on how disturbances or other processes interfere with the exclusion of species that occurs as communities attain equilibrium.

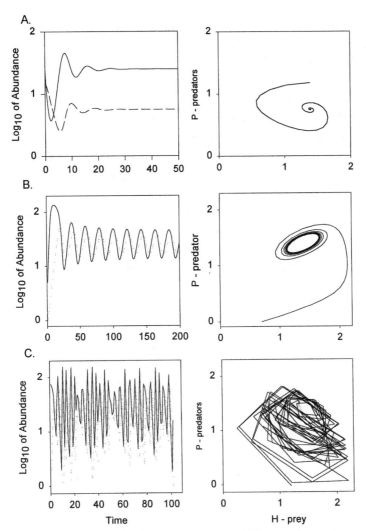

FIGURE 12.1. *Examples of different kinds of equilibrium and nonequilibrium behaviors generated by simple predator-prey models. Each behavior is shown both as a time series (left) and as a trajectory of population dynamics in phase space (right). The time series show prey and predator dynamics as solid and dashed lines, respectively. (A) Stable equilibrium. (B) Limit cycles. (C) Chaos.*

The controversy surrounding the relative importance of equilibrium and nonequilibrium explanations for diversity comes from a basic uncertainty about whether natural communities are near or far from equilibrium. The evidence that can be brought to bear on whether populations within communities are near any stable equilibrium is surprisingly scant and indirect. Recall that in simple models of population dynamics, a stable equilibrium refers to the tendency of a population to return to a population size at which $dN/dt = 0$ following any increase or decrease in population size. The evidence needed to assess whether natural populations exhibit stable equilibrium behavior is difficult to obtain (Connell and Sousa 1983).

Are natural communities ever in equilibrium?

First, one must show that following a perturbation, populations will return to a density (presumably the equilibrium density) seen before the perturbation. A test for equilibrium behavior should therefore first perturb populations by changing the abundance of one or more species and then follow populations for a sufficient number of generations to see whether they return to the presumed equilibrium density (Figure 12.2). Such direct perturbations rarely happen by design, and when they do, their consequences are often not followed long enough to assess whether the system returns to its previous state. Instead, ecologists often make inferences about stable equilibrium behavior from the long-term dynamics of unmanipulated populations. Even then, because long-term data are available primarily for long-lived organisms, the observed dynamics occur over time periods that are too short to yield

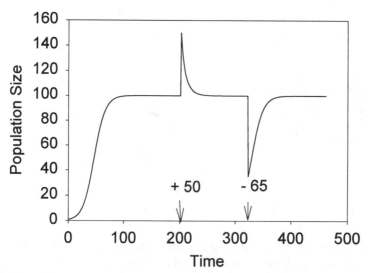

FIGURE 12.2. *An example of how a model population with a stable equilibrium responds over time to perturbations that increase (by 50 individuals) or decrease (by 65 individuals) population size. The model assumes logistic growth (r = 0.5) and a stable equilibrium size of k = 100.*

much useful information about stability. For example, observations of little change in population size over intervals spanning less than a complete generation, or a complete population turnover, say little about stability.

The apparent constancy of populations in many communities, such as long-lived forest trees or slowly growing reef-building corals, may suggest unchanging communities with stable equilibria, but perhaps reflect only the nearly imperceptible responses of long-lived organisms to gradually changing surroundings (Frank 1968; Connell and Sousa 1983; Davis 1986). These communities change at rates that are difficult for the average observer to recognize. For example, to a short-lived (relative to the average tree) human observer, North American temperate forests appear to consist of relatively constant associations of tree species. However, paleoecological data show that many of these associations are recent and transient consequences of the northward expansion of species' ranges following the retreat of the Pleistocene glaciers (Davis 1986; Figure 12.3). The same data show that each species seems to have recovered from the last glaciation in an individualistic fashion, suggesting that the forests in question have not expanded as tightly integrated communities with a consistent species composition. Information about the persistence of populations and communities must be collected over timescales that span many generations of organisms to infer anything about stability. Unfortunately, because of the extraordinary time and effort involved, such data seldom exist.

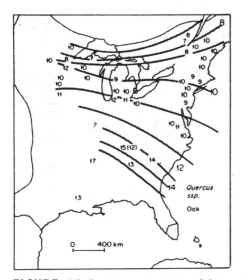
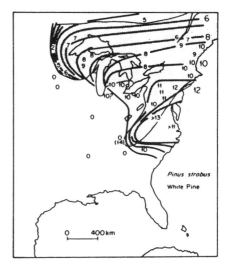

FIGURE 12.3. *Reconstructions of the northward progression of common forest trees after the last Pleistocene glaciation, showing different rates of movement by different species. Numbered lines indicate approximate northern limits of species ranges in thousands of years before the present. (Reprinted with permission from Forest Succession:* Concepts and Application, *Quaternary history and the stability of forest communities. M. B. Davis. Figure 10.8, p. 144; Figure 10.9, p. 145. © 1981 Springer-Verlag.)*

What can surveys of population dynamics tell us about the stability of populations and communities? If natural populations fall into two groups corresponding to equilibrium and nonequilibrium situations, surveys might describe a bimodal distribution of temporal variation in population size.

 Natural populations exhibit a broad range of variation in dynamics.

Equilibrium populations might vary little in size, whereas nonequilibrium populations could vary considerably. These expectations also assume that both kinds of populations would experience similar kinds of perturbations. Surveys show that organisms display a continuous range of temporal variation in population size, a pattern that offers little support for distinct classes of populations with either equilibrium or nonequilibrium dynamics (Connell and Sousa 1983). There appears to be a continuum of temporal variation in population size, ranging from relative constancy to extensive fluctuations (Figure 12.4), but comparisons are complicated by the fact that different studies measure population dynamics over different temporal and spatial scales, which can influence various statistics used to quantify temporal variation (McArdle et al. 1990). Unless these factors are also controlled when

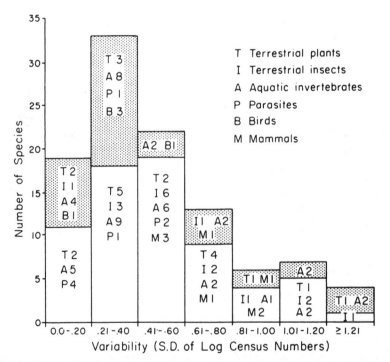

FIGURE 12.4. *The frequency distribution of one measure of variation in abundance over time (standard deviation of the log of abundance over time) for the species reviewed by Connell and Sousa (1983). (Reprinted from Connell and Sousa, 1983, with permission of the University of Chicago Press.)*

comparing measures of temporal variability among different species or different locations, observed differences may not indicate differences in population dynamics. This limitation suggests that it may be difficult to conclude much about the commonness of stable equilibrium situations from Connell and Sousa's (1983) literature survey of population dynamics.

Of course, simple models also exhibit a range of behaviors that depart from strictly defined stable equilibria, but which nonetheless correspond to species persistence for prolonged periods of time. Stable limit cycles, and the situations that produce them, provide one example of oscillatory dynamics that are not strictly stable but which could correspond to the fluctuating dynamics of organisms in nature. In models with stable limit cycles, any perturbation away from an equilibrium causes population sizes to cycle with a period and amplitude determined by the parameters of the model.

Other kinds of nonequilibrium behavior exist, but those behaviors are difficult to document in natural systems with any certainty. One example is chaos, which can arise in populations with time lags, or in systems of three or more interacting species. Hastings and Powell (1991) show that models of three species arranged in simple linear food chains can exhibit chaos under biologically realistic circumstances (Figure 12.5). McCann and Yodzis (1994) also describe chaotic systems that persist for long periods of time and then suddenly go extinct. Extinctions happen when chaotically fluctuating populations move too close to an alternate equilibrium point, which corresponds to the loss of one or more species from the system. These alternate outcomes provide a theoretical example of phenomena that have been termed **multiple basins of attraction**, or **alternate stable equilibria**, within a range of dynamic behavior (Figure 12.6). Of course, both of these examples are purely theoretical. In fact, so far there is only one convincing experimental demonstration of chaotic behavior in a population of real organisms (Costantino et al. 1997), and that is in a highly stage-structured insect population under carefully contrived laboratory conditions. Other observational studies have found patterns that are consistent with chaotic population dynamics (Ellner and Turchin 1995), but in practice it can be very difficult to distinguish between chaotic and random population fluctuations. Other populations appear to fluctuate randomly within upper and lower bounds. Bounded stochastic fluctuations, which often appear to depend only weakly on population density, also occur in some natural populations (Chesson 1978; Wiens 1983).

EXPERIMENTAL STUDIES OF COMMUNITY DYNAMICS

Relatively few studies address whether collections of multiple species exhibit stable compositions that might correspond to stable equilibria in

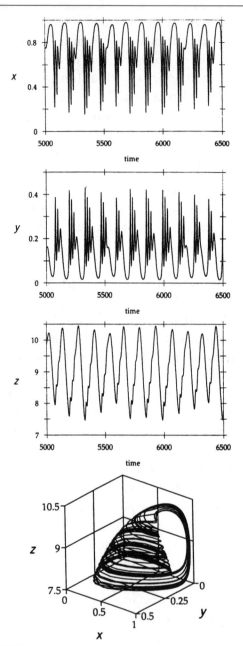

FIGURE 12.5. *An example of chaotic population dynamics generated by interactions among three hypothetical species (x, y, and z) in a three-level linear food chain. (Reprinted from Hastings and Powell, 1991, with permission of the Ecological Society of America.)*

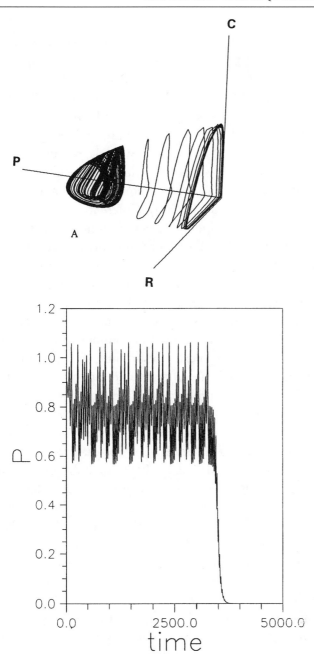

FIGURE 12.6. *An example of alternate outcomes in a simple model system in which initially chaotic fluctuations lead to extinctions after populations come too close to an alternate equilibrium point. (Reprinted from McCann and Yodzis, 1994, with permission of the University of Chicago Press.)*

mathematical models. Most studies of community stability suffer from short-comings imposed by the limited duration of observations relative to the life spans of important organisms, and the conclusions that can be drawn are correspondingly limited (Sutherland 1981; Connell and Sousa 1983).

John Sutherland (1974, 1981) studied assemblages of sessile marine organisms (sometimes called fouling organisms) on hard substrates to answer basic questions about temporal patterns of community development and composition. Sutherland found it useful

 Perturbations of different intensity may produce different community responses.

to distinguish among three different kinds of perturbations that might produce qualitatively different community-level responses. Some minor perturbations occur often but have no detectable effect on community composition. The ability of a community to buffer the effects of such perturbations is sometimes called **resistance**. Other, stronger perturbations create transient changes in the abundance of one or more species, but the community eventually recovers and returns to its predisturbance composition. The strongest perturbations create potentially permanent changes in community composition, which persist until an equally strong perturbation shifts the composition to another state. Sutherland likened the different community compositions before and after these strongest perturbations to alternate stable states.

Lewontin (1969) had suggested that alternate or multiple stable states might exist for some ecological communities, but unequivocal examples of alternate states in natural systems remain elusive. For example, Barkai and McQuaid (1988) describe one possible

Do communities exhibit multiple stable states?

example of alternate stable states in a marine system, where initial establishment of high densities of either predatory whelks or rock lobsters precludes the subsequent establishment of the other dominant predator species. Transplant experiments show that lobsters are rapidly overwhelmed by high densities of whelks, whereas lobsters can consume whelks under some circumstances.

Evidence for alternate community states in Sutherland's system came from observations of communities of fouling organisms that developed on ceramic tiles submerged below the low-tide line. Very different communities sometimes developed on the tiles, despite close physical proximity and exposure to a common species pool of potential colonists. Within an annual cycle of community development, tiles dominated by the tunicate *Styela* or the hydroid *Hydractinia* tended to resist invasion by other species of settling larvae. Sutherland suggested that these two different communities constituted alternate states that appeared relatively stable within an annual sequence of community development. Over longer periods of time the stability of these patterns seems questionable. At the end of each annual cycle of settlement and growth, organisms die and slough off of the tiles, opening up space and creating new sites where organisms can settle. Over periods of several years, *Styela* tends to recruit more consistently than other species and therefore eventually predominates, even though it cannot directly displace *Hydractinia* within an annual cycle of recruitment.

Connell and Sousa (1983) disagreed with Sutherland's interpretation of these differences in fouling community composition as alternate stable states. First, they suggested that differences in composition did not persist long enough to be considered really stable. Since the organisms that Sutherland studied have approximately annual life cycles, differences in species composition would have to persist for multiple years to be considered stable. As Sutherland noted, the initial differences among communities with a predominance of *Styela* or *Hydractinia* eventually disappeared as *Styela* dominated more sites after successive years of colonization.

Connell and Sousa (1983) also argued that other examples of alternate stable community states were unconvincing, either because the studies were too short relative to organism generation times or because differences in species composition between two sites were confounded with other factors, such as differences in the physical characteristics of the sites. For example, alternate stable states should not be the result of different environmental conditions acting on a species pool, but should instead correspond to different possible outcomes of species interactions in comparable environments.

The requirement for similar physical habitats in examples of alternate stable states may be unrealistically stringent in that it excludes situations in which organisms act as ecosystem engineers, modifying their physical habitats in important ways that in turn promote the development of different sets of species (Peterson 1984; Jones et al. 1994, 1997). Peterson (1984) used another example of marine community patterns to make this point. Two infaunal organisms, the burrowing ghost shrimp, *Callianasa*, and the bivalve *Sanguinolaria*, tend to occur in mutually exclusive patches in sandy substrates along portions of the California coast. The ghost shrimp extensively modifies the substrate by excavating burrows where it lives. Burrowing by shrimp produces substrates with more coarse particles than in areas without shrimp. Substrates with finer particles predominate in areas with *Sanguinolaria*. Neither *Callianasa* nor *Sanguinolaria* seems to invade patches where the other species is abundant, which suggests that the two types of habitat patches are stable. However, removal of ghost shrimp by disrupting their burrows leads to a gradual reversion to a finer substrate, and that reversion is further enhanced by the successful settlement of *Sanguinolaria*. This result led Peterson to suggest that the alternate states characterized by *Callianasa* and *Sanguinolaria* could not be separated from physical changes in the habitat wrought by each species.

Examples of Stable Community Patterns

A few long-term studies of community patterns suggest that some long-lasting community patterns are relatively stable. Lawton and Gaston (1989) studied an assemblage of 18 to 20 herbivorous insects living on bracken fern over a period of seven years at two sites. Since the insects have annual life cycles, observations of community composition over seven years would correspond to at least seven complete turnovers, or replacements, of each insect population. Over time, both the taxonomic composition of the arthropod assemblage and the relative abundances of different species remained similar. Rare species remained rare, common species remained common, and the seasonal phenologies of species also remained similar over time. These constant patterns suggest that the insects feeding on bracken remained highly predictable despite the natural perturbations that almost certainly affected the community.

Other studies of relatively constant long-term patterns of community composition in short-lived organisms include patterns of odonate community composition over 10 years in a temperate North American lake (Crowley and Johnson 1992), zooplankton abundance in the central Pacific Ocean (Fager and McGowan 1963; McGowan and Walker 1979), and the composition of vegetation in managed grasslands in England (Silvertown 1987). All these studies describe community patterns that persist much longer than the generation times of the dominant organisms. However, community responses to perturbations of known magnitude typically remain unmeasured, although the Rothamsted park grass experiments (Silvertown 1987) have been subjected to regular episodes of mowing and nutrient amendments for over 100 years. Consequently, it is unclear whether such long-term examples of constant community composition are examples of stable equilibria or examples of constancy in the absence of significant perturbations.

EQUILIBRIUM EXPLANATIONS FOR DIVERSITY

Equilibrium explanations for species diversity assume that coexisting species occur in stable equilibrium configurations (Connell 1978) and emphasize special circumstances that enhance the number of species that can stably coexist. The majority of proposed equilibrium mechanisms operate by preventing any species from obtaining a competitive monopoly. Proposed equilibrium mechanisms can operate via 1) increased specialization of resource use, sometimes called niche diversification, 2) intransitive networks of competitive interactions that involve different competitive mechanisms, and 3)

 Equilibrium-based explanations for diversity.

stabilization of otherwise unstable competitive interactions by predation or other sources of mortality that fall with differential severity on competitively dominant species.

Niche Diversification

Niche diversification (or, equivalently, resource specialization) can allow more species to coexist along a particular resource spectrum by packing more but narrower niches into a given range of resources (Figure 12.7). The extent to which this actually happens is controversial (Pianka 1966; Connell 1978), especially for organisms such as tropical trees and corals that seem to be relatively unspecialized with respect to resource use. For other organisms, such as herbivores specialized for feeding on particular species of plants, niche diversification seems at least plausible, although it begs the question of what

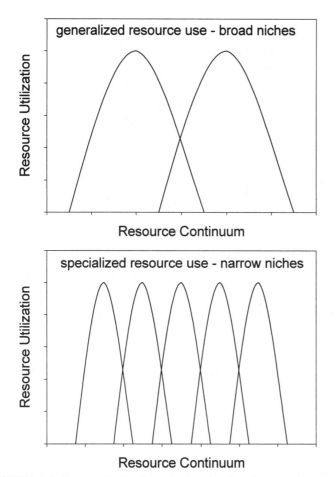

FIGURE 12.7. *An idealized graph showing how increased resource specialization (niche compression) might permit a greater number of species to pack into a resource spectrum.*

causes the initial diversity of resource species (MacArthur 1972; Diamond 1975).

Intransitive Competitive Networks

Intransitive competitive networks occur in situations in which three competing species interact so that A competitively excludes B, B competitively excludes C, and C excludes A. Such situations can arise when the mechanism of competition between A and B and B and C differs from the mechanism of competition between C and A. Examples of this sort of interaction occur in the encrusting sessile organisms found in some marine communities (Jackson and Buss 1975). In general, intransitive competitive interactions seem infrequent, and there is little evidence for such interactions beyond the communities of encrusting reef-dwelling organisms where they were originally described.

Compensatory Mortality

Compensatory mortality occurs in situations in which a competitively dominant species suffers disproportionately greater mortality than the competitively inferior species that it would otherwise tend to exclude. The mortality can come from either biological or physical sources. The enhancement of prey diversity by *Pisaster* in rocky intertidal systems (Paine 1966, 1974) and the similar effects of grazing by *Littorina* on tide pool algae (Lubchenco 1978) occur when consumers selectively prey on competitively dominant species. Connell (1978) points out other examples in which mortality from physical factors can fall with greater severity on superior competitors. For instance, competitively dominant corals suffer greater mortality from storm-related wave damage than competitively inferior species. In this particular case, the competitively superior corals have upright branching growth forms that make some corals superior over competitors but also predisposes them to wave damage. The common theme in all these examples is that compensatory mortality promotes diversity by enhancing local species coexistence. Predators can also act in a rather different, nonequilibrium fashion described previously (Caswell 1978) and reviewed below.

NONEQUILIBRIUM EXPLANATIONS FOR DIVERSITY

Although communities are often assumed to be equilibrium assemblages of coexisting species, the preceding discussion suggests that there is really very little compelling evidence that communities ever reach an equilibrium. There are, however, some interesting speculations about mechanisms that might maintain

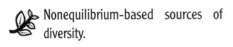 Nonequilibrium-based sources of diversity.

diversity in systems that spend much of their time far away from equilibrium.

Gradually Changing Environments and the Paradox of the Plankton

G. E. Hutchinson (1961) proposed one of the earliest ideas about how non-equilibrium mechanisms might promote high diversity in the planktonic algae of many freshwater lakes. Hutchinson suggested that the apparent coexistence of 30 to 40 species of planktonic algae in temperate lakes was at odds with the basic predictions of equilibrium-based competition theory (see Figure 9.2A). Most algae compete for the same array of resources, including carbon dioxide, nitrogen, phosphorus, sulfur, trace elements, and vitamins. Given such similar resource needs, how can a large number of algal species manage to coexist in a structurally simple environment? Competition theory suggests that the best competitor should exclude weaker competitors. The apparent failure of the competitive exclusion principle is the so-called paradox of the plankton.

> Interactions in gradually changing environments.

Hutchinson suggested that the impact of gradually changing physical conditions within lakes on competition among algae might provide a resolution of the paradox. Environmental conditions may change sufficiently over time so that no single species remains competitively superior long enough to exclude other species. This idea can be formalized by defining two terms, t_c, the length of time required for one species to competitively exclude another, and t_e, the length of time required for the environment to change sufficiently so that the outcome of competition between two species will reverse. The relative durations of t_c and t_e then define three different situations in which competitive exclusion is or is not expected.

1. When t_c is much less than t_e, competitive exclusion occurs rapidly and the system reaches a competitive equilibrium long before the environment changes sufficiently to alter the outcome of competition. This kind of situation occurs in laboratory studies of organisms with short generation times, where environmental conditions are held relatively constant.

2. When t_c is approximately equal to t_e, no competitive equilibrium is ever reached, because the environmental conditions affecting the outcome of competition change on about the same timescale as is required for competitive exclusion to occur. This situation might apply to organisms with relatively short generation times, ranging from algae to various invertebrates, in systems where environmental changes are relatively frequent.

3. When t_c is much greater than t_e, competitive exclusion again becomes possible, as large, long-lived organisms simply integrate over short-term

environmental fluctuations. This situation should hold for organisms that live for at least several years, such as birds, mammals, and perennial plants.

Hutchinson suggested that differences in the relative duration of t_c and t_e could account for apparent differences in the importance of interspecific competition in different systems. For example, laboratory studies of short-lived organisms or field studies of long-lived organisms should provide situations in which competition might lead to exclusion, whereas studies of short-lived organisms in rapidly changing seasonal environments might not.

Recent theoretical work suggests that fluctuating environmental conditions alone are not sufficient to maintain diversity within a group of ecologically identical organisms (Chesson and Huntly 1997). Chesson and Huntly suggest that species must differ in their responses to environmental change for fluctuating environments to maintain or enhance species diversity. Although this idea seems implicit in Hutchinson's original scenario, it emphasizes that species must differ in some way for diversity to be maintained by fluctuating environments.

The Storage Effect

Other situations similar to the ones described by Hutchinson can also promote the nonequilibrium maintenance of species diversity (Warner and Chesson 1985; Chesson 1990). These conditions can result in a phenomenon called the **storage effect**, since variable conditions that are only occasionally favorable for reproduction by long-lived organisms are effectively "stored" as long-lived adults or dormant stages until conditions again become favorable. For the storage effect to enhance species diversity, two conditions must hold. First, the environment must vary in such a way that each species encounters favorable and unfavorable conditions for reproduction, and conditions favorable for one species must be unfavorable for others. Second, the organisms must be long lived, either as adults or as dormant propagules, so that they can endure unfavorable conditions. The storage effect also works best if established adults do not compete strongly, since strong competition might drive reproductively mature organisms extinct before favorable environmental conditions recur. The storage effect might explain the coexistence of high diversities of organisms such as annual desert plants (with dormant seeds), zooplankton (with diapausing eggs), and many long-lived organisms that average over long periods of environmental fluctuations.

 The storage effect.

Lottery Models

The storage effect builds on a group of lottery models developed to explain the maintenance of diversity in other systems, such as coral reef fish (Sale 1977; Chesson and Warner 1981). Briefly, lottery models account for diversity by assuming that openings (such as empty territories for fish) are filled at random by recruits from a large pool of potential colonists. The species composition of this pool should not be tightly linked to the composition of the local community; otherwise, the most abundant local species will have an advantage in filling any open sites. The conditions fostering coexistence in lottery models are similar to those promoting the storage effect. Species should have low adult mortality, high fecundity, and environmentally dependent recruitment rates, so that they can persist until opportunities for recruitment occur. Lottery models work best where different species have similar average reproductive rates and similar competitive abilities, so that no one species gains the upper hand. As in the storage effect, the species should respond differently to environmental variation. The pattern of environmental variation should have little effect on adults, but exert strong stochastic mortality on juveniles. It is easy to imagine the kinds of high-diversity communities in which lottery models might operate. These include trees in tropical forests, corals, coral reef fish, and amphibians and other organisms that exploit temporary pools.

Nonequilibrium Predator-Mediated Coexistence

The spatially subdivided predator-prey interaction modeled by Caswell (1978) and described in Chapter 11 provides another example of a nonequilibrium mechanism that can maintain diversity. Since this example has already been described in detail, here we will only emphasize how it differs from the corresponding equilibrium models of predator-mediated coexistence. In the nonequilibrium Caswell model, predators and prey do not coexist locally for any length of time, and neither do competing prey species, within any conveniently defined habitat unit, such as a predator's home range. That failure to coexist defines the nonequilibrium nature of the model and the mechanisms producing diversity. In contrast, equilibrium models of predator-mediated coexistence (Roughgarden and Feldman 1975; Cramer and May 1972; Comins and Hassell 1976) focus on conditions in which predators enhance the ability of competing prey to coexist locally. Interestingly, both equilibrium and nonequilibrium models have been invoked to explain the same empirical examples of predator-mediated coexistence, such as the impact of *Pisaster* on prey diversity in the rocky intertidal zone. Clearly, decisions about which model best explains the phenomenon depend on whether prey and predators coexist locally. In most cases, we lack the necessary information to make that decision.

The Intermediate Disturbance Hypothesis

Other nonequilibrium hypotheses draw on Hutchinson's original ideas about the importance of disturbance in maintaining diversity. The **intermediate disturbance hypothesis** (Connell 1978) focuses on the fact that both the frequency and intensity of various kinds of abiotic disturbances would affect patterns of diversity. The basic idea is that disturbance maintains diversity by preventing competitively dominant species from excluding others. However, diversity is thought to be maximized by disturbances of intermediate frequency and intensity (Figure 12.8). Weak or infrequent disturbances are not sufficient to alter the progress of competitive exclusion, and diversity consequently declines. Intense or frequent disturbances so disrupt the community that species are actively excluded, leading to reduced diversity through the loss of species that are particularly sensitive to disturbance.

The intermediate disturbance hypothesis.

The intermediate disturbance hypothesis has been invoked to explain high diversities of coexisting coral species and tropical forest trees (Connell 1978). One of the more convincing experimental demonstrations of the role of intermediate disturbance focused on patterns of algal diversity on intertidal boulders in California (Sousa 1979b). The main disturbance in this system occurs when waves overturn algae-covered boulders, exposing new substrate for colonization and smothering and killing algae on the underside of the rocks. The frequency of disturbance depends on boulder size, since small boulders are frequently rolled by waves whereas larger boulders are overturned only in exceptional storms. Sousa described the diversity of algae found on boulders of various size and found maximal numbers of species on substrates that presumably experienced intermediate amounts of disturbance (Figure 12.9). Of course, the boulders also differed with respect to many other factors, perhaps most importantly in size, so it was difficult to attribute the pattern entirely to differences in disturbance frequency.

Sousa directly manipulated disturbance frequency by immobilizing sets of small, similarly sized boulders, and compared the community that developed on these boulders with the algae on other rocks of similar size that were free to roll in the surf. The results showed that frequent disturbance created low-diversity assemblages dominated by early-successional algal species, particularly the green alga *Ulva*, whereas undisturbed substrates acquired a more diverse algal assemblage (see Figure 12.9). Other work (Sousa 1979a) in the same system shows that a much longer period of study would be required to

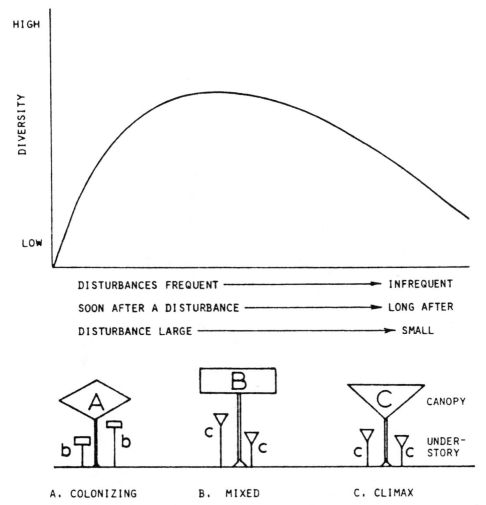

FIGURE 12.8. *The intermediate disturbance hypothesis (Connell 1978). Disturbances of intermediate frequency and intensity promote diversity by preventing competitive exclusion by potentially dominant species. (Reprinted with permission from Connell, 1978. © 1978 American Association for the Advancement of Science.)*

see the eventual domination of undisturbed artificial boulders by *Gigartina*, which is the dominant late-successional species at these sites.

STABILITY AND COMPLEXITY

Quite separate from the discussion of factors influencing diversity is the prolonged controversy over possible relations between species diversity and various aspects of population, community, and ecosystem stability. In this context, diversity and complexity are often used synonymously, but, as previously discussed, stability tends to mean rather different things. In some cases,

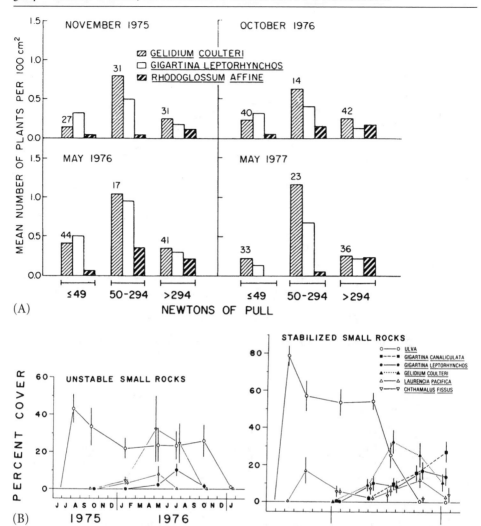

FIGURE 12.9. *(A) Relations between boulder size and algal species composition in the rocky intertidal system studied by Sousa (1979b). Boulder size is inversely related to disturbance frequency. (B) Effects of the presence or absence of disturbance on boulders of similar size. (Reprinted from Sousa, 1979b, with permission of the Ecological Society of America.)*

> Are complex systems more or less stable than simple systems?

stability refers specifically to the dynamics of a single population, whereas in other cases it describes the tendency of multiple populations (the entire community of interest) to remain unchanged in the face of various perturbations. These rather different meanings of stability have in turn caused a fair bit of confusion, since the stability of single-species populations and that of community composition are quite different things, and the existence of one need not imply the other. Other studies focus on the apparent stability of various ecosystem processes and collective attributes, such as biomass (e.g., Hurd et al. 1971; McNaughton 1977; Tilman and Downing 1994; Tilman 1996). The stability of ecosystem properties may not be tightly linked to the population dynamics of individual species, which raises the possibility that a system may have rather unstable populations while exhibiting quite stable ecosystem properties (King and Pimm 1983; Tilman 1996).

Robert MacArthur (1955) and Charles Elton (1958) suggested several reasons why more-complex communities might be more stable than simple ones. Here stability means a range of things, but can refer to both the tendency for populations to persist while showing low levels of temporal variation and to the tendency for community composition to remain unchanged. MacArthur's essentially graphical argument (Figure 12.10) considered the stability of a population of top predators feeding on single or multiple prey species. The obvious advantage gained by feeding on more than one prey is that alternate food sources are available if one prey suffers a population crash. MacArthur's argument ignores whether the entire system of interacting populations is more or less likely to be stable as the complexity of feeding links increases. Other work by Robert May (1972), described previously in Chapter 6, suggested that increasing complexity should decrease the stability of the entire system of interacting populations in randomly connected model food webs.

Elton's argument was based on several different lines of evidence, ranging from the behavior of simple mathematical models to observations of natural history. The ideas focus primarily on the apparent instability of low-diversity systems, which by extension implies a greater stability of more complex systems. For example, Elton noted that both simple models of two-species predator-prey interactions and laboratory systems containing two species

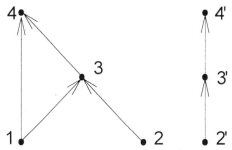

FIGURE 12.10. *Possible effects of food web complexity on the stability of a top predator. A predator with access to more alternate pathways of energy flow (species 4) should be less dependent on fluctuations in any single prey species and better buffered against the consequences of prey extinction than a predator that relies on a single source of energy (species 4'). (After MacArthur, 1955.)*

tend to oscillate. Inclusion of cover in the latter, an aspect of complexity, tends to reduce the observed population fluctuations. Elton also suggested that the relatively simple communities found on small islands tended to be more vulnerable to invasions by new species than the more complex communities found in mainland situations, although this seems to be more of an assertion than a pattern supported by data. There is no shortage of examples of invasions by exotic species in mainland communities. Elton also noted the tendency for simple agricultural communities to experience invasions by weedy species and outbreaks of insect pests, both indicators of reduced stability. In contrast, more diverse tropical forests seem much less prone to insect outbreaks than do less diverse temperate forests, which are also prone to invasions by exotic pests (e.g., gypsy moths, *Lymantria dispar*, in North America). Finally, Elton noted that artificial reductions in the diversity of already simple systems, such as occur following the control of insect pests in orchards by chemical pesticides, frequently result in insect outbreaks. These outbreaks probably result because pesticides eliminate the natural enemies of the pests. This collective argument was widely accepted by ecologists and became the conventional wisdom until it was challenged by more recent empirical and theoretical work.

Hairston et al. (1968) provided one of the first experimental tests of relations between diversity and stability, using an array of simple communities assembled from bacteria, bacterivorous protists (*Paramecium*), and their predators (*Didinium* and *Woodruffia*). Hairston et al. manipulated diversity in each of three trophic levels, specifically examining 1) how enhanced bacterial diversity (one to three species) affected the stability of bacterivores, 2) how increasing bacterivore diversity (one to three species) affected bacterivore stability, and 3) how increasing diversity of predators (zero to two species) affected the entire system. Results based on both the frequency of population extinctions and the dynamics of persisting populations provided inconsistent support for positive effects of diversity on stability. For example, increasing bacterial diversity did tend to enhance the stability of species that consumed bacteria, whereas increasing either the diversity of bacterivores or their predators tended to increase extinctions. Apparently, the top predators used in this experiment failed to act in a keystone fashion and did not promote prey coexistence. This is not surprising for *Didinium*, which is well known for its ability to overexploit *Paramecium* under some circumstances (Gause 1934), but other work indicates that *Woodruffia* (Salt 1967) should coexist with prey for prolonged periods of time.

 Laboratory studies suggest that populations may be less stable in more complex systems.

Sharon Lawler's (1993b) study of the effects of increasing complexity on stability, described previously in Chapter 6, also found that extinctions became more frequent as diversity increased. In her study, this happened even though the predator-prey pairs used to assemble increasingly complex communities were themselves stable.

The findings of both Hairston et al. (1968) and Lawler (1993b) are consistent with the predictions of simple mathematical models of randomly connected food webs studied by Robert May (1973). Those models suggest that increases in either species richness, connectance, or the average strength of interspecific interactions will tend to decrease the stability of an entire system of equations. This makes it likely that some species will drop out of the system, but it says nothing about the stability of the remaining species. Other theoretical work by King and Pimm

 However, aggregate properties of communities and ecosystems may be more stable in more complex settings.

(1983) shows that increasing complexity in model food webs can decrease the stability of individual populations within the web while increasing the stability of aggregate properties of the entire assemblage, such as biomass, by providing opportunities for compensation among competing species. This result points to the need to identify whether the property whose stability is of interest resides at the population, community, or ecosystem level of ecological organization. Clearly, the stability of population dynamics and community or ecosystem attributes need not be tightly linked.

EFFECTS OF BIODIVERSITY ON ECOSYSTEM FUNCTION

Recent studies have renewed interest in questions concerning the effect of community complexity on various measures of ecosystem function. The complexity measure used is typically species richness, and its effects on a range of community and ecosystem

 Diversity is related to some measure of ecosystem function in grasslands.

processes have been explored. David Tilman and John Downing (1994) noticed an interesting effect of plant species richness on how grassland assemblages responded to a drought of unusual severity. A previous series of nutrient manipulations (additions of nitrogen) produced a gradient of plant species richness running from 1 to 25 species. Tilman and Downing noticed that plots containing more species appeared more resistant to the perturbation imposed by the drought, in the sense that total plant biomass decreased less in plots with many species (Figure 12.11). After the drought ended, plots with more species also seemed more resilient, since plant biomass tended to recover more rapidly in species-rich plots. Relations between plant species richness and drought resistance or recovery were curvilinear, displaying a rapid increase with initial increases in species

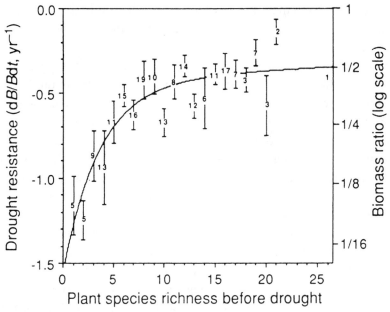

FIGURE 12.11. *Relations between plant species richness before a natural disturbance (drought) and plant biomass after the disturbance. Communities with more species display greater resistance to disturbance. (Reprinted with permission from* Nature *367: 363–365, D. Tilman and J. A. Downing. © 1994 Macmillan Magazines Limited.)*

richness and then leveling off when plots contained more than 10 to 12 species. This kind of curvilinear ecosystem response to increasing species richness would be expected if species tended to be somewhat functionally redundant (Lawton and Brown 1993), in the sense that different species may play functionally similar roles in the community. Where some functional redundancy exists, a few species can be lost without noticeable degradation in ecosystem performance, since other species with similar functional roles remain. However, with greater species losses it becomes increasingly likely that all members of a functional group will be missing, with levels of the corresponding function suffering accordingly.

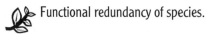 Functional redundancy of species.

Tilman and Downing's conclusions met with some skepticism (Givnish 1994; Huston 1997), in part because their species richness gradient may have been confounded with other species properties; for example, perhaps species that tended to occur in low-richness plots were also more susceptible to drought. To rigorously test the effects of species richness on ecosystem properties, Tilman et al. (1996) created a gradient of plant species richness by randomly selecting sets of species from a pool of over 20 species. This study confirmed the nonlinear effects of species richness on several ecosystem properties, including the accumulation of biomass and nutrient retention. Consequently, whether the gradient in plant species richness occurred as an accidental consequence of other manipulations or by design, the effects on ecosystem properties seem comparable.

Simple models indicate how increasing diversity might lead to increases in biomass and greater reductions in free nutrients in terrestrial systems (Tilman et al. 1997). The simplest of these models assumes that species differ in R^*, the equilibrium concentration of a limiting resource. Within a species pool, the values of R^* will range from R^*_{min} to R^*_{max}, and the species with R^*_{min} should yield the greatest biomass since a greater amount of resource can be assimilated and converted into biomass. This relationship can be formalized by approximating plant biomass as

$$B = a\,Q(S - R^*_i)$$

where B is biomass, S is a resource supply rate, R^*_i is the value for species i, Q is a coefficient that converts the resource used into biomass, and a is a rate of resource mineralization (loss by the organism). For a set of N species, Tilman et al. show that the expected value of plant biomass is

$$B_{(N)} = a\,Q\left\{S - \left[R^*_{min} + \frac{(R^*_{max} - R^*_{min})}{N+1}\right]\right\}$$

This relation reproduces the increase in biomass and decline in free resource levels seen in Tilman et al. (1996; Figure 12.12). More-complex models considering competition among species for more than one resource produce similar patterns. The operating principle that generates this pattern is essentially a statistical sampling effect. As species richness increases, it becomes more likely that the community will contain a species with a particular functional attribute, such as maximal productivity.

 The sampling effect.

Other studies have manipulated diversity in multiple trophic levels in terrestrial and aquatic communities and found similar effects of species richness on a range of processes. Naeem et al. (1994, 1995) created three levels of species richness within multilevel terrestrial food webs assembled in controlled laboratory growth chambers. They found that carbon dioxide uptake and two measures of plant productivity increased with increasing diversity, a result similar to that seen in the study by Tilman et al. (1996). Treatments affected other processes, including decomposition rates and nutrient retention, but not in a way that suggested simple increases or decreases in these processes with diversity. The findings suggest that species loss could have deleterious effects on various ecosystem services, including the ability of

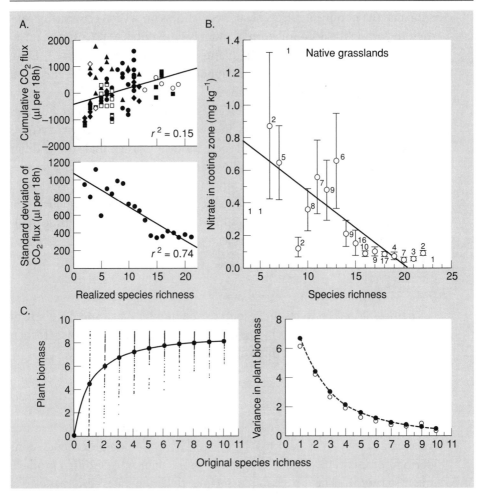

FIGURE 12.12. *Examples of relations between species richness and patterns of ecosystem properties. (A) Carbon dioxide flux in aquatic microcosms. (Reprinted with permission from* Nature *390: 162–165, J. McGrady-Steed et al. © 1997 Macmillan Magazines Limited.) (B) Plant biomass in reconstructed grassland vegetation. (Adapted with permission from* Nature *379: 718–720. D. Tilman et al. © 1996 Macmillan Magazines Limited.) (C) Predicted pattern of variation caused by the sampling effect in model communities. (Adapted from Tilman et al., 1997. © 1997 National Academy of Sciences, U.S.A.)*

terrestrial communities to absorb carbon dioxide, which has been implicated as a possible agent in global climate change.

McGrady-Steed et al. (1997) created aquatic communities from assemblages of bacteria, protists, and small metazoans to test whether the effects of species richness noted in terrestrial communities held for aquatic systems. Like Naeem et al., they manipulated diversity across all levels in an entire food web, rather than varying the species richness of only the primary producers.

Unlike the situation seen in terrestrial systems, carbon dioxide uptake, a measure of productivity, decreased with increasing diversity, to the extent that diverse assemblages were net producers of carbon dioxide. Another ecosystem function, decomposition, increased nonlinearly with species richness, suggesting some degree of functional redundancy for this ecosystem process. Invasion resistance also increased with increasing diversity, but the effect could be accounted for by variation in the abundance of particular species rather than by species richness itself. Similar effects of species richness on invasion resistance have been described by Tilman (1997) for terrestrial vegetation.

One feature noted in all these studies is that variation in several processes, either among replicate communities or within replicates over time, decreases as diversity increases (McGrady-Steed et al. 1997). Naeem and Li (1997) describe a similar pattern in aquatic microcosms, in which variability in abundances of bacteria and primary producers decreases with increases in the number of species per functional group. This pattern

> The predictability or reliability of community processes may increase with diversity.

means that in some respects, the functional properties of ecosystems become more predictable as diversity increases. One source of this decreasing variability may be the sampling effect noted by Tilman et al. (1997), since it produces a similar pattern in model communities. However, this pattern also materializes in studies in which species composition within a particular diversity level remains constant (Naeem et al. 1994, 1995; McGrady-Steed et al. 1997), which means that it cannot be driven by random draws from the species pool.

Another possibility is that the key species driving ecosystem processes may simply have more-variable population dynamics in less-diverse communities. If variation in ecosystem processes closely tracks population dynamics, that could produce the observed result. Tilman's (1996) observations of plant population dynamics suggest precisely the opposite trend. He shows that as species richness increases, population dynamics tend to become more variable, and aggregate community properties, such as total biomass, become less variable. One possible mechanism that could increase population variability while stabilizing community attributes would be compensatory increases by some species in response to declines by their competitors.

One other possibility is that the reduced variation observed in more-diverse communities is a simple consequence of the statistical properties of the sum of an increasing number of variables. Community biomass, the sum of the individual biomass of each species in a system, is a real-world example of this kind of measure. Doak et al. (1998) have modeled how the variability of sums of variables declines as the sum incorporates increasing numbers of variables. They find that a decrease in variability emerges as a simple statistical consequence of summing more and more variables in more-diverse communities.

The question of whether the increasing predictability of more-diverse communities is a consequence of interesting biological interactions, the statistical properties of sums of random variables, or some mixture of the two remains unanswered. Clever experimental tests to separate the contribution of these two mechanisms will be required to settle the debate. At this writing, the best design for those studies is far from clear.

RELATIONS BETWEEN DIVERSITY AND PRODUCTIVITY

Although manipulative studies of terrestrial communities suggest that productivity can increase with species richness in a particular location (Tilman et al. 1996; Naeem et al. 1994, 1996), surveys incorporating many locations distributed across a wide range of physical conditions suggest a more complex unimodal relation between productivity and species richness (Figure 12.13; Tilman and Pacala 1993; Leibold 1996; Leibold et al. 1997). These surveys are not entirely comparable to the experiments linking species richness and productivity in single sites. Productivity and diversity tend to be measured in different ways in different studies. An even greater complication is that the range of productivity occurring within a single experimental study is likely to be considerably smaller than the range seen in observations distributed across many geographic locations.

> Diversity and productivity show a curvilinear relation over a broad range of sites.

There are several competing explanations for unimodal relations between species richness and productivity. The simplest explanation invokes the paradox of enrichment (Rosenzweig 1971), which suggests that model systems of predators and prey will become less stable as productivity increases. If the paradox of enrichment accurately describes the situation in real communities, food chains should begin to fall apart as increased productivity destabilizes predator-prey interactions, with diversity declining accordingly.

Another explanation invokes resource-ratio competition theory (Tilman 1982) to predict a gradual change in species composition as the ratio of resource supply rates changes with increasing productivity. Obviously, if resource supply ratios do not change with increasing productivity, this explanation cannot account for observed changes in species richness.

Mathew Leibold (1996) has suggested that unimodal productivity-diversity relations could be driven by changing intensities of keystone

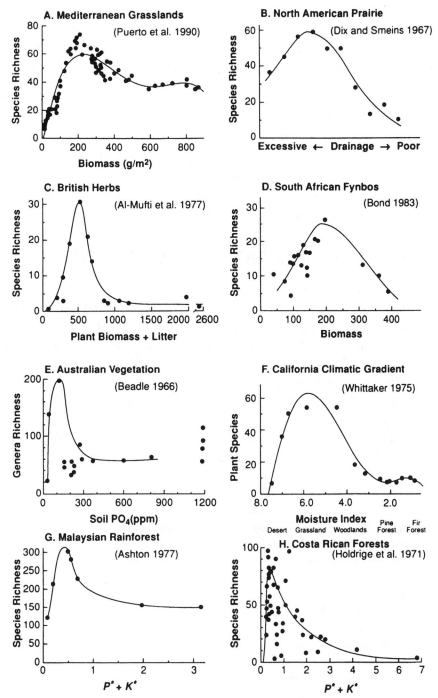

FIGURE 12.13. *Relations between productivity (or other variables known to be associated with productivity) and measures of species diversity in a variety of terrestrial ecosystems. (Reprinted from Tilman and Pacala, 1993, with permission of the University of Chicago Press.)*

predation along a productivity gradient. At low productivity, predators are too rare to mediate competition among prey, and diversity is low because the best competitors predominate. Intermediate values of productivity support higher predator densities, and prey diversity increases as competitively superior prey coexist with competitively inferior but predation-resistant species. At the highest levels of productivity, high predator densities eliminate all but the most predation-resistant species from the community, producing a decline in diversity. These alternate hypotheses make testable predictions about population dynamics, or the attributes of populations that predominate in different portions of the productivity gradient. So far, there are no definitive experimental tests of these competing explanations for nonlinear relations between productivity and diversity.

LATITUDINAL GRADIENTS IN SPECIES DIVERSITY

One of the most striking species diversity patterns remains essentially unexplained. As one goes from the poles toward the equator, the species richness of most groups of organisms increases dramatically (Pianka 1966, 1988; Currie 1991; Currie and Paquin 1987). One example of this pattern is shown in Figure 12.14, which describes the pattern of latitudinal variation in woody plants in North America. There are at least 10 hypotheses that have been suggested to explain this pattern (Table 12-1). The hypotheses are neither mutually exclusive nor necessarily testable by experiments. Rather than simply perpetuate this list of correlated and potentially interacting factors, it makes sense to discuss related ideas together.

TABLE 12-1.   Hypotheses proposed to explain latitudinal gradients in species diversity.

1. Evolutionary time
2. Ecological time
3. Climatic stability
4. Climatic predictability
5. Spatial heterogeneity
6. Productivity
7. Stability of primary production
8. Competition
9. Disturbance
10. Predation

Source: Pianka (1988).

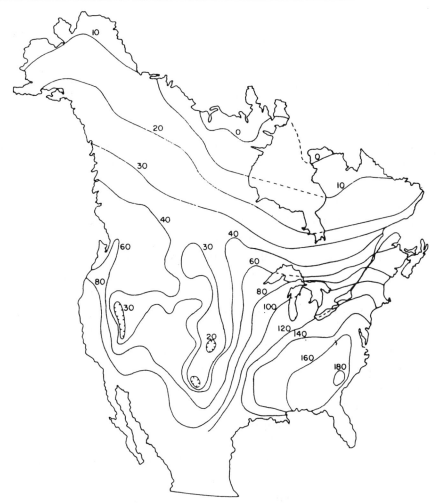

FIGURE 12.14. *Patterns of geographic variation in the species richness of woody plants in North America. The latitudinal gradient is most pronounced along the Atlantic Coast of North America. (Reprinted with permission from* Nature *329: 326–327, D. J. Currie and V. Paquin. © 1987 Macmillan Magazines Limited.)*

Climate and Its Effects on Productivity

One suggestion is that the tropics support more species than other regions because they are more productive, in the sense that more carbon is fixed there annually per unit area. This idea requires species richness to increase with primary production. Of course, the productivity-diversity relations described previously indicate that on smaller geographic scales, maximal species richness does not always occur in the most productive environments. And although lush tropical forests may appear highly productive, in fact the large standing crop of biomass in some tropical communities results from the slow accumulation of biomass in relatively unproductive habitats. Similarly, other spectacularly diverse tropic systems, such as coral reef communities, are relatively unproductive.

Other analyses do make a convincing case for relations between primary productivity and diversity of particular groups of organisms. Currie and Paquin (1987) found that geographic variation in the species richness of woody plants in North America and Europe was tightly correlated with annual evapotranspiration, which is a surrogate for productivity. Patterns of diversity for a number of other groups of animals are not so tightly linked to productivity (Currie 1991). Although latitudinal variation in productivity appears to account for latitudinal variation in the species richness of some groups, particularly plants, it does not appear to provide a general explanation for the pattern in other groups.

The seasonal distribution or predictability of productivity may be as important as its annual amount in promoting diversity. The idea is that organisms in strongly seasonal environments must have relatively general (unspecialized) patterns of resource use if they are to successfully average over seasonal variation in resource availability that is driven by variation in productivity. The broad niches of these generalized species restrict the number of species that can be packed into a given resource gradient. In contrast, species in less variable environments can specialize on temporally reliable resources. This greater specialization allows more species to pack into a particular resource spectrum, thereby promoting diversity.

Other variations on this theme stress the greater predictability or reduced seasonality of aspects of climate, such as rainfall or temperature, that are likely to be correlated with productivity. These ideas again suggest that more-predictable environments can support a greater diversity of organisms, either because greater specialization of resource use is possible or because there are more opportunities for species to use those environments.

The Relative Ages of Tropical and Temperate Habitats and Opportunities for Speciation

This idea suggests that if tropical habitats are geologically much older than temperate or polar ones, then the greater diversity of organisms in tropical systems could simply result from speciation occurring over longer periods of time. The problem with this idea is that paleoecological evidence shows that the whole range of polar to tropical conditions have existed for long periods of time, and their location and areal extent has varied with the expansion and contraction of ice sheets during past episodes of glaciation. A related idea is that the tropics may have seen higher rates of speciation precisely because

 Are tropical habitats older than temperate ones?

past contractions of a previously widespread tropical habitat into subdivided refugia may have fostered many opportunities for allopatric speciation.

Related ideas suggest that processes operating on shorter ecological timescales may account for differences in tropical and temperate diversity. For example, current high levels of tropical diversity may reflect the fact that ecological interactions among many recently generated tropical taxa have not yet proceeded long enough for competition to reduce diversity to levels seen in temperate communities. Or, conversely, species found in the tropics and capable of living in temperate systems have not yet managed to disperse to those sites. This could explain patterns for relatively sedentary organisms with weak powers of dispersal, but seems unlikely for mobile organisms such as birds and mammals that readily migrate across continents. It also ignores the fact that some organisms are physiologically unable to cope with harsher environments.

Latitudinal Differences in Ecological Interactions

Other ideas suggest that the same kinds of ecological interactions that generate differences in diversity on local scales also operate on larger latitudinal scales. For example, a greater intensity of predation in tropical systems might lead to situations in which more prey species are able to coexist, via keystone predation. Alternatively, greater intensities of competition could select for narrower niches, which in turn allow more species to pack into an available resource gradient. There also appear to be more spectacular examples of mutualisms in tropical communities than in other settings, but the net effect of these interactions on expected patterns of species diversity has received little theoretical attention.

Disturbance

This hypothesis draws on Connell's intermediate disturbance hypothesis (Connell 1978) to suggest that greater diversity in the tropics reflects a disturbance regime of intermediate frequency and intensity compared with other latitudes. Evidence for the relative intensity of disturbance in temperate and tropical systems is anecdotal at best (Connell 1978). Some disturbances, such as tropical storms (hurricanes and typhoons), are certainly more frequent in the tropics. However, other kinds of disturbances, including fire, may be more frequent in temperate regions. The problem in applying these ideas to natural disturbance gradients comes from the difficulty in deciding where a particular community or location falls along the frequency or intensity gradient.

Spatial Heterogeneity

This idea suggests that tropical communities have more spatial heterogeneity than others. Examples of this include the greater development of canopy vegetation in tropical forests and the complex spatial matrix created by reef-building corals. This hypothesis obviously must rely on other explanations to account for the initial diversity of organisms that creates the heterogeneous spatial scaffolding on which the remaining community depends. Given that the spatial heterogeneity exists, the assumption is that it creates a greater range of niches for other tropical species to use.

It seems likely that no single explanation can account for the spectacular diversity of tropical communities, and that latitudinal patterns may reflect a synergistic interaction of several underlying mechanisms. Given the long timescale on which some of the proposed mechanisms operate, it also seems unlikely that definitive experimental tests of some hypotheses (such as greater speciation rates) are even possible. Consequently, any general theory that accounts for one of the most fundamental patterns in community ecology will almost certainly require a new synthesis of the ideas described above. Recent efforts to explain large-scale patterns in species diversity suggest some of the forms that these syntheses might take (Rosenzweig 1995; Brown 1995; Huston 1994).

CONCLUSIONS

Competing explanations for the maintenance of species diversity in natural communities make different assumptions about the dynamics of communities, specifically in whether communities are usually near an equilibrium composition, or are frequently perturbed far from equilibrium. Equilibrium and nonequilibrium theories for the maintenance of species diversity probably apply with equal validity to communities that experience different frequencies and intensities of disturbance. Recent studies suggest that more diverse communities are better buffered against disturbances, or have greater rates of key processes, such as production. Other studies suggest that more diverse communities are also less variable, or more predictable, with respect to certain processes. Explanations for differences in diversity along gradients of productivity or latitude invoke interactions among multiple factors that remain largely untested.

CHAPTER 13

Succession

OVERVIEW

Ecological succession is the process of temporal change in community composition. This chapter begins by reviewing the historical development of ideas about succession. Early studies inferred repeatable patterns of community change from comparisons of communities of different age, but stopped short of directly investigating the mechanisms responsible for species replacements. These early ideas differed in whether succession was seen as an integrative process operating at the community level or as simply the natural consequence of life history differences among species. Succession has been alternately viewed as a directional process that maximizes various ecosystem processes or as the simple outcome of various interspecific interactions that only coincidentally cause changes in ecosystem properties.

Three mechanisms—facilitation, tolerance, and inhibition—probably describe species-by-species replacements during succession. Recent conceptual models of succession integrate the importance of multiple factors, which operate hierarchically, in driving temporal patterns of community composition. Two quantitative models of succession (one descriptive, one mechanistic) show how temporal changes in community composition can be rigorously described. Finally, case studies of succession in different field settings provide insights into the relative merits of different models and point to the need for studies that directly test the mechanisms thought to drive successional patterns.

SUCCESSION

Succession is the nearly universal phenomenon of temporal change in species composition following natural or anthropogenic disturbances. Succession is a conspicuous feature of plant assemblages that develop on barren sites created by geological events (volcanism, glaciation) or other disturbances. Although the study of succession was pioneered by plant ecologists who traditionally focused on vegetation change, succession results in corresponding changes in the community composition of animals, fungi, bacteria, and protists.

 Succession is the process of community change through time.

Early studies by plant ecologists identified striking temporal changes in plant species composition (Cowles 1899; Clements 1916; Keever 1950; Bard 1952). These studies emphasized sequences of species replacements that appeared to be typical in particular locations. For example, after agricultural fields in the North Carolina piedmont are abandoned, there is a gradual transition from horseweed (*Leptilon*) to asters (*Aster*) to broomsedge (*Andropogon*) to coniferous trees such as loblolly pine (*Pinus taeda*). Eventually a forest dominated by mixed deciduous hardwood species becomes established (Keever 1950). Analogous patterns involving different species occur at other sites. In the New Jersey piedmont, some 500 kilometers north of North Carolina, abandoned fields undergo a transition from annuals such as ragweed (*Ambrosia*) and biennials such as evening primrose (*Oenothera*) to perennial herbs such as goldenrod (*Solidago*) and then to short-lived conifers (*Juniperus*). Eventually another mixed-hardwood forest dominated by oaks and hickories occurs (Bard 1952). Although the dominant species vary between these two locations, the common theme is a transition from annuals and biennials to perennial herbs and then to a coniferous tree. And although the patterns are not as frequently studied, succession in the identity of dominant plant species is accompanied by changes in animals and fungi.

Some ecologists suggest that the phrase *plant succession* should be replaced by *vegetation dynamics* to emphasize that the population dynamics of interacting organisms are ultimately responsible for successional patterns (Pickett and McDonnell 1989). Similarly, the term *community dynamics* could describe the overall process of temporal change in community composition, thereby emphasizing that succession involves the dynamics of the entire complement of species interacting in communities, not just the vegetation. The point is that temporal changes in community composition are a natural consequence of interactions among species with different life history strategies, rather than some special process operating only at the community level.

Successional patterns are conspicuous, site-specific, and influenced by many factors (Pickett and McDonnell 1989). Historically, ecologists have distinguished between **primary succession** on sites without existing vegetation and **secondary succession** on sites with

 Primary vs. secondary succession.

established vegetation. Primary succession occurs on the sterile inorganic substrates generated by volcanism or glaciation. During primary succession species arrive from other, sometimes distant, locations. The process may proceed slowly because early colonists often must transform the environment before other species can become established. For example, soil formation requires the breakdown of rocks, accumulation of dead organic material, and the gradual establishment of soil microorganisms. Secondary succession occurs after disturbances disrupt established communities without completely eliminating all life. Storms, fires, clear cuts, mining, and agricultural clearings all provide the

 Autogenic vs. allogenic succession.

kinds of disturbances that set the stage for secondary succession. Ecologists also sometimes distinguish between **autogenic succession**, which is driven by processes operating within a particular location, and **allogenic succession**, which is driven by factors outside a particular site.

By understanding succession, it is possible to predict and perhaps accelerate rates of community change after natural disturbances such as fires or storms. Optimal ecological restoration of sites disturbed by human activity clearly depends on knowledge of the factors that promote rapid or otherwise desirable successional changes. Succession provides a conceptual framework for integrating the many diverse processes that affect natural community patterns and can also inform the application of sound ecological principles when restoring degraded ecosystems.

Other kinds of temporal change in community composition are sometimes referred to as different kinds of succession, although they differ from primary and secondary succession in that they are not initiated by disturbance. **Seasonal succession** refers to a

Seasonal succession.

regular annual phenology of abundance or activity that occurs without the permanent loss or addition of species from the community. Examples could include the sequence of flowering by woodland plants (Poole and Rathcke 1979), seasonal variation in insect reproduction

(Morin 1984b), or seasonal variation in the activity and abundance of aquatic microorganisms (Finlay et al. 1997). **Cyclic succession** occurs in special circumstances in which a small number of species tends to replace each other over time. A classic example of cyclic succession involves the heather, *Calluna*, and the bracken fern, *Pteridium* (Watt 1947, 1955). *Calluna* can invade stands of *Pteridium* under some situations. However, as the *Calluna* plants age and senesce, they can in turn be replaced by *Pteridium*. The cycle may take over 25 years to complete.

 Cyclic succession.

A BRIEF HISTORY OF SUCCESSION

Henry Chandler Cowles (1899) was one of the first individuals to recognize succession as an ecological phenomenon. Cowles used differences in the vegetation occurring on sand dunes of different age to infer successional patterns along the southern coast of Lake Michigan. Moving inland from the lake margin, dunes increase in age and differ in the species composition of the dominant plants. Observations of communities of known age in different locations is a common approach, sometimes called **space for time substitution** (see Pickett 1989) or **chronoseries**, used to infer successional patterns. The approach has the advantage that community changes can be inferred without long-term observations of a single site, which might take decades to describe. However, because differences in "comparable" sites are used to infer successional patterns, sites must be similar with respect to their species pool, environmental conditions, and any other factors that might influence succession. In practice, it is often difficult to know how much of the difference in species composition among different-aged sites is caused by conditions unique to each site, such as disturbance history, the species pool of available colonists, and so on.

Indirect ways of reconstructing successional patterns.

Early controversies about the process of succession involved some of the more bitter debates in community ecology (see Kingsland 1985; McIntosh 1985). The controversy was driven largely by the very different views of Frederick Clements (1916) and Henry Gleason (1917). Clements promoted the idea that ecological communities were analogous to superorganisms, with different species interacting in mutually supportive and often altruistic ways to promote a directed pattern of community development. Clements went so far as to compare primary succession with embryonic development, and sec-

Are communities superorganisms . . .

ondary succession with wound healing in individual organisms. For example, Clements (1916) wrote about the plant community, which he called a formation, in the following way:

> The developmental study of vegetation necessarily rests upon the assumption that the unit or climax formation is an organic entity. As an organism the formation arises, grows, matures, and dies. Its response to the habitat is shown in processes or functions and in structures which are the record as well as the result of these functions. Furthermore, each climax formation is able to reproduce itself, repeating with essential fidelity the stages of its development. The life-history of a formation is a complex but definite process, comparable in its chief features with the life-history of an individual plant. (p. 16)

Clements developed a complex and cumbersome terminology to describe succession through a series of intermediate stages, called **seres**, until a stable endpoint, called the **climax**, was reached. Clements argued that different climax communities, or **formations**, were the product of different environmental regimes and that failure to reach these typical climax communities was the consequence of various disturbances, such as fire, which maintained communities in a state of disclimax. For many years, Clements' terminology for various successional seres was used by ecologists as a standard way to describe the particular communities that they studied.

The main dispute between Clements and Gleason concerned whether communities developed as tightly integrated sets of species, with some species selflessly paving the way for others, or whether communities simply developed as a consequence of the individualistic responses of each species to the particular set of physiological constraints imposed by a particular location. Gleason argued that species tended to increase or decrease independent of one another through successional time, rather than occurring in mutually beneficial associations. This individualistic view of community development sees the community as nothing more than the collection of species whose individual physiological requirements allow them to exploit a particular location. In Gleason's (1926) own words:

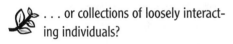 . . . or collections of loosely interacting individuals?

> In conclusion, it may be said that every species of plant is a law unto itself, the distribution of which in space depends upon its individual peculiarities of migration and environmental requirements. Its disseminules migrate everywhere, and grow wherever they find favorable conditions. The species disappears from areas where the environment is no longer endurable. It grows in company with any other species of similar environmental require-

ments, irrespective of their normal associational affiliations. The behavior of the plant offers in itself no reason at all for the segregation of definite communities. Plant associations, the most conspicuous illustration of the space relation of plants, depend solely on the coincidence of environmental selection and migration over an area of recognizable extent and usually for a time of considerable duration. A rigid definition of the scope or extent of the association is impossible, and a logical classification of associations into larger groups, or into successional series, has not yet been achieved. (p. 26)

In other words, the community is simply the sum of the species living in a particular place, and species do not cooperate to generate special community attributes.

In the short run, Clements' views were widely accepted and promoted, and Gleason left ecology. In retrospect, Gleason's ideas have proven to be a better description of the process of plant succession. The disagreement between Clements and Gleason over the nature of plant succession has resurfaced in slightly different guises during the subsequent development of community ecology. For example, Whittaker's (1956, 1975) conclusions about patterns of changing community composition along environmental gradients clearly had much in common with Gleason's individualistic view of plant assemblages. Subsequent controversies about whether communities consist of arbitrary assemblages of organisms or groups of species that are highly structured by deterministic interspecific interactions (Strong et al. 1984) echoed the contentious exchange between Clements and Gleason.

Frank Egler (1952) provided the next important conceptual development in succession to follow the prolonged debate between Clements and Gleason. Egler promoted an idea that became known as the **initial floristic composition hypothesis**. This idea specifically concerns patterns of secondary succession and holds that succession at a site is determined largely by the species composition of plant propagules already present when the site is disturbed. Subsequent changes in community composition can be attributed to the fact that some species live longer than others, but grow slowly and take time to become dominant features of the community. Short-lived species are gradually replaced by longer-lived ones to create a gradual transition in the identity of dominant species.

 Successional patterns may be strongly affected by the initial composition of species.

Eugene Odum (1969) viewed succession as an orderly (i.e., predictable) pattern of community development that produced significant changes in a variety of ecosystem attributes (Table 13-1). The orderly nature of the process was thought to result from modifications of the environment driven by the developing community. In Odum's view, succession culminated in a stabilized

TABLE 13-1.   Suggested changes in a variety of ecosystem properties during ecological succession.

Variable	Early Succession	Late Succession
GPP/Respiration	> or <1	~1
GPP/Biomass	High	Low
Biomass/Energy	Low	High
Yield = NPP	High	Low
Food chains	Linear	Web-like
Total organic matter	Small	Large
Nutrients	Extrabiotic	Intrabiotic
Species richness	Low	High
Species evenness	Low	High
Biochemical diversity	Low	High
Stratification and pattern	Poorly organized	Well organized
Niche specialization	Broad	Narrow
Size	Small	Large
Life cycles	Short, simple	Long, complex
Mineral cycles	Open	Closed
Nutrient exchange	Rapid	Slow
Role of detritus	Unimportant	Important
Selection on growth form	r-Selection	k-Selection
Selection on production	Quantity	Quality
Symbiosis	Undeveloped	Developed
Nutrient conservation	Poor	Good
Stability	Low	High
Entropy	High	Low
Information content	Low	High

GPP = gross primary production = total photosynthesis.
NPP = net primary production = total photosynthesis-respiration.
Modified and reprinted with permission from Odum, 1969. © 1969
American Association for the Advancement of Science.

ecosystem in which biomass and levels of symbioses were maximized per unit of energy flow into the system. Odum outlined a large list of ecosystem properties and attributes that might change during succession (see Table 13-1). Although many of these observations are accurate, others have dismissed the trends as little more than truisms. Many of the patterns are simple consequences of the fact that communities gradually acquire species over time, and as those species grow, biomass tends to accumulate in long-lived species (Connell and Slatyer 1977). Recent models also suggest that some properties, such as increased productivity, biomass, and nutrient cycling over successional time, are a simple and predictable consequence of competition for resources (Loreau 1998).

Drury and Nisbet (1973) presented a somewhat more complex multifactor conceptual framework that presented succession as a consequence of the differential growth, survival, and colonizing ability of species along various environmental gradients. The basic idea is that all communities fall along gradients in soil conditions, stress, and other abiotic factors. Different species have different life history characteristics that make them specialized to exploit different sets of conditions along these environmental gradients. Once established at particular sites, individual plants have a competitive advantage over seedlings and immigrants of other species that attempt to become established. This notion clearly is at odds with Clements' earlier ideas, in that early arrivals do not pave the way for later colonists. Other life history characteristics, such as stress tolerance, are thought to be correlated with high dispersal ability, which creates a situation in which the first species to arrive at stressful early-successional sites are the ones that can also best handle stress. The final trade-off is that colonizing ability and somatic growth rates are inversely correlated with longevity and size at maturity. This life history trade-off potentially explains the transition from rapidly growing, weedy early colonists to slow-growing, large, dominant species like trees.

Connell and Slatyer (1977) viewed succession from a different perspective than Drury and Nisbet (1973) and emphasized the different mechanisms that might be involved in species-by-species replacements as communities develop over time. They were concerned that the absence of direct experimental studies of succession probably led to an overemphasis of the role of competition among plant species in generating successional patterns. A second concern was that many of the correlates of succession suggested by Odum (1969) were a tautological consequence of the simple growth of organisms or other gradual community changes through time. They pointed out that any transition involving species that arrive in the community at different times could be the result of three basic kinds of interactions between species that arrive early and later in succession:

 Mechanisms of species-by-species replacements.

1. Facilitation, in which early species enhance establishment of later species.

2. Tolerance (equals no interaction), in which early species have no effect on later ones.

3. Inhibition, in which early species actively inhibit establishment of later ones.

These ideas were already discussed in relation to priority effects, but they clearly play a role in community processes operating on long successional timescales.

In all these interactions, it is assumed that early species cannot invade and grow once the site is fully occupied, that conditions favoring establishment of later species depend on the particular mechanism, and that early species are eliminated either by competition with later ones (under facilitation and tolerance) or by some local disturbance (inhibition). How often do these different mechanisms operate? There is some evidence for facilitation, particularly among autotrophs during primary succession, and for the succession of groups of heterotrophs in various kinds of decomposing organisms. Connell and Slatyer noted little evidence for tolerance, whereas there are many examples of inhibition (e.g., Sousa 1979a).

Walker and Chapin (1987) expanded the notion of succession as a temporal gradient in the importance of various interspecific interactions and other events. They describe the relative importance of various processes in driving changes in species composition during three stages of succession that they call colonization, maturation, and senescence. They also consider how these processes might differ in importance between severe and favorable environments and in their contribution to primary and secondary succession. The factors that they consider include seed dispersal, availability of propagules on site, the importance of stochastic events, facilitation, competition, maximum growth rates, longevity, mycorrhizae, and herbivory by insects, pathogens, and mammals (Figure 13.1). Their suggestions emphasize that most successional sequences will defy simple generalizations about temporal variation in the importance of particular factors and will require qualifications concerning the harshness of the environment or whether primary or secondary succession is happening.

Pickett and McDonnell (1989) expanded on the multifactor approach of Walker and Chapin (1987) by pointing out that the whole process of vegetation dynamics is the end result of a hierarchy of interacting factors (Figure 13.2). They emphasize that temporal patterns depend first on the availability of sites, the species pool of potential colonists, and factors affecting species performance. Site availability depends on the size, severity, and spatial dispersion of the disturbance that initiates succession. Species availability depends on dispersal and the presence of a propagule pool. To understand resulting successional patterns, one needs to know about the initial disturbance, the composition of the species pool, and the ways in which species interact.

 Succession results from a hierarchy of interacting factors.

Current views hold that succession is not a single simple process. Rather, succession is a consequence of complex interactions initiated by disturbances that create opportunities for establishment. Life history characteristics and interspecific interactions combine to create repeatable changes in community composition over time. The next section shows how succession can be

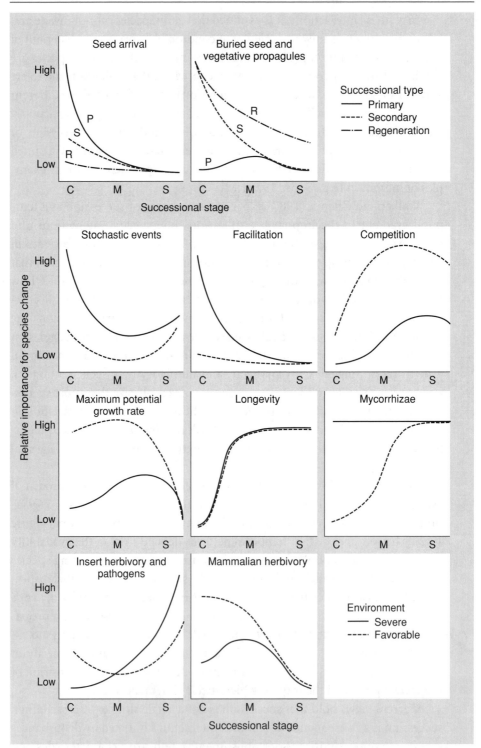

FIGURE 13.1. *Suggested changes in the importance of different factors operating early (C), midway (M), and late (S) in primary and secondary succession. (Adapted from Walker and Chapin, 1987, with permission of Oikos.)*

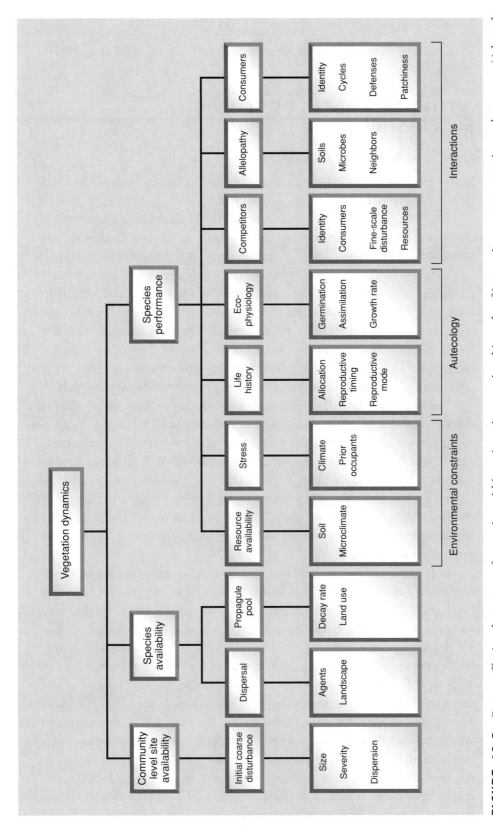

FIGURE 13.2. *Factors affecting the pattern of succession within a given site occur in a hierarchy of interacting events, properties, and processes. (Adapted from* Trends in Ecology and Evolution, *Vol. 4, Pickett, S. T. A., and M. J. McDonnell, Changing perspectives in community dynamics: a theory of successional forces, pages 241–245, copyright 1989, with permission from Elsevier Science.)*

modeled and considers some mechanisms that might drive successional patterns.

QUANTITATIVE MODELS OF ECOLOGICAL SUCCESSION

Markov Models of Species Transitions

Henry Horn (1974, 1975) used a simple mathematical framework to model transitions from early- to late-successional species in eastern deciduous forests of the United States. The approach

 Matrix models provide one way to simulate forest succession.

involves some simple matrix algebra, and models species replacements as transitions from one species to another on a particular site. A forest is assumed to represent a honeycomb of sites, with each site, or cell, in the honeycomb occupied by a single mature tree. Each tree is assumed to have a probability of being replaced by a tree of the same or different species with a probability that depends on the biology of the interacting species. The sequence of species transitions at each tree-occupied site is referred to as a **Markov chain**, and the models are termed **Markov models**.

The approach uses the following formalism. Species composition is described by a vector of abundances of each species, $c_0 = (N_1, N_2, N_3, \ldots, N_s)$, summed over all of the available sites at a particular starting point in time. The change in species composition after a period of time corresponding to the death and replacement of every tree in the forest is calculated by multiplying this initial community composition vector by a matrix of transition probabilities. Each probability, or entry in the matrix, describes the probability that an individual of some species will be replaced by the same species or by some other species.

In a simplified example based on Horn's work, assume there are only four important species: gray birch (GB), black gum (BG), red maple (RM), and beech (BE). Assume that you can also estimate the probability of transition to the same or different species. In practice, you might do this by assuming that the replacement of an existing tree depends on the relative abundance of different species of seedlings beneath that tree. By counting the number of seedlings of each tree species beneath each mature tree, the relative abundance of each species in the seedling pool beneath each tree species provides an estimate of the transition probabilities. For example, if there are 100 seedlings beneath a sample of gray birch trees, and only 5 are gray birch seedlings, the probability of gray birch replacing gray birch is 5/100 = 0.05. For the same sample, if 50 seedlings are of red maple, the probability that maple replaces birch is 0.50. A simple example of a four-species transition matrix is shown in Table 13-2. Elements in each row describe the probability

TABLE 13-2.   Example of the use of transition matrices to model changes in species composition in a simple forest stand containing four tree species.

Assume that a plot of forest contains an initial composition consisting of 100 gray birch trees and 0 of the remaining three species, black gum, red maple, and beech. The initial species composition vector would be

$c = (100 \quad 0 \quad 0 \quad 0)$

The matrix of species transition probabilities would be given by the matrix S, where each element corresponds to the probability that a species (identified by different rows) at time 1 is replaced by a species (identified by different columns) at time 2. Here, rows from top to bottom correspond to gray birch, black gum, red maple, and beech. Columns, from left to right, correspond to the same sequence of species.

$$S = \begin{array}{c} \\ GB \\ BG \\ RM \\ BE \end{array} \begin{array}{cccc} GB & BG & RM & BE \\ \begin{bmatrix} 0.05 & 0.36 & 0.50 & 0.09 \\ 0.01 & 0.57 & 0.25 & 0.17 \\ 0.00 & 0.14 & 0.55 & 0.31 \\ 0.00 & 0.01 & 0.03 & 0.96 \end{bmatrix} \end{array}$$

The species composition vector after one round of species replacement is given by

$c \times S = (5 \quad 36 \quad 50 \quad 9)$

Composition after two rounds of replacement is

$c \times S \times S = (0.61 \quad 29.41 \quad 39.27 \quad 30.71)$

By extension, composition after N rounds of replacement is

$c \times S^N$

Plotting the values of each species over time, as N goes from 1 to 30, we see the following pattern:

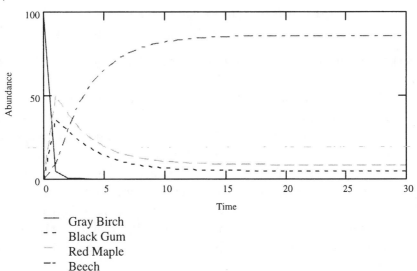

Source: Horn (1975).

that a tree of the species listed in that row will be replaced by the species corresponding to each column. So, for GB, the first row, the probability of replacement by GB is 0.05, by BG is 0.36, by RM is 0.50, and by BE is 0.09. This means that seedlings of BG and RM are much more common under GB than conspecifics or seedlings of BE.

To find the relative composition of the forest in the next time period, the species composition vector, c_0, is post-multiplied by the transition matrix using standard rules of matrix algebra. In this case, the first element of the new species composition vector is given by the product of the species composition vector and the first column of S. The vector product is obtained by summing the products of the corresponding vector elements. For this example, assuming we start with a pure stand of 100 GB, that vector product is

$$(100, 0, 0, 0) \times (0.05, 0.01, 0.0, 0.0)'$$
$$= (100 \times 0.05) + (0 \times 0.01) + (0 \times 0.0) + (0 \times 0.0) = 5$$

For the transition to BG, the corresponding vector product is

$$(100, 0, 0, 0) \times (0.36, 0.57, 0.14, 0.01)'$$
$$= (100 \times 0.36) + (0 \times 0.57) + (0 \times 0.14) + (0 \times 0.01) = 36$$

For the transition to RM, the corresponding vector product is

$$(100, 0, 0, 0) \times (0.50, 0.25, 0.55, 0.03)'$$
$$= (100 \times 0.50) + (0 \times 0.25) + (0 \times 0.55) + (0 \times 0.03) = 50$$

For the transition to BE, the corresponding vector product is

$$(100, 0, 0, 0) \times (0.09, 0.17, 0.31, 0.96)'$$
$$= (100 \times 0.09) + (0 \times 0.17) + (0 \times 0.31) + (0 \times 0.96) = 9$$

Therefore, the species composition vector in the next generation is (5, 36, 50, 9).

Notice that the forest contains the same number of trees (occupied sites). What has changed is the relative distribution of species among those sites. If we take the new species composition vector, c_1, and post-multiply it by the transition matrix, the next species composition vector is given by $c_2 = (1, 29, 39, 31)$. After many iterations, the community attains a stable composition consisting of (0, 5, 9, 86), indicating that the forest consists mostly of BE, with a few BG and RM. This end result does not depend on the initial value of the composition vector used to describe the starting conditions. The final community composition depends only on the transition probabilities given in the matrix, S.

This idealized model of succession captures the interesting biology involved in species transitions in the probabilities in the matrix S. For instance, failure to become established in the shade of other trees would appear as low probabilities for transitions to shade-intolerant species. In contrast, shade-tolerant superior competitors should have larger transition probabilities. The final species composition is always fixed for a given S, regardless of initial values of c_0. The approach can be varied somewhat by including different transition matrices to describe altered probabilities associated with different environmental conditions (say, alternations of harsh and benign climate, represented by the transition matrices D and S). Succession over two sequential favorable transitions followed by one harsh period is given by the product of $c_0 \times S \times S \times D$.

Horn's matrix approach makes many simplifying assumptions about how species replacements occur during succession. One of the more unrealistic assumptions is that the replacement probabilities are density independent; that is, they do not depend on the abundances given in the species composition vector. The approach also assumes that species transition probabilities remain constant over time unless different transition matrices are included in the successional sequence to account for possible changes. Despite these oversimplifications, the late-successional composition predicted by the model comes fairly close to the patterns seen in old wood lots in New Jersey.

Other, much more complex, simulation models have been developed to explore patterns of vegetation change in different well-studied systems. The models typically approach forest succession as a tree-by-tree replacement process and keep track of the growth and survival of a large number of individual trees within a simulated plot of forest. Early simulation models that are known by the acronyms FORET (Shugart and West 1977, 1980) and JABOWA (Botkin et al. 1972) provide reasonably good agreement between simulated patterns of forest succession and patterns reconstructed from historical information. More recent efforts such as the SORTIE model of Pacala et al. (1996) use information about seed input, light-dependent growth, and growth-dependent mortality to explore successional patterns in forests. The SORTIE model does a reasonable job of predicting the general successional trends expected in forests of the northeastern United States and has the advantage of being based on a series of basic ecological attributes of tree species that can be readily measured in the field. An example of the kind of succession predicted by SORTIE is shown in Figure 13.3.

 Complex simulation models include more biological details, and make fairly realistic predictions.

The Resource Ratio Model of Succession

Tilman (1985) has extended his mechanistic model for resource competition among plants to describe how vegetation might change in an orderly fashion during succession. The model assumes that plants compete for two limiting

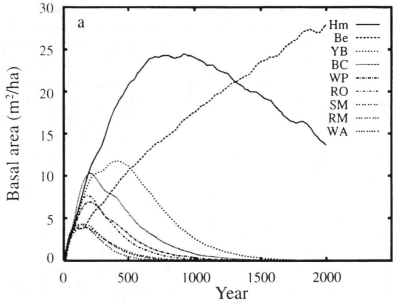

FIGURE 13.3. *Patterns of forest succession predicted by the SORTIE model. The model predicts replacement of shade-intolerant species by the shade-tolerant species eastern hemlock and American beech. Key to species: Hm = eastern hemlock, Be = American beech, YB = yellow birch, BC = black cherry, WP = white pine, RO = red oak, SM = sugar maple, RM = red maple, WA = white ash. (Reprinted from Pacala et al., 1996, with permission of the Ecological Society of America.)*

 Succession as a consequence of changing resource ratios through time.

resources. The model also assumes that resource supply rates change in some orderly fashion during succession, as a result of consumption, biogeochemical processes, or disturbance. Finally, the model assumes that competition for these resources is what drives the replacement process in communities. Each plant species is assumed to be a superior competitor at a particular ratio of limiting resources.

Tilman argues that random or positively correlated changes in resource supply rates should not produce the orderly changes in species composition that are usually associated with succession. The reason is that neither of these patterns leads to a consistent trajectory of resource supply rates across the regions corresponding to different competitive outcomes (Figure 13.4). In contrast, negative correlations among resource supply rates over succession will produce trajectories that cut across regions corresponding to dominance or coexistence by multiple species (Figure 13.5). Negative correlations might arise as simple consequences of plants competing for resources such as light and nutrients.

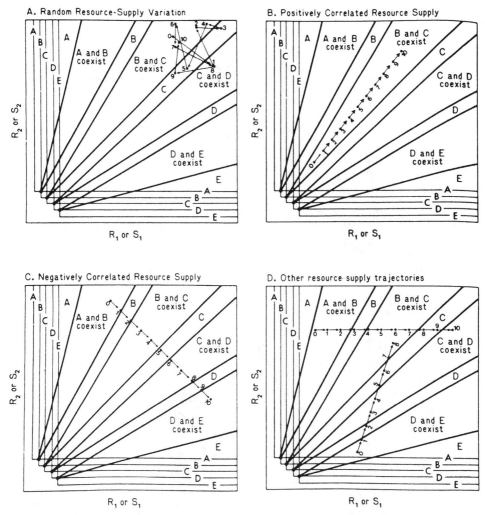

FIGURE 13.4. *Tilman's (1985) resource ratio framework for ecological succession. Plants are assumed to compete for two resources. Succession occurs when there is a change in the supply rates of both resources through time such that supply rates are negatively correlated. Random temporal changes or positively correlated changes in resource supply rates do not produce a predictable sequence of species transitions over time. (Reprinted from Tilman, 1985, with permission of the University of Chicago Press.)*

Observational studies of patterns of light and nutrient availability in old agricultural fields of different ages support a negative correlation between light and nutrients (Inouye et al. 1987). Early in primary succession, as plants first become established, light is abundant but nutrients are in short supply. As plants become abundant and nutrients accumulate, light at the soil surface decreases and nutrients increase. This process could create the kind of negative correlation that would lead to an orderly transition of species during suc-

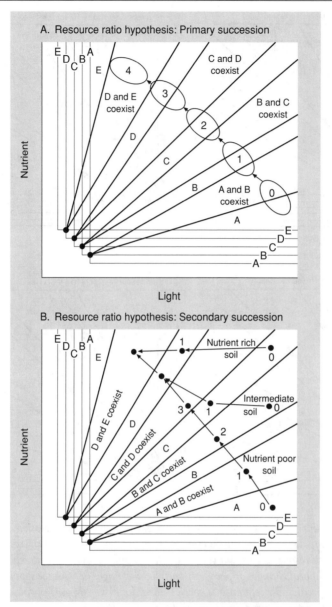

FIGURE 13.5. *Possible differences in successional pathways during primary and secondary succession that result from different initial values and rates of change for resource supply rates. (Adapted from Tilman, 1985, with permission of the University of Chicago Press.)*

cession. Tilman suggests that primary succession and secondary succession on nutrient-poor soils should show similar patterns, whereas succession on nutrient-rich soils should be rapid and should involve fewer species (see Figure 13.5). Rapid succession on nutrient-rich soils might happen because starting conditions occur at a location in resource supply space that allows fewer possible transitions in dominant species.

Experimental tests of the resource ratio succession hypothesis (Tilman 1987) focus mainly on the consequences of nutrient additions for plant species composition in different situations. For example, Tilman (1987) added nitrogen fertilizer at nine different rates, ranging from 0 to $27.2 \, \text{g} \, \text{m}^{-2} \text{yr}^{-1}$, to fields of four different successional ages and followed the patterns of community change for four years. Basic responses of the community to nitrogen addition showed that plant biomass increased as light declined, supporting a negative correlation between the supply rates of these two resources. Plant species richness also tended to decline over time, and the decline was most rapid in plots with the highest rate of nutrient supply. This result is qualitatively similar to the predictions of the graphical mechanistic theory.

CASE STUDIES OF SUCCESSION IN DIFFERENT KINDS OF HABITATS

Although the majority of concepts and systems considered in this chapter have focused on succession in temperate terrestrial plant communities, succession happens in any situation in which a disturbance creates opportunities for establishment and subsequent species transitions. One of the more illuminating studies of species replacement mechanisms comes from the rocky intertidal zone of California (Sousa 1979a). After a disturbance creates a patch of bare rock in the intertidal zone, a succession of different algal species occupy the site. Because the transitions are relatively rapid, occurring over just a few years, it is possible to observe how changes in the abundance of early-successional species influence the establishment of later ones.

Sousa created opportunities for algal succession by removing all algae from small patches of rock. He was then able to observe how removals of early-successional species influenced the establishment of late-successional species. Removals of early-successional species tended to enhance establishment of late-successional species (Figure 13.6). This result is consistent with inhibition of species replacement by established species (Connell and Slatyer 1977), but not with either facilitation or tolerance. Sousa concluded that algal succession in his system was accelerated when disturbances or herbivores removed early-successional species and allowed late-successional species to become established.

Tilman (1983, 1984, 1987) showed that differences in nutrient availability have major effects on early patterns of succession. In his system, additions of nitrogen increase rates of successional change while decreasing the number of coexisting species (Figure 13.7). One fascinating complication in this study is

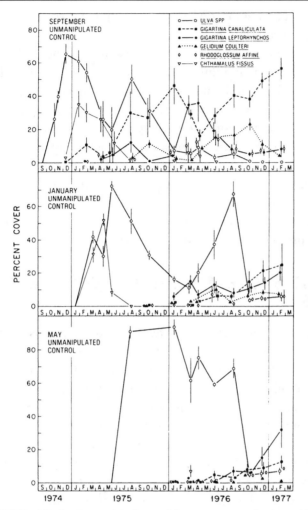

FIGURE 13.6. *Successional transitions among rocky intertidal algae that colonize open substrate at different times of the year. Early-successional species such as* Ulva *tend to slow recruitment by late-successional species. (Reprinted from Sousa, 1979a, with permission of the Ecological Society of America.)*

that herbivores, in this case subterranean-feeding pocket gophers, differentially attacked plots with high levels of nitrogen addition and high standing stocks of biomass. Unlike the situation in marine systems, herbivory seems to slow rather than speed succession in terrestrial situations (Tansley and Adamson 1925; Hope-Simpson 1940). There is abundant anecdotal evidence that strong herbivory on woody seedlings is sufficient to slow or arrest the establishment of woody species, especially in situations in which natural enemies of the herbivores are rare or absent. This is the current situation in portions of the northeastern United States, where abundant populations of white-tailed deer inhibit the establishment of woody species or select for par-

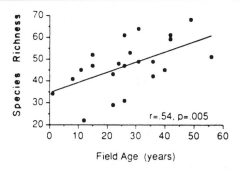

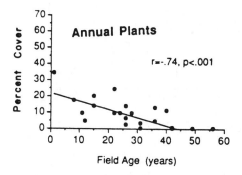

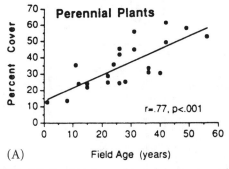

(A)   Field Age (years)

FIGURE 13.7. *(A) General trends in species richness and species composition in a chronosequence of old field communities. (Reprinted from Inouye et al., 1987, with permission of the Ecological Society of America.) (B) Effects of nitrogen addition on patterns of plant species composition in old fields in Minnesota. (Reprinted from Tilman, 1987, with permission of the Ecological Society of America.)*

ticularly unpalatable ones. It is surprising how few careful studies of the effects of herbivory on terrestrial succession have been done.

Christopher Uhl (1987) has used factorial experiments to study the causes of successional patterns in tropical forests. Succession has been little studied in highly diverse tropical systems, and consequently we know little about factors that might be manipulated to accelerate the restoration of tropical forests that have been devastated by slash-and-burn agriculture. Uhl's

Old Field Plant Abundances

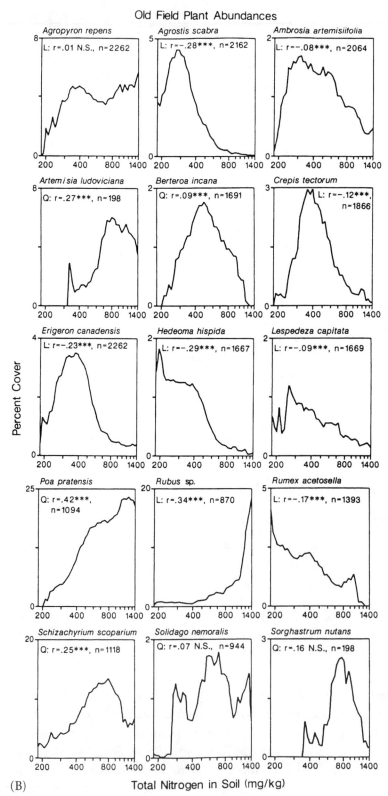

(B)

Total Nitrogen in Soil (mg/kg)

FIGURE 13.7. *(Continued)*

studies show that establishment of woody tropical forest species depends strongly on factors affecting propagule dispersal and herbivory.

Uhl explored the importance of herbivory and site characteristics on the persistence of plant propagules placed in either a natural forest gap or an abandoned farm site. For most species considered, few propagules survived longer than one month in the former farm site, whereas survival was considerably higher in the natural forest gap (Table 13-3). Apparently, animals that feed on seeds and fruits took a much larger toll in postagricultural sites, either because the diaspores were easier to locate or because more consumers frequented these sites. Larger propagules also tended to survive longer, suggesting that removals were the work of relatively small consumers. The implication is that restoration of forest species in these locations requires measures to reduce predation on plant propagules. However, simply reducing predation may be insufficient, since postagricultural sites also appear to be physiologically stressful to any propagules that do become established as seedlings.

Uhl also transplanted seedlings of rain forest trees into shaded and exposed sites in farmland that had been abandoned for different amounts of time. Most seedlings transplanted into open unshaded fields died within two days, due to the effects of intense tropical sunlight. In contrast, seedling survival in shady older fields was closer to 90% over the same time period. This result clearly suggests a positive effect of early colonists on the establishment of later ones. Similarly, after one year of growth, woody stems were approximately one order of magnitude more abundant in shady sites compared with exposed sunny sites. Uhl's work on tropical secondary succession suggests an important role for facilitation of later colonists by pioneering species that has not been documented in most studies of temperate terrestrial succession. It also emphasizes the striking effects of herbivores, which can greatly slow or limit the establishment of primary forest species if those species manage to disperse to postagricultural sites.

TABLE 13-3. Mortality of diaspores (fruits and seeds) of different primary forest tree species placed in either an open abandoned agricultural field or a natural forest gap.

Species	*Ocotea* (small)	*Jessenia* (small)	*Aldinla* (large)
Time	1 month	1 month	4 months
Farm	100%	100%	35%
Gap	88%	68%	6%

Source: Data from Uhl (1987).

Most studies of ecological succession have emphasized description of patterns without directly measuring the mechanisms responsible for species replacements or establishment. Consequently, the expected patterns of changing community composition are well known for a variety of sites, but the reasons for differences among sites, or for the replacement series seen within each site, remain poorly understood. Much important work remains to be done, particularly in understanding how interactions with other organisms, especially consumers and mutualists, alter the rate and pattern of community change over time. This information is crucial for situations in which it is desirable to reestablish particular kinds of communities after various kinds of natural or anthropogenic disturbances.

EFFECTS OF PLANT SUCCESSION ON ANIMAL ASSEMBLAGES

Well-known associations among certain kinds of vegetation and particular animal species are usually cited to suggest that successional changes in vegetation should generate differences in animal species composition (Lack and Venables 1939; Kendeigh 1948; Odum 1950; Johnston and Odum 1956). However, most explicit studies of succession focus on plants and pay little attention to concordant changes in other groups of organisms. Studies that attribute changes in animal species composition to successional changes in plant communities often do so indirectly and for very limited groups of species, primarily birds. In the eastern United States, declines in some bird species, such as eastern bluebirds that frequent agricultural fields, have been ascribed to the loss of extensive early-successional communities through secondary succession to woodlands. Other species, such as the red-cockaded woodpecker, are restricted to old-growth *Pinus plaustris* forests because the birds only nest in older longleaf pine trees that have been infected by a particular fungus that attacks the heartwood of the tree (Jackson 1977, 1986). Reasons for associations between particular plants typical of certain successional communities and other animal species are probably as numerous as the factors affecting the distributions of any organisms, and include physiological constraints, feeding preferences, and habitat selection driven by predator avoidance.

Studies of bird species composition in communities of different successional age do show important changes in species composition and species richness that accompany changes in plant community structure. In the piedmont of Georgia (United States), both bird species richness and the density of breeding pairs per unit area tend to increase through successional time (Johnston and Odum 1956; Figure 13.8; Table 13-4). These changes reflect some turnovers of species that occur in very early successional communities,

A.

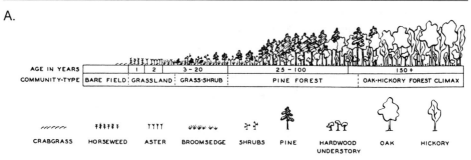

B.

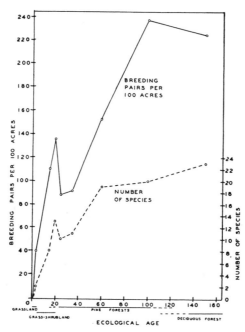

FIGURE 13.8. *(A) Schematic diagram of plant succession in the Georgia (United States) piedmont. (B) Changes in bird species richness and total numbers of bird breeding pairs in sites of different successional age in the piedmont of Georgia. (Reprinted from Johnston and Odum, 1956, with permission of the Ecological Society of America.)*

but not in later ones, as well as the addition of many species that occur only in forest communities.

It is unclear whether other groups of organisms, which are much more difficult to sample, display similar patterns along successional gradients. Other organisms, such as arthropods, sometimes show associations with plant species that are likely to translate into differences in animal community composition as plant communities change through time (Whittaker 1952), although the fidelity of those associations varies greatly among species (Futuyma and Gould 1979).

TABLE 13-4.   Patterns in the density of different species of breeding birds occurring in different secondary successional sites in the piedmont of Georgia (United States). Numbers represent territories or pairs per 100 acres.

| | Successional Stage and Age (in years) | | | | | | | | |
| | Grass-Forb | | Grass-Shrub | | Pine Forest | | | | Oak-Hickory |
Species	1	3	15	20	25	35	60	100	150+
Grasshopper sparrow	10	30	25						
Eastern meadowlark	5	10	15	2					
Field sparrow			35	48	25	8	3		
Yellowthroat			15	18					
Yellow-breasted chat			5	16					
Cardinal			5	4	9	10	14	20	23
Eastern towhee			5	8	13	10	15	15	
Bachman's sparrow				8	6	4			
Prairie warbler				6	6				
White-eyed vireo				8		4	5		
Pine warbler					16	34	43	55	
Summer tanager					6	13	13	15	10
Carolina wren						4	5	20	10
Carolina chickadee						2	5	5	5
Blue-gray gnatcatcher						2	13		13
Brown-headed nuthatch							2	5	
Blue jay							3	10	5
Eastern wood pewee							10	1	3
Ruby-throated hummingbird							9	10	10
Tufted titmouse							6	10	15
Yellow-throated vireo							3	5	7
Hooded warbler							3	30	11
Red-eyed vireo							3	10	43
Hairy woodpecker							1	3	5
Downy woodpecker							1	2	5
Crested flycatcher							1	10	6
Wood thrush							1	5	23
Yellow-billed cuckoo								1	9
Black-and-white warbler									8
Kentucky warbler									5
Acadian flycatcher									5

CONCLUSIONS

Succession is a community-level phenomenon that results from the full panoply of interspecific interactions, historical effects, and spatial dynamics that operate in developing communities. The history of the study of succession is one of often heated debate about the best way to describe the phenomenon and the mechanisms of community change. Simple models can mimic the patterns of species replacement over time, with or without the inclusion of explicit mechanisms of species interactions. Most studies of succession have been largely descriptive, which means that the actual mechanisms of species replacement in most natural communities remain highly speculative.

CHAPTER 14

Applied Community Ecology

OVERVIEW

Insights obtained from community ecology can be used to solve important applied problems concerning the management of natural, altered, or reconstructed communities. Historically, many attempts to manage populations, introduced or natural, ignored the many possible direct and indirect interactions of those populations with other species that are at the core of community ecology. The application of community ecology to emerging problems in human-dominated ecosystems is still in its infancy, but there are many problems that might benefit from a community ecology perspective. These problems range from management strategies for important diseases transmitted by animals, to the restoration and reconstruction of viable communities. Such applications provide important opportunities for research in community ecology, since the current limits of knowledge often compromise the effectiveness of applied community ecology.

MOST of this book has focused on the study of community ecology for its own sake, building on the assumption that understanding processes that drive patterns in biological diversity has its own intrinsic merits, just like any other field of study. This final chapter departs from that emphasis to highlight a few of the many possible ways that knowledge of community ecology can be used to address problems in the real

TABLE 14-1.   Examples of problems that might be solved through the imaginative application of insights obtained from community ecology.

Epidemiology of animal-borne diseases
Restoration of community composition and function
Biological control of invasive species
Biomanipulation of water quality
Management of multispecies fisheries
Optimal design of nature preserves
Predicting and managing responses to global environmental change
Maximization of yield in mixed-species agricultural systems
Assembly of viable communities in novel environments

world. In some cases, the applications of community ecology are real and ongoing. In other cases the value of those possible applications is just beginning to be realized, and operational applications are still a long way off.

Important problems can be traced to the increasing domination of natural communities and ecosystems by human activities (Vitousek et al. 1997). Humans have altered natural systems in several important ways. Some changes are the consequence of overharvesting natural populations, such as the crashes in fish populations that have almost invariably accompanied human exploitation of fish stocks. In other cases, introductions of species into novel communities without natural enemies have led to outbreaks of "pests" that vastly alter natural or agricultural systems. Other changes take the form of large-scale transformations of natural ecosystems through deforestation, mining, grazing, agriculture, and other forms of development that replace natural communities with very different, and usually much less diverse, systems. All of these changes create problems and challenges that could be profitably addressed by the creative application of community ecology. Table 14-1 lists a few such problems, but there are many others. The following examples briefly outline ways in which community ecology can be put to work.

EPIDEMIOLOGY OF ANIMAL-BORNE DISEASES

One fascinating example of the importance of insights drawn from community ecology concerns the web of strong interactions in forests of the northeastern United States that may influence the prevalence of both an introduced insect pest and an increasingly important human pathogen (Jones et al. 1998). The web of interactions in this system is outlined in Figure 14.1. The driving force in this system appears to be the periodic production of large

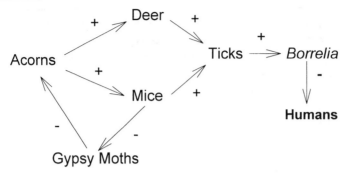

FIGURE 14.1. *Interaction web for the primary effects of mast fruiting by oaks on outbreaks of ticks (*Ixodes, *transmitters of* Borrelia*) and* Lymantria *(gypsy moths).*

crops of acorns by oaks (*Quercus* sp.) at intervals of two to five years. Such mast crops of acorns have positive effects on the densities of two mammals, white-tailed deer (*Odocoileus virginianus*) and the white-footed mouse (*Peromyscus leucopus*). Deer simply aggregate in areas of high acorn abundance, whereas mice respond numerically via enhanced survival and reproduction. Deer are the hosts for adult black-legged ticks, *Ixodes scapularis*, which feed on the blood of mammals and can carry the spirochete bacterium *Borrelia burgdorferi*. In humans, infection by *Borrelia* transmitted through the bite of an infected tick causes the debilitating condition known as Lyme disease. The juvenile ticks feed and mature on *Peromyscus*, which also serve as an important reservoir for the spirochete. As shown in Figure 14.1, high densities of acorns have positive effects on deer and mice, which in turn increase the number of ticks found on the mice. It is unclear whether the incidence of *Borrelia* in ticks also increases when mice and ticks are more abundant. The causal effects were demonstrated by manipulating densities of acorns in experimental woodland plots and monitoring changes in the densities of deer, mice, and ticks.

The community ecology of disease transmission.

A second chain reaction involving the consequences of acorn mast crops concerns interactions with the introduced gypsy moth, *Lymantria dispar*, which undergoes outbreaks that can defoliate oaks and many other trees. The high densities of *Peromyscus* that follow large acorn crops have negative effects on *Lymantria*, primarily through predation by *Peromyscus* on *Lymantria* pupae. Predation by *Peromyscus* is probably not sufficient in itself to regulate *Lymantria* populations, but may contribute to the prevention of outbreaks. In contrast, *Lymantria* outbreaks, through their negative effects on oak reproduction and mast production, conceivably depress *Peromyscus* populations, creating a positive feedback that may promote additional *Lymantria* outbreaks.

Jones et al. (1998) point out that management of oaks for reduced tick abundance (fewer acorns and fewer mice) would lead to a greater likelihood of gypsy moth outbreaks, whereas management for reduced moth outbreaks (more acorns and more mice) would lead to greater numbers of ticks, with possible consequences for increased Lyme disease transmission. It may not be possible to manipulate the system so that insect outbreaks and tick abundances are both minimized. The reforestation of the northeastern United States following the widespread abandonment of small farms has in turn created situations in which all elements of the interaction chain leading from oaks to *Borrelia* have prospered.

RESTORATION OF COMMUNITY COMPOSITION AND FUNCTION

Restoration ecology is an important new field that can potentially use ecological principles, many derived from com-

 Ecological restoration is an application of successional theory.

munity ecology, to establish and accelerate the development of communities in degraded or otherwise altered ecosystems (Montalvo et al. 1997; Palmer et al. 1997; Parker 1997). Dobson et al. (1997) outline the rationale, goals, and important case studies in this important developing field. In some cases, restoration efforts parallel the process of primary succession, in that the goal is to establish a functional community where one did not previously exist (Parker 1997). Abandoned mine tailings and closed landfill sites are examples of such situations. In other situations, the goal is to create new systems as replacements in kind for natural communities, such as freshwater wetlands, that have been lost to development. Still other cases involve attempts to restore existing but greatly altered communities to their natural state, typically by removals of nonnative invasive species and by reintroductions of native species that may have been lost during the process of community alteration.

Left to their own devices, most sites will gradually acquire a community of one sort or another, unless the site is so physiologically challenging, due to toxins or other stresses, that species cannot gain a foothold. The goals of restoration ecology are consequently focused on making sites physiologically tolerable, where needed, and then accelerating the processes of primary or secondary succession. Changes in the physical environment can be engineered by important dominant organisms (Bertness and Callaway 1994; Jones et al. 1994, 1997), which then permit the establishment of others. Establishment of some species can require the joint introduction of mutualists that are essential for growth and reproduction, such as the mycorrhizal fungi and pollinators associated with many higher plants (Montalvo et al. 1997). In some cases, species may fail to return to restored sites because of problems associated with limited opportunities for dispersal, and special efforts must be made to provide opportunities for reintroduction.

Obviously, enlightened restoration efforts could profitably incorporate what is known about assembly rules (remarkably little), priority effects (almost as little), and successional pathways (well described in many systems, but often not mechanistically understood). Palmer et al. (1997) review the various ways that community theory can potentially be used to guide and assess the success of restoration efforts. Ecological restoration efforts also provide important large-scale opportunities for experimental studies of the process of community development.

BIOLOGICAL CONTROL OF INVASIVE SPECIES

Examples of the successful biological control of invasive species, such as the ones outlined in Chapter 4, illustrate how community-level interactions, including the effects of predators and

 Community ecology can be used to devise effective geological control strategies.

pathogens, can limit or reverse the spread of unwanted species. The successful cases typically involve the reestablishment of a key fragment of a naturally occurring food web that was missing during the initial introduction and subsequent population expansion of the invader. Some properties of specialized predator-prey interactions, which are desirable from the standpoint of limiting the impact of natural enemies on targeted invasive species, may also tend to make these systems less stable and more prone to outbreaks (MacArthur 1955; Elton 1958). Although biological control efforts clearly involve the engineering of persisting food web fragments, there is as yet little indication that food web theory is consciously applied to biological control programs (Crowder et al. 1996; Ehler 1996).

One example of the possible application of community ecology to biological control comes from the work by Karban et al. (1994) discussed in Chapter 8 on indirect effects. Recall that the effectiveness of a biological control agent, in this case a predatory mite, was enhanced when it was introduced on grapevines together with an alternate prey species, the Willamette mite. Predatory mites reduced the target species, the Pacific mite, to a greater extent when the Willamette mites were also present than when alone, perhaps via apparent competition. This example shows how insights into community ecology might lead to a very different management strategy (introduction of a predator together with an alternate prey) than the standard biological control approach (introduction of one or more species of specialized predators).

BIOMANIPULATION OF WATER QUALITY

One strategy that is conceptually related to biological control is the biomanipulation of aquatic systems to alter aspects of water quality (Carpenter et al. 1985; Carpenter and Kitchell 1988). The idea builds on the notion that changes in food chain length, or changes in the abundance of top predators, may generate trophic cascades that have desirable effects on lake ecosystems. Typically, the desire is to improve water clarity in systems that have become somewhat eutrophic after a history of elevated nutrient inputs, either from agricultural runoff or sewage inputs.

Biomanipulation strategies use trophic cascades to increase herbivory by zooplankton on phytoplankton in lakes. The basic idea is outlined in Figure 14.2. Lakes with algal blooms are assumed to have essentially three trophic levels: phytoplankton, zooplankton, and zooplanktivores, usually small fish. Addition or increased abundance of a fourth trophic level, piscivorous fish that consume planktivorous fish, should generate a trophic cascade, leading to an increase in zooplankton and a decrease in phytoplankton. In practice, the manipulation would involve stocking the biomanipulated lake with high densities of piscivores, which have the added advantage of usually being a desirable target for sport fishing. The actual efficacy of biomanipulation as a

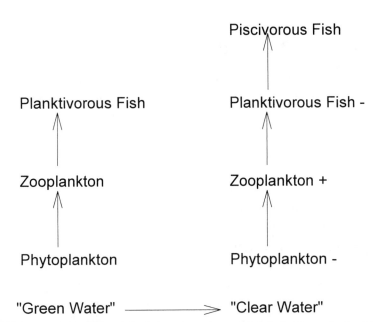

FIGURE 14.2. *Idealized food chains in a eutrophic lake before and after biomanipulation. Augmenting the top trophic level should generate a trophic cascade resulting in the reduction of phytoplankton in the water column. Plus and minus signs indicate changes in population sizes or densities relative to the unmanipulated three-level food chain.*

strategy for lake management remains somewhat controversial (DeMelo et al. 1992; Carpenter and Kitchell 1992).

MANAGEMENT OF MULTISPECIES FISHERIES

Most natural exploited fish populations are components of complex food webs (Yodzis 1994). Despite this, fish populations have historically been managed by assuming that they are single populations subjected to a particular level of additional mortality imposed by harvesting (Ricker 1975; Larkin 1978). One consequence of this simplistic management is the tendency for exploited fisheries, and other heavily harvested resources, to be overexploited to the point of collapse (Hillborn et al. 1995). A contrasting community ecology perspective would emphasize that changes in the abundance of species in a complex interconnected system may produce changes in the abundance of other species, resources, competitors, or predators, which may alter subsequent patterns of recovery from exploitation (Botsford et al. 1997). For example, changes in a heavily exploited food web may make it difficult for a heavily harvested species that relied on those resources to subsequently recover from overexploitation. A key unresolved question in applying community ecology to manage multispecies fisheries involves the number of food web members that must be considered in any management strategy (Yodzis 1994).

OPTIMAL DESIGN OF NATURE PRESERVES

The application of ecological theory to the design of nature preserves has prompted a surprising amount of controversy. Some of the concern stems from the uncritical application of island biogeography theory to preserve maximal numbers of species (Simberloff 1988) in cases where the benefits of particular reserve designs are based on theoretical assumptions rather than hard facts.

Several aspects of community theory impinge on the design of nature preserves. Preserves that focus on the protection of single species of endangered status must still be designed so that the larger community in which that species is embedded maintains its structural and functional integrity. Where species persist largely through nonequilibrium mechanisms in patchy habitats, multiple habitat patches, a natural disturbance regime, and opportunities for dispersal among patches must also be preserved. Pickett and Thompson (1978) describe some of the ways in which patch dynamics should influence the design of nature preserves. That said, it is important to point out that the most convincing demonstrations of the importance of metapopulation dynamics come from studies of laboratory systems rather than natural communities.

Recent work on relations between biodiversity and the level and predictability of ecosystem function (Tilman et al. 1996; McGrady-Steed et al. 1997; Naeem and Li 1997) suggests that systems containing larger numbers of species will support higher and more predictable levels of some ecosystem processes. The extent to which these findings apply to the full range of processes needed to establish a persisting community within the confines of a reserve is far from clear and requires much further study. In most cases, we know very little about the extent of functional redundancy within communities, or how that functional redundancy changes as communities are restricted to smaller fragmented systems that correspond to a patchwork of reserves.

PREDICTING AND MANAGING RESPONSES TO GLOBAL ENVIRONMENTAL CHANGE

As responses of communities to the Pleistocene glaciations show (Davis 1981), the ranges of species shift, expand, and contract in response to changing regimes of temperature and rainfall. One problem in understanding possible responses to environmental change, such as global warming, is that the remaining habitat available for shifting species distributions is sometimes highly fragmented. Species restricted to small disconnected reserves may not be able to migrate or shift their distributions to track changing physical conditions as they might have in the past.

A larger issue is whether the so-called climate envelope approach, which uses current correlations between climate and distribution to predict how future species distributions will shift with changes in climate, will accurately predict shifts in species ranges. The climate envelope approach ignores the impact of community-level interactions on species distributions. Recent experiments show that such interactions can strongly influence the distributions of competing species along environmental gradients. Davis et al. (1998) have experimentally explored how interspecific interactions among three *Drosophila* species affect their distributions among a series of connected population cages subjected to a laboratory temperature gradient. The gradient consisted of four different temperatures: 10°, 15°, 20°, and 25°C. Flies could move through tubing that connected the cages. Comparisons of the distributions of flies established in single-species or three-species communities showed that competition altered the distributions of all three species relative to the patterns of apparent temperature preference seen in the absence of competitors. In one case, *Drosophila melanogaster* occurred along the entire temperature gradient without competitors, but was restricted to the three warmest levels when competing with *D. simulans* and *D. subobscura*. The key point is that predictions based solely on the physiology and behavior of the individual species would not have accurately predicted where interacting species would occur along a temperature gradient. These experimental results were also prefigured by some theoretical work done by Ives and Gilchrist (1993), who predicted that directional climate change should have interesting effects on the relative abundances of competitors.

MAXIMIZATION OF YIELD IN MIXED-SPECIES AGRICULTURAL SYSTEMS

Modern agricultural ecosystems tend to be dominated by monocultures of desirable species (Matson et al. 1997), even though polycultures may be the most productive systems at a given site. It might be argued that evidence for increased production in more diverse communities (Tilman and Downing 1994; Naeem et al. 1994) is nothing more than a nonapplied example of the well-known agricultural phenomenon of overyielding in species mixtures (Harper 1977). Although overyielding is typically attributed to differences in resource utilization among species, species mixtures that are more productive than monocultures are typically discovered by experimental trial and error rather than by prior consideration of differences in resource use that could promote maximal production (as in Tilman et al. 1997). Species mixtures may also be constructed to take advantage of associational defenses against natural enemies (Matson et al. 1997), such as those described in Chapter 7. Standard large-scale agricultural practices make little effort to assemble stable polyculture communities as opposed to the usual monocultures, which typically require large inputs of pesticides and nutrients to support high yields (Matson et al. 1997). Sustainable agricultural practices that build on the insights obtained from community ecology are still in the early stages of development and provide interesting opportunities for experimental studies of community structure and function.

ASSEMBLY OF VIABLE COMMUNITIES IN NOVEL ENVIRONMENTS

One common grade-school exercise in community ecology is the creation of a balanced aquarium, a system in which producers, consumers, and decomposers generate and recycle all of the oxygen, carbon dioxide, and nutrients needed to sustain the enclosed community as long as light from an outside source is provided to drive photosynthesis (Beyers and Odum 1992). Such systems are typically much simpler than natural aquatic communities, but they embody the basic properties and problems encountered in the assembly of any new community in a novel, self-contained environment. One future application of community ecology is the scaling up of simple self-contained systems such as the balanced aquarium to create functioning communities in novel human-engineered environments. The ultimate goal of such efforts could be the establishment of ecological communities in spacecraft, in other isolated settings, or even on other planets.

The Biosphere 2 project (see Beyers and Odum 1992; Cohen and Tilman 1996) was an initial large-scale effort to assemble organisms from several biomes in a closed environment to ascertain whether such systems would persist without exchanges of material with the outside biosphere. As originally conceived, the Biosphere 2 project was unreplicated, and thus lacked one of the key features of any well-designed ecological study. Even in the absence of replication, it became obvious that the enclosed communities and their encapsulated ecosystem failed to function as hoped. Many species went extinct, and concentrations of carbon dioxide and oxygen departed substantially from natural levels. The lesson was that simply putting a group of organisms in a large controlled environmental facility would not necessarily lead to a persistent, self-sustaining, diverse community. Some of these problems might have been anticipated from what community ecologists have learned about species-area relations and the complex contingency of the community assembly process.

CONCLUSIONS

It is sobering that we still know so little about ecological communities that we cannot reassemble them with anything approaching real success. Despite the accumulated insights obtained from nearly a century of research on the description and function of natural communities, any ecologist would be hard pressed to recommend a plan for building a community from scratch that might have any chance of resembling an intended endpoint, let alone supporting natural levels of key ecosystem services. The current limits of knowledge, coupled with the daunting complexity of ecological systems, make it all the more imperative to study and preserve the natural communities that support the basic ecosystem functions on which we depend. There is much left to learn about the structure and function of communities, and that potential knowledge will be lost forever if we allow natural communities to disappear before we learn their secrets.

APPENDIX

Stability Analysis

T HE following information is provided as a "cookbook" example of the steps involved in a stability analysis of a simple mathematical model of a community-level interaction. Those craving a deeper understanding of the underlying mathematics—why it works, rather than how it is done— are encouraged to consult the many excellent texts that describe the process in much greater depth. Good choices include Edelstein-Keshet (1988), Bulmer (1994), and Hastings (1997).

The example considered here involves a pair of differential equations that describe the interaction between a predator and prey, where the prey population exhibits density-dependent population regulation via the inclusion of a logistic-style term. This is the same system of equations described in Chapter 5 as Equations 5.3 and 5.4. The equations, which are also two functions F_1 and F_2 of H and P, are

$$dH/dt = bH(1 - H/K) - PaH = F_1(H, P) \qquad (A.1)$$

and

$$dP/dt = e(PaH) - sP = F_2(H, P) \qquad (A.2)$$

where H and P are respectively the prey and predator population sizes, b is the per capita birth rate of the prey, a is a per capita attack rate, aH is the per capita consumption rate, or functional response, of predators on a given density of prey, e is the conversion efficiency of consumed prey into new predators, and $-s$ is the per capita rate at which predators die in the absence of prey.

The basic steps in the stability analysis are simply listed first to provide a general roadmap of the process and are then described in slightly greater detail. The sequence of steps involves the following operations:

1. Solving the equations for the equilibrium values of H and P, denoted H^* and P^*.

2. Creating a matrix of the partial derivatives for both equations with respect to H and P, called the Jacobian matrix.

376

3. Substituting the equilibrium values of H^* and P^* into the partial derivatives in the Jacobian matrix.

4. Conducting an eigenanalysis of the Jacobian matrix to determine whether the values of the model parameters will yield eigenvalues, or characteristic roots, of the matrix with negative real parts.

That said, a few words are in order to explain what these things are, and why they are useful.

Before conducting a stability analysis, it is first necessary to establish whether equilibrium values of H^* and P^* exist such that $dH/dt = 0$ and $dP/dt = 0$. These values will take the form of various combinations of the parameters included in the equations, and describe when population growth of each species is zero. Trivial cases in which H and/or P equals zero are not of interest, since one or both populations have gone extinct. We focus on the case where both populations have equilibrium population sizes that are greater than zero, which corresponds to one way in which the populations might coexist. Nonequilibrium coexistence is a possibility not considered by this analysis. When we conduct a stability analysis, we ask whether a small change in the values of either H^* or P^* will lead to an eventual return to the equilibrium values H^* and P^*. If so, the system is stable. The answer to that question depends on the properties of the Jacobian matrix.

Solving for the values of H^* and P^* simply involves a bit of algebra. Factoring H out of the prey equation and P out of the predator equation and setting both equations equal to 0 yields the following:

$$dH/dt = H(b - Hb/K - Pa) = 0 \qquad (A.3)$$

and

$$dP/dt = P(eaH - s) = 0 \qquad (A.4)$$

Since we are not interested in the case where $H = 0$ or $P = 0$, we want to solve for values of H and P such that

$$(b - Hb/K - Pa) = 0 \qquad (A.5)$$

and

$$(eaH - s) = 0 \qquad (A.6)$$

These equations describe the zero-growth isoclines for H and P, that is, the combinations of values of H and P that produce a net population growth of zero for each species. If and where the lines intersect, population growth rates of both species are simultaneously zero, and an equilibrium exists. Figure A.1 shows these isoclines for a particular set of parameter values. Solving Equa-

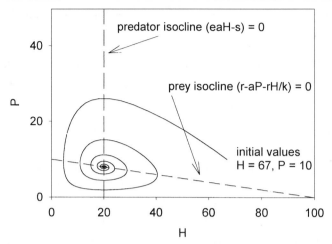

FIGURE A.1. *Zero-growth isoclines and population dynamics for Equations A.1 and A.2.*

tion A.6 for H gives $H^* = s/ea$. Solving Equation A.5 for P gives $P^* = b/[a(1 - H/k)]$. Substituting s/ea for H gives $P^* = b/[a(1 - s/eak)]$.

The Jacobian matrix consists of the partial derivatives of each equation with respect to H and P, in the two-variable, two-equation case. Referral to your freshman-year calculus book will remind you that when one takes partial derivatives with respect to a particular variable, all terms in the equation that do not contain that variable are treated like constants. Consequently, each partial derivative describes how a change in the variable of interest will change the value of the function. In this case, that description means how a change in H or P will affect the population growth rates of H or P. Since we are interested in the case where H and P are at or near H^* and P^*, we substitute the values obtained for H^* and P^* into the partial derivatives.

Whether a small change in H or P will continue to grow (unstable) or gradually decay (stable) so that values return to H^* and P^* is determined by a complex function of the elements of the Jacobian matrix. The values that function takes, called eigenvalues or characteristic roots, provide the criterion for determining whether the system is stable. The eigenvalues can be complex numbers, with real and imaginary parts. For the system to be stable, its eigenvalues must have negative real parts. If so, any perturbation away from H^* and P^* will decrease over time due to the net effects of intraspecific and interspecific interactions described by the elements of the Jacobian. The imaginary parts of the eigenvalues, if present, describe the tendency of the system to oscillate. If the eigenvalues consist solely of imaginary numbers, the system will oscillate without any tendency for the oscillations to increase or decrease in amplitude.

In the two-species, two-equation case, the eigenvalues of the Jacobian matrix , J, are given by the following relation. Where the Jacobian matrix,

$$J = \begin{pmatrix} a & b \\ c & d \end{pmatrix}$$

and a, b, c, and d are the numerical values of the partial derivatives of F_1 and F_2 evaluated at equilibrium, the eigenvalues of J, denoted by λ, are given by solving the equation

$$(a - \lambda)(d - \lambda) - bc = 0 \qquad (A.7)$$

which can be rewritten as

$$\lambda^2 - (a + d)\lambda + ad - bc = 0 \qquad (A.8)$$

Using the formula for the solution of a quadratic equation, we get

$$\lambda = \left\{ (a + d)\left[(a + d)^2 - 4(ad - bc) \right]^{0.5} \right\} \Big/ 2 \qquad (A.9)$$

However, in practice, we'll typically use the mathematical software package of our choice to painlessly do the same thing. The example below was solved using Mathcad. An example worked out for numerical values of the model parameters follows.

Consider the case of Equations A.1 and A. 2 above, where $b = 0.5$, $k = 100$, $a = 0.05$, $e = 0.5$, and $s = 0.5$. These values of the parameters yield equilibrium values of $H^* = 20$ and $P^* = 8$, since $H^* = s/ea$ and $P^* = b/[a(1 - s/eak)]$. The zero-growth isoclines are shown in Figure A.1. The Jacobian matrix of partial derivatives of Equations A.1 and A.2 is

$$\begin{pmatrix} b - aP - 2bH/k & -aH \\ eaP & eaH - s \end{pmatrix} = \begin{pmatrix} \partial F_1/\partial H & \partial F_1/\partial P \\ \partial F_2/\partial H & \partial F_2/\partial P \end{pmatrix}$$

which, for these parameter values and substituting values of H^* and P^* for H and P, becomes

$$\begin{pmatrix} -0.1 & -1.0 \\ 0.2 & 0.0 \end{pmatrix}$$

The eigenvalues of this matrix are $-0.05 + 0.444i$ and $-0.05 - 0.444i$. Since the real parts of both eigenvalues are negative, the system is stable. The tendency for the system to oscillate as it returns to equilibrium is indicated by the presence of imaginary parts of the eigenvalues. The damped oscillatory return to equilibrium is shown in the phase space trace of population dynamics in Figure A.1.

Bibliography

Abrams, P. A. 1987. Indirect interactions between species that share a predator: varieties of indirect effects. Pages 38–54 in W. C. Kerfoot and A. Sih (eds.), *Predation: Direct and Indirect Impacts on Aquatic Communities.* University Press of New England, Hanover, NH.

Abrams, P. A. 1987. On classifying interactions between populations. Oecologia 73: 272–281.

Abrams, P. A. 1993. Effect of increased productivity on the abundances of trophic levels. Amer. Natur. 141: 351–371.

Addicott, J. F. 1974. Predation and prey community structure: an experimental study of the effect of mosquito larvae on the protozoan communities of pitcher plants. Ecology 55: 475–492.

Adler, F. R. and W. F. Morris. 1994. A general test for interaction modification. Ecology 75: 1552–1559.

Aker, C. L. and D. Udovic. 1981. Oviposition and pollination behavior of the Yucca Moth, *Tegeticula maculata* (Lepidoptera: Prodoxidae), and its relation to the reproductive biology of *Yucca whipplei* (Agavaceae). Oecologia 49: 96–101.

Alford, R. A. 1989. Variation in predator phenology affects predator performance and prey community composition. Ecology 70: 206–219.

Allan, J. D. 1982. The effects of reduction in trout density on the invertebrate community of a mountain stream. Ecology 63: 1444–1455.

Allen, M. F. 1991. *The Ecology of Mycorrhizae.* Cambridge University Press, Cambridge.

Altmann, S. A. and J. Altmann. 1970. *Baboon Ecology, African Field Research.* University of Chicago Press, Chicago.

Antonovics, J. and D. A. Levin. 1980. The ecological and genetic consequences of density-dependent regulation in plants. Ann. Rev. Ecol. Syst. 11: 411–452.

Atkinson, W. D. and B. Shorrocks. 1981. Competition on a divided and ephemeral resource: a simulation model. J. Anim. Ecol. 50: 461–471.

Atsatt, P. R. and D. J. O'Dowd. 1976. Plant defense guilds. Science 193: 24–29.

Bacon, C. W., J. K. Porter, and J. D. Robbins. 1975. Toxicity and occurrence of *Balansia* on grasses from toxic fescue pastures. Appl. Microbiol. 29: 553–556.

Bard, G. E. 1952. Secondary succession on the piedmont of New Jersey. Ecological Monographs 22: 195–215.

Barkai, A. and C. McQuaid. 1988. Predator-prey reversal in a marine benthic ecosystem. Science 242: 62–64.

Barnard, C. J. and D. B. A. Thompson. 1985. *Gulls and Plovers: The Ecology and Behaviour of Mixed Species Foraging Groups.* Croom Helm, London.

Beattie, A. J. 1985. *The Evolutionary Ecology of Ant-Plant Mutualisms.* Cambridge University Press.

Bender, E. A., T. J. Case, and M. E. Gilpin. 1984. Perturbation experiments in community ecology: theory and practice. Ecology 65: 1–13.

Bengtsson, J. 1989. Interspecific competition increases local extinction rate in a metapopulation system. Nature 340: 713–715.

Benke, A. C. 1978. Interactions among coexisting predators: a field experiment with dragonfly larvae. J. Anim. Ecol. 47: 335–350.

Benke, A. C. and S. S. Benke. 1975. Comparative dynamics and life histories of coexisting dragonfly populations. Ecology 56: 302–317.

Bergelson, J. 1990. Life after death: site pre-emption by the remains of *Poa annua.* Ecology 71: 2157–2165.

Berger, J. 1980. Feeding behaviour of *Didinium nasutum* on *Paramecium bursaria* with normal and apochlorotic zoochlorellae. J. Gen. Microbiol. 118: 397–404.

Bertness, M. D. 1992. The ecology of a New England salt marsh. Amer. Sci. 80: 260–268.

Bertness, M. D. and R. Callaway. 1994. Positive interactions in communities. TREE 9: 191–193.

Bertness, M. D. and S. W. Shumway. 1993. Competition and facilitation in marsh plants. Amer. Natur. 142: 718–724.

Beyers, R. J. and H. T. Odum. 1992. *Ecological Microcosms.* Springer-Verlag, New York.

Bierregaard, R. O., Jr., T. E. Lovejoy, V. Kapos, A. A. dos Santos, and R. W. Hutchings. 1992. The biological dynamics of tropical forest fragments. Bioscience 42: 859–866.

Billick, I. and T. J. Case. 1994. Higher order interactions in ecological communities: what are they and how can they be detected. Ecology 75: 1529–1543.

Borg-Karlson, A.-K. 1990. Chemical and ethological studies of pollination in the genus *Ophrys* (Orchidaceae). Phytochemistry 29: 1359–1387.

Botsford, L. W., J. C. Castilla, and C. H. Peterson. 1997. The management of fisheries and marine ecosystems. Science 277: 509–515.

Botkin, D. B., J. F. Janak, and J. R. Wallis. 1972. Some ecological consequences of a computer model of forest growth. J. Ecology 60: 849–872.

Boucher, D. H. 1985. The idea of mutualism, past and future. Pages 1–28 in D. H. Boucher (ed.), *The Biology of Mutualism.* Oxford University Press.

Boucher, D. H., S. James, and K. H. Keeler. 1982. The ecology of mutualism. Annual Review of Ecology and Systematics 13: 315–347.

Briand, F. 1983. Environmental control of food web structure. Ecology 64: 253–263.

Briand, F. and J. E. Cohen. 1984. Community food webs have scale-invariant structure. Nature 307: 264–267.

Briand, F. and J. E. Cohen. 1987. Environmental correlates of food chain length. Science 238: 956–960.

Bronstein, J. L. 1994. Our current understanding of mutualism. The Quarterly Review of Biology 69: 31–51.

Brooks, J. L. and S. I. Dodson. 1965. Predation, body size, and composition of plankton. Science 150: 28–35.

Brown, J. H. 1971. Mammals on mountaintops: nonequilibrium insular biogeography. The American Naturalist 105: 467–478.

Brown, J. H. 1995. *Macroecology.* University of Chicago Press, Chicago.

Brown, J. H. and D. W. Davidson. 1977. Competition between seed-eating rodents and ants in desert ecosystems. Science 196: 880–882.

Brown, V. K. 1985. Insect herbivores and plant succession. Oikos 44: 17–22.

Brown, W. L., Jr. and E. O. Wilson. 1956. Character displacement. Syst. Zool. 5: 49–64.

Bulmer, M. G. 1994. *Theoretical Evolutionary Biology.* Sinauer, Sunderland.

Burkholder, P. R. 1952. Cooperation and conflict among primitive organisms. American Scientist 40: 601–631.

Buss, L. 1986. Competition and community organization on hard surfaces in the sea. Pages 517–536, In J. Diamond and T. J. Case (eds.), *Community Ecology,* Harper and Row, New York.

Carpenter, S. R., J. F. Kitchell, and J. R. Hodgson. 1985. Cascading trophic interactions and lake productivity. BioScience 35: 634–639.

Carpenter, S. R. and J. F. Kitchell. 1988. Consumer control of lake productivity. Bioscience 38: 764–769.

Carpenter, S. R. and J. F. Kitchell. 1992. Trophic cascade and biomanipulation: interface of research and management. Limnology and Oceanography 37: 208–213.

Carpenter, S. R., J. F. Kitchell, J. R. Hodgson, P. A. Cochran, J. J. Elser, M. M. Elser, D. M. Lodge, D. Kretchmer, X. He, and C. N. von Ende. 1987. Regulation of lake primary productivity by food web structure. Ecology 68: 1863–1876.

Carson, W. P. 1993. *The Influence of Phytophagous Insects on Successional Plant Communities.* Ph.D. Dissertation. Cornell University, Ithaca, NY.

Case, T. J. and E. A. Bender. 1981. Testing for higher order interactions. Amer. Natur. 118: 920–929.

Caswell, H. 1978. Predator mediated coexistence: a nonequilibrium model. Amer. Natur. 112: 127–154.

Chapman, H. H. 1945. The effect of overhead shade on the survival of loblolly pine seedlings. Ecology 26: 274–282.

Chesson, P. 1978. Predator-prey theory and variability. Ann. Rev. Ecol. Syst. 9: 323–347.

Chesson, P. 1990. Geometry, heterogeneity and competition in variable environments. Phil. Trans. Roy. Soc. Lond. B 330: 165–173.

Chesson, P. and N. Huntly. 1997. The roles of harsh and fluctuating conditions in the dynamics of ecological communities. Amer. Natur. 150: 519–533.

Chesson, P. L. and R. R. Warner. 1981. Environmental variability promotes coexistence in lottery competitive systems. Amer. Natur. 117: 923–943.

Clatworthy, J. N. and J. L. Harper. 1962. The comparative biology of closely related species living in the same area. V. Inter- and intraspecific interference within cultures of *Lemna* spp. and *Salvia natans.* J. Exp. Botany 13: 307–324.

Clay, K. 1990. Fungal endophytes of grasses. Annu. Rev. Ecol. Syst. 21: 275–297.

Clay, K., T. N. Hardy, and A. M. Hammond Jr. 1985. Fungal endophytes of grasses and their effects on an insect herbivore. Oecologia 66: 1–6.

Clements, F. E. 1916. *Plant Succession.* Carnegie Inst. Wash. Publ. 242. 512 pp.

Cohen, J. E. 1968. Alternate derivations of a species-abundance relation. Amer. Natur. 102: 165–172.

Cohen, J. E. 1977. Ratio of prey to predators in community food webs. Nature 270: 165–167.

Cohen, J. E. 1978. *Food Webs and Niche Space.* Princeton University Press, Princeton, NJ.

Cohen, J. E. and F. Briand. 1984. Trophic links of community food webs. Proc. National Acad. Sci. USA 81: 4105–4109.

Cohen, J. E. and C. M. Newman. 1985. A stochastic theory of community food webs. I. Models and aggregated data. Proc. R. Soc. London B. 224: 421–448.

Cohen, J. E. and D. Tilman. 1996. Biosphere 2 and biodiversity: the lessons so far. Science 274: 1150–1151.

Cohen, J. E., F. Briand, and C. M. Newman. 1986. A stochastic theory of community food webs. III. Predicted and observed lengths of food chains. Proc. R. Soc. London B. 228: 317–353.

Cohen, J. E., C. M. Newman, and F. Briand. 1985. A stochastic theory of community food webs. II. Individual webs. Proc. R. Soc. London B. 224: 449–461.

Coley, P. D. 1986. Cost and benefits of defense by tannins in a neotropical tree. Oecologia 70: 238–241.

Colwell, R. K. and D. W. Winkler. 1984. A null model for null models in biogeography. Pages 344–359 in D. R. Strong, Jr., D. Simberloff, L. G. Abele, and A. B. Thistle (eds.), *Ecological Communities: Conceptual Issues and the Evidence,* Princeton University Press.

Comins, H. N. and M. P. Hassell. 1976. Predation in multi-prey communities. J. Theor. Biol. 62: 93–114.

Comins, H. N., M. P. Hassell, and R. M. May. 1992. The spatial dynamics of host-parasitoid systems. J. Anim. Ecol. 61: 735–748.

Connell, J. H. 1961. The influence of interspecific competition and other factors on the distribution of the barnacle *Chthamalus stellatus.* Ecology 42: 710–723.

Connell, J. H. 1975. Some mechanisms producing structure in natural communities: a model and evidence from field experiments, pages 460–490 in M. L. Cody and J. Diamond (eds.), *Ecology and Evolution of Communities.* Harvard University Press, Cambridge, Mass.

Connell, J. H. 1978. Diversity in tropical rainforests and coral reefs. Science 199: 1302–1310.

Connell, J. H. 1979. Tropical rain forest and coral reefs as open nonequilibrium systems. Pages 141–163 in R. M. Anderson, B. D. Turner, and L. R. Taylor (eds.), *Population Dynamics.* Blackwell, Oxford.

Connell, J. H. 1983. On the prevalence and relative importance of interspecific competition: evidence from field experiments. Amer. Natur. 122: 661–696.

Connell, J. H. and R. O. Slatyer. 1977. Mechanisms of succession in natural communities and their role in community stability and organization. Amer. Natur. 111: 1119–1144.

Connell, J. H. and W. P. Sousa. 1983. On the evidence needed to judge ecological stability or persistence. Amer. Natur. 121: 789–824.

Connor, E. F. and D. Simberloff. 1978. Species number and compositional similarity of the Galapagos flora and avifauna. Ecological Monographs 48: 219–248.

Connor, E. F. and D. Simberloff. 1979. The assembly of species communities: chance or competition. Ecology 60: 1132–1140.

Connor, R. C. 1995. The benefits of mutualism: a conceptual framework. Biol. Rev. 70: 427–457.

Costantino, R. F., R. A. Desharnais, J. M. Cushing, B. Dennis. 1997. Chaotic dynamics in an insect population. Science 275: 389–391.

Cowles, H. C. 1899. The ecological relations of the vegetation on the sand dunes of lake Michigan. The Botanical Gazette 27: 95–117, 167–202, 281–308, 361–391.

Cramer, N. F. and R. M. May. 1972. Interspecific competition, predation, and species diversity: a comment. J. Theor. Biol. 34: 289–293.

Crawley, M. J. 1983. *Herbivory: The Dynamics of Animal-Plant Interactions.* The University of California Press, Berkeley, CA.

Crowder, L. B. and W. E. Cooper. 1982. Habitat structural complexity and the interaction between bluegills and their prey. Ecology 63: 1802–1813.

Crowder, L. B., D. P. Reagan, and D. W. Freckman. 1996. Food web dynamics and applied problems. Pages 327–336 in G. A. Polis and K. Winemiller (eds.), *Food Webs: Integration of Patterns & Dynamics.* Chapman and Hall, London.

Crowley, P. H. and D. M. Johnson. 1992. Variability and stability of a dragonfly assemblage. Oecologia 90: 260–269.

Cruden, R. W. 1972. Pollination biology of *Nemophilla meziesii* (Hydrophyllaceae) with comments on the evolution of oligolectic bees. Evolution 26: 373–389.

Currie, D. J. 1991. Energy and large-scale patterns of animal- and plant-species richness. Amer. Natur. 137: 27–49.

Currie, D. J. and V. Paquin. 1987. Large-scale biogeographical patterns of species richness in trees. Nature 329: 326–327.

Darwin, C. 1859. *On the Origin of Species.* Murray, London.

Davidson, D. W., R. S. Inouye, and J. H. Brown. 1984. Granivory in a desert ecosystem: Experimental evidence for indirect facilitation of ants by rodents. Ecology 65: 1780–1786.

Davis, A. J., L. S. Jenkinson, J. H. Lawton, B. Shorrocks, and S. Wood. 1998. Making mistakes when predicting shifts in species range in response to global warming. Nature 391: 783–786.

Davis, M. B. 1981. Quaternary history and the stability of forest communities. Pages 132–153 in D. C. West, H. H. Shugart, and D. B. Botkin (eds.), *Forest Succession: Concepts and Application.* Springer-Verlag, New York.

Davis, M. B. 1986. Climatic instability, time lags, and community disequilibrium. Pages 269–284 in J. Diamond and T. J. Case (eds.), *Community Ecology,* Harper and Row.

Dayan, T., D. Simberloff, E. Tchernov, and Y. Yom-Tov. 1990. Feline canines: community-wide character displacement among the small cats of Israel. Amer. Natur. 136: 39–60.

Dayton, P. K. 1971. Competition, disturbance, and community organization: the provision and subsequent utilization of space in a rocky intertidal community. Ecol. Monogr. 41: 351–389.

Dayton, P. K. 1975. Experimental evaluation of ecological dominance in a rocky intertidal algal community. Ecol. Monogr. 45: 137–159.

Dean, A. M. 1983. A simple model of mutualism. Amer. Natur. 121: 409–417.

Dean, T. A. and L. E. Hurd. 1980. Development in an estuarine community: the influence of early colonists on later arrivals. Oecologia 46: 295–301.

DeAngelis, D. L. 1975. Stability and connectance in food web models. Ecology 56: 238–243.

DeAngelis, D. L., P. J. Mulholland, A. V. Palumbo, A. D. Steinman, M. A. Huston, and J. W. Elwood. 1989. Nutrient dynamics and food-web stability. Ann. Rev. Ecol. Syst. 20: 71–95.

DeMelo, R., R. France, and D. J. McQueen. 1992. Biomanipulation: Hit or myth? Limnology and Oceanography 37: 192–207.

Dethier, M. N. and D. O. Duggins. 1984. An "indirect commensalism" between marine herbivores and the importance of competitive hierarchies. Amer. Natur. 124: 205–219.

Diamond, J. M. 1975. Assembly of species communities. Pages 342–444 in M. L. Cody and J. M. Diamond (eds.), *Ecology and Evolution of Communities.* Harvard University Press, Cambridge, Mass.

Diamond, J. M. 1986. Overview: Laboratory experiments, field experiments, and natural experiments. Pages 3–22 in J. Diamond and T. J. Case (eds.), *Community Ecology,* Harper and Row.

Diamond, J. M. and M. E. Gilpin. 1982. Examination of the "null" model of Connor and Simberloff for species co-occurrences on islands. Oecologia 52: 64–74.

Doak, D. F., D. Bigger, E. K. Harding, M. A. Marvier, R. E. O'Malley, and D. Thomson. 1998. The statistical inevitability of stability-diversity relationships in community ecology. Amer. Natur. 151: 264–276.

Dobson, A. P., A. D. Bradshaw, and A. J. M. Baker. 1997. Hopes for the future: restoration ecology and conservation biology. Science 277: 515–522.

Docters van Leeuwen, W. M. 1936. Krakatau, 1883 to 1933. Ann. Jard. Botan. Buitenzorg 56–57: 1–506.

Dodd, A. P. 1959. The biological control of prickly pear in Australia. Pages 565–577 in A. Keast, R. L. Crocker, and C. S. Christian (eds.), *Biogeography and Ecology in Australia,* Monographiae Biologicae VIII, Dr. W. Junk, The Hague.

Dodson, S. I. 1970. Complementary feeding niches sustained by size-selective predation. Limnol. Oceanogr. 15: 131–137.

Dodson, S. I. 1974. Zooplankton competition and predation: an experimental test of the size-efficiency hypothesis. Ecology 55: 605–613.

Drake, J. A. 1990. Communities as assembled structures: do rules govern pattern? TREE 5: 159–164.

Drake, J. A. 1991. Community-assembly mechanics and the structure of an experimental species ensemble. Amer. Natur. 137: 1–26.

Drury, W. H. and I. C. T. Nisbet. 1973. Succession. J. Arnold Arboretum 54: 331–368.

Dunham, A. E. 1980. An experimental study of interspecific competition between the iguanid lizards *Sceloporus merriami* and *Urosaurus ornatus*. Ecol. Monog. 50: 309–330.

Edelstein-Keshet, L. 1988. *Mathematical Models in Biology*. Random House, New York.

Egler, F. 1952. Vegetation science concepts. I. Initial floristic composition a factor in old-field vegetation development. Vegetation 4: 412–417.

Ehler, L. E. 1996. Structure and impact of natural enemy guilds in biological control of insect pests. Pages 337–342 in G. A. Polis and K. Winemiller (eds.), *Food Webs: Integration of Patterns & Dynamics*. Chapman and Hall, London.

Ehrlich, P. R. and L. C. Birch. 1967. The "balance of nature" and "population control." Amer. Natur. 101: 97–107.

Eisner, T. 1970. Chemical defense against predation in arthropods. Pages 157–217 in E. Sondheimer and J. B. Simeone (eds.), *Chemical Ecology*. Academic Press, NY.

Ellner, S. and P. Turchin. 1995. Chaos in a noisy world: new methods and evidence from time-series analysis. Amer. Natur. 145: 343–375.

Elton, C. 1927. *Animal Ecology*. Sidgwick and Jackson, London. 207 pages.

Elton, C. S. 1958. *The Ecology of Invasions by Animals and Plants*. Chapman & Hall, London.

Elton, C. S. 1966. *The Pattern of Animal Communities*. Chapman & Hall, London.

Emlen, J. M. 1977. *Ecology: An Evolutionary Approach*. Addison-Wesley, Reading, Massachusetts.

Engelmann, G. 1872. The flower of *Yucca* and its fertilization. Bull. Torrey Bot. Club 3: 37.

Errington, P. L. 1946. Predation and vertebrate populations. Quart. Rev. Biol. 21: 144–177.

Erwin, T. L. 1982. Tropical forests: their richness in Coleoptera and other Arthropod species. Coleopt. Bull. 36: 74–75.

Facelli, J. M. 1994. Multiple indirect effects of plant litter affect the establishment of woody seedlings in old fields. Ecology 75: 1727–1735.

Fairweather, P. G. 1988. Consequences of supply-side ecology: manipulating the recruitment of intertidal barnacles affects the intensity of predation upon them. Biol. Bull. 175: 349–354.

Fager, E. W. and J. A. McGowan. 1963. Zooplankton species groups in the North Pacific. Science 140: 453–460.

Feeney, P. P. 1976. Plant apparency and chemical defense. Pages 1–40 in J. W. Wallace and R. L. Mansell (eds.), *Biochemical Interaction between Plants and Insects*. Plenum, NY.

Fenner, F. 1983. Biological control, as exemplified by smallpox eradication and myxomatosis. Proc. Royal Soc. London, B. 218: 259–285.

Finlay, B. J., S. C. Maberly, and J. I. Cooper. 1997. Microbial diversity and ecosystem function. Oikos 80: 209–213.

Flaspohler, D. J. and M. S. Laska. 1994. Nest site selection by birds in Acacia trees in a Costa Rican dry deciduous forest. Wilson Bull. 106: 162–165.

Fowler, N. L. 1981. Competition and coexistence in a North Carolina grassland II. The effects of the experimental removal of species. Journal of Ecology 69: 843–854.

Frank, P. W. 1968. Life histories and community stability. Ecology 49: 355–357.

Fretwell, S. 1977. The regulation of plant communities by the food chains exploiting them. Perspect. Biol. Med. 20: 169–185.

Futuyma, D. J. and F. Gould. 1979. Associations of plants and insects in a deciduous forest. Ecological Monographs 49: 33–50.

Gaines, S. and J. Roughgarden. 1985. Larval settlement rate: a leading determinant of structure in an ecological community of the marine intertidal zone. Proc. Nat. Acad. Sci. USA 82: 3707–3711.

Gaines, S. and J. Roughgarden. 1987. Fish in offshore kelp forests affect recruitment to intertidal barnacle populations. Science 235: 479–481.

Gauch, H. G., Jr. 1982. *Multivariate Analysis in Community Ecology.* Cambridge University Press, Cambridge.

Gaudet, C. L. and P. A. Keddy. 1988. A comparative approach to predicting competitive ability from plant traits. Nature 334: 242–243.

Gause, G. F. 1934. *The Struggle for Existence.* Williams & Wilkins, Baltimore.

Gause, G. F. and A. A. Witt. 1935. Behavior of mixed populations and the problem of natural selection. Amer. Natur. 69: 596–609.

Giguere, L. 1979. An experimental test of Dodson's hypothesis that *Ambystoma* (a salamander) and *Chaoborus* (a phantom midge) have complementary feeding niches. Can. J. Zool. 57: 1091–1097.

Gill, D. E. and N. G. Hairston. 1972. The dynamics of a natural population of *Paramecium* and the role of interspecific competition in community structure. J. Anim. Ecol. 41: 137–151.

Gilpin, M. E. and J. M. Diamond. 1982. Factors contributing to nonrandomness in species co-occurrences on islands. Oecologia 52: 75–84.

Gilpin, M. E. and J. M. Diamond. 1984. Are species co-occurrences on islands non-random, and are null hypotheses useful in community ecology? Pages 297–315, in D. R. Strong, Jr., D. Simberloff, L. G. Abele, and A. B. Thistle (eds.). *Ecological Communities: Conceptual Issues and the Evidence,* Princeton University Press, Princeton, NJ.

Gilpin, M. E., M. P. Carpenter, and M. J. Pomerantz. 1986. The assembly of a laboratory community: multispecies competition in *Drosophila.* Pages 23–40, in, J. Diamond and T. J. Case (eds.), *Community Ecology,* Harper and Row.

Ginzberg, L. R. and H. R. Akcakaya. 1992. Consequences of ratio-dependent predation for steady-state properties of ecosystems. Ecology 73: 1536–1543.

Givnish, T. J. 1994. Does diversity beget stability? Nature 371: 113–114.

Gleason, H. A. 1917. The structure and development of the plant association. Bull. Torrey Bot. Club 44: 463–481.

Gleason, H. A. 1926. The individualistic concept of the plant association. Bull. Torrey Bot. Club 53: 7–26.

Gliwicz, Z. M. 1990. Food thresholds and body size in cladocerans. Nature 343: 638–640.

Goldberg, D. E. and A. M. Barton. 1992. Patterns and consequences of interspecific competition in natural communities: a review of field experiments with plants. Amer. Natur. 139: 771–801.

Goulden, C. E., L. L. Henry, and A. J. Tessier. 1982. Body size, energy reserves, and competitive ability in three species of caldocera. Ecology 63: 1780–1789.

Grant, P. R. 1986. *Ecology and Evolution of Darwin's Finches.* Princeton University Press, Princeton, NJ.

Griffiths, R. A., J. Denton, A. L.-C. Wong. 1993. The effect of food level on competition in tadpoles: interference mediated by protothecan algae? J. Anim. Ecol. 274–279.

Grime, J. P., J. M. L. Mackey, S. H. Hillier, and D. J. Read. 1987. Floristic diversity in a model system using experimental microcosms. Nature 328: 420–422.

Grinnell, J. 1914. An account of the mammals and birds of the Lower Colorado Valley with especial reference to the distributional problems presented. University of Colorado Publication in Zoology 12: 51–294.

Grosberg, R. 1981. Competitive ability influences habitat choice in marine invertebrates. Nature 290: 700–702.

Gurevitch, J. 1986. Competition and the local distribution of the grass *Stipa neomexicana*. Ecology 67: 46–57.

Gurevitch, J., L. L. Morrow, A. Wallace, and J. S. Walsch. 1992. A meta-analysis of competition in field experiments. Amer. Natur. 140: 539–572.

Hairston, N. G. 1949. The local distribution and ecology of the plethodontid salamanders of the southern Appalachians. Ecological Monographs 19: 47–63.

Hairston, N. G. 1980a. The experimental test of an analysis of field distributions: competition in terrestrial salamanders. Ecology 61: 817–826.

Hairston, N. G. 1980b. Evolution under interspecific competition. Field experiments on terrestrial salamanders. Evolution 34: 409–420.

Hairston, N. G. 1981. An experimental test of a guild. Ecology. 62: 65–72.

Hairston, N. G. 1987. *Community Ecology and Salamander Guilds.* Cambridge University Press, Cambridge.

Hairston, N. G. 1989. *Ecological Experiments.* Cambridge University Press, Cambridge.

Hairston, N. G., J. D. Allan, R. K. Colwell, D. J. Futuyma, J. Howell, M. D. Lubin, J. Mathais, and J. H. Vandermeer. 1968. The relationship between species diversity and stability: and experimental approach with protozoa and bacteria. Ecology 49: 1091–1101.

Hairston, N. G., Jr., and N. G. Hairston, Sr. 1993. Cause-effect relationships in energy flow, trophic structure, and interspecific interactions. Amer. Natur. 142: 379–411.

Hairston, N. G., F. E. Smith, and L. B. Slobodkin. 1960. Community structure, population control, and competition. Amer. Natur. 94: 421–425.

Hall, D. J., W. E. Cooper, and E. E. Werner. 1970. An experimental approach to the production dynamics and structure of freshwater animal communities. Limnology and Oceanography 15: 839–928.

Handel, S. N. 1978. The competitive relationship of three woodland sedges and its bearing on the evolution of ant-dispersal of *Carex pedunculata*. Evolution 32: 151–163.

Hanzawa, F. M., A. J. Beattie and D. C. Culver. 1988. Directed dispersal: demographic analysis of an ant-seed mutualism. Amer. Natur. 131: 1–13.

Hardin, G. 1960. The competitive exclusion principle. Science 131: 1292–1297.

Harper, J. L. 1961. Approaches to the study of plant competition. Symposia of the Society for Experimental Biology 15: 1–39.

Harper, J. L. 1977. *Population Biology of Plants.* Academic Press.

Harrison, G. W. 1995. Comparing predator-prey models to Luckinbill's experiment with *Didinium* and *Paramecium.* Ecology 76: 357–374.

Harvey, P. H. and P. J. Greenwood. 1978. Anti-predator defence strategies: some evolutionary problems. Pages 129–151 in J. R. Krebs and N. B. Davies (eds.), *Behavioural Ecology: an Evolutionary Approach,* Sinauer, Sunderland, MA.

Hassell, M. P. 1978. *The Dynamics of Arthropod Predator-prey Systems.* Princeton University Press, Princeton, NJ. 237.

Hassell, M. P., H. N. Comins, and R. M. May. 1991a. Spatial structure and chaos in insect population dynamics. Nature 353: 255–258.

Hassell, M. P., R. M. May, S. W. Pacala, and P. L. Chesson. 1991b. The persistence of host-parasite associations in patchy environments. 1. A general criterion. Amer. Natur. 138: 568–583.

Hastings, A. 1997. *Population Biology: Concepts and Models.* Springer-Verlag, New York.

Hastings, A. and T. Powell. 1991. Chaos in a three-species food chain. Ecology 72: 896–903.

Hay, M. E. 1986. Associational plant defenses and the maintenance of species diversity: turning competitors into accomplices. Amer. Natur. 128: 617–641.

Heske, E. J., J. H. Brown, and S. Mistry. 1994. Long-term experimental study of a Chihuahuan desert rodent community: 13 years of competition. Ecology 75: 438–445.

Hillborn, R., C. J. Walters, and D. Ludwig. 1995. Sustainable exploitation of renewable resources. Annu. Rev. Ecol. Syst. 26: 45–67.

Holdridge, L. R. 1947. Determination of world plant formations from simple climatic data. Science 105: 367–368.

Holldobler, B. and E. O. Wilson. 1990. *The Ants.* Harvard University Press, Cambridge, Massachusetts.

Holling, C. S. 1965. The functional response of predators to prey density and its role in mimicry and population regulation. Mem. Ent. Soc. Canada 45: 3–60.

Holmes, R. T., R. E. Bonney, Jr., and S. W. Pacala. 1979a. Guild structure of the Hubbard Brook bird community: a multivariate approach. Ecology 60: 512–520.

Holmes, R. T., J. C. Schultz, and P. Nothnagle. 1979b. Bird predation on forest insects: an exclosure experiment. Science 206: 462–463.

Holomuzki, J. R. 1986. Predator avoidance and diel patterns of microhabitat use by larval tiger salamanders. Ecology 67: 737–748.

Holt, R. D. 1977. Predation, apparent competition, and the structure of prey communities. Theoretical Population Biology 12: 197–229.

Holt, R. D., J. Grover, and D. Tilman. 1994. Simple rules for interspecific dominance in systems with exploitative and apparent competition. American Naturalist 144: 741–771.

Holt, R. D. and J. H. Lawton. 1994. The ecological consequences of shared natural enemies. Annu. Rev. Ecol. Syst. 25: 495–520.

Holyoak, M. and S. P. Lawler. 1996a. Persistence of an extinction-prone predator-prey interaction through metapopulation dynamics. Ecology 77: 1867–1879.

Holyoak, M. and S. P. Lawler. 1996b. The role of dispersal in predator-prey metapopulation dynamics. Journal of Animal Ecology 65: 640–652.

Hope-Simpson, J. F. 1940. Studies of the vegetation of the English chalk. VI. Late stages in succession leading to chalk grassland. J. Ecology 28: 386–402.

Horn, H. S. 1974. The Ecology of Secondary Succession. Ann. Rev. Ecol. Syst. 5: 25–37.

Horn, H. S. 1975. Markovian properties of forest succession. Pages 196–211 in M. L. Cody and J. Diamond, eds. *Ecology and Evolution of Communities*. Belknap Press, Cambridge.

Horn, H. S. and R. M. May. 1977. Limits to similarity among coexisting competitors. Nature 270: 660–661.

Howe, H. F. and J. Smallwood. 1982. Ecology of seed dispersal. Ann. Rev. Ecol. Syst. 13: 201–228.

Hrbacek, J., M. Dvorakova, V. Korinek, and L. Prochazkova. 1961. Demonstration of the effect of the fish stock on the species composition of zooplankton and the intensity of metabolism of the whole plankton association. Verh. Internat. Verein. Limnol. 14: 192–195.

Hubbell, S. P. 1979. Tree dispersion, abundance, and diversity in a tropical dry forest. Science 203: 1299–1309.

Hubbell, S. P. and R. B. Foster. 1983. Diversity of canopy trees in a neotropical forest and implications for conservation. Pages 25–41 in S. Sutton, T. C. Whitmore, and A. Chadwick (eds.), *Tropical Rain Forest: Ecology and Management*. Blackwell, Oxford.

Huffaker, C. B. 1958. Experimental studies on predation: dispersion factors and predator-prey oscillations. Hilgardia 27: 343–383.

Huffaker, C. B. and C. E. Kennett. 1959. A ten year study of vegetational changes associated with biological control of klamath weed. J. Range Management 12: 69–82.

Hurd, L. E., M. V. Mellinger, L. L. Wolf, and S. J. McNaughton. 1971. Stability and diversity at three levels in terrestrial successional ecosystems. Science 173: 1134–1136.

Hurlbert, S. H. 1971. The non-concept of species diversity: a critique and alternate parameters. Ecology 52: 577–586.

Huston, M. A. 1994. *Biological Diversity: The Coexistence of Species on Changing Landscapes*. Cambridge University Press, Cambridge.

Huston, M. A. 1997. Hidden treatments in ecological experiments: re-evaluating the ecosystem function of biodiversity. Oecologia 110: 449–460.

Hutchinson, G. E. 1957. Concluding remarks. Cold Spring Harbor Symposia on Quantitative Biology 22: 415–427.

Hutchinson, G. E. 1959. Homage to Santa Rosalia, or why are there so many different kinds of animals? Amer. Natur. 93: 145–159.

Hutchinson, G. E. 1961. The paradox of the plankton. Amer. Natur. 95: 137–145.

Hutchinson, G. E. 1967. *A Treatise on Limnology, Vol. 2. Introduction to Lake Biology and the Limnoplankton.* Wiley, New York.

Hutchinson, G. E. 1978. *An Introduction to Population Ecology.* Yale University Press, New Haven.

Inger, R. F. and R. K. Colwell. 1977. Organization of contiguous communities of amphibians and reptiles in Thailand. Ecological Monographs 47: 229–253.

Inouye, R. S., N. J. Huntly, D. Tilman, J. R. Tester, M. Stillwell, and K. C. Zinnel. 1987. Old-field succession on a Minnesota sand plain. Ecology 68: 12–26.

Istock, C. A. 1973. Population characteristics of a species ensemble of water boatmen. Ecology 54: 535–544.

Ives, A. R. 1991. Aggregation and coexistence in a carrion fly community. Ecological Monographs 61: 75–94.

Ives, A. R. and G. Gilchrist. 1993. Climate change and ecological interactions. Pages 120–146 in P. Kareiva, J. Kingsolver, and R. Huey (eds.), *Biotic Interactions and Global Change.* Sinauer, Sunderland, MA.

Ives, A. R. and R. M. May. 1985. Competition within and between species in a patchy environment: relations between microscopic and macroscopic models. J. Theor. Biol. 115: 65–92.

Jackson, J. A. 1977. Red-cockaded woodpeckers and pine red heart disease. Auk 94: 160–163.

Jackson, J. A. 1986. Biopolitics, management of federal lands, and the conservation of the red-cockaded woodpecker. Amer. Birds 40: 1162–1168.

Jackson, J. B. C. and L. Buss. 1975. Allelopathy and spatial competition among coral reef invertebrates. Proc. Natl. Acad. Sci. USA 72: 5160–5163.

Janzen, D. H. 1966. Coevolution between ants and acacias in Central America. Evolution 20: 249–275.

Janzen, D. H. 1971a. Euglossine bees as long-distance pollinators of tropical plants. Science 171: 203–205.

Janzen, D. H. 1971b. Escape of *Cassia grandis* L. beans from predators in time and space. Ecology 52: 964–979.

Janzen, D. H. and P. S. Martin. 1982. Neotropical anachronisms: the fruits the Gomphotheres ate. Science 215: 19–27.

Jenkins, B., R. L. Kitching, and S. L. Pimm. 1992. Productivity, disturbance, and food web structure at a local spatial scale in experimental container habitats. Oikos 65: 249–255.

Jeffries, M. J. and J. H. Lawton. 1984. Enemy free space and the structure of ecological communities. Biol. J. of the Linnean Soc. 23: 269–286.

Jones, C. G., J. H. Lawton, and M. Shachak. 1994. Organisms as ecosystem engineers. Oikos 69: 373–386.

Jones, C. G., J. H. Lawton, and M. Shachak. 1997. Positive and negative effects of organisms as physical ecosystem engineers. Ecology 78: 1946–1957.

Jones, C. G., R. S. Ostfeld, M. P. Richard, E. M. Schauber, and J. O. Wolff. 1998. Chain reactions linking acorns to gypsy moth outbreaks and Lyme disease risk. Science 279: 1023–1026.

John-Alder, H. B., P. J. Morin, and S. P. Lawler. 1988. Thermal physiology, phenology, and distribution of tree frogs. Amer. Natur. 132: 506–520.

Johnson, L. W. and R. D. Riess. 1982. *Numerical Analysis*. Addison-Wesley, Reading.

Johnston, D. W. and E. P. Odum. 1956. Breeding bird populations in relation to plant succession on the piedmont of Georgia. Ecology 37: 50–62.

Juliano, S. A. and J. H. Lawton. 1990a. The relationship between competition and morphology. I. Morphological patterns among co-occurring dytiscid beetles. J. Anim. Ecol. 59: 403–419.

Juliano, S. A. and J. H. Lawton. 1990b. The relationship between competition and morphology. II. Experiments on co-occurring dytiscid beetles. J. Anim. Ecol. 59: 813–848.

Karban, R., D. Hougen-Eitzmann, and G. English-Loeb. 1994. Predator-mediated apparent competition between two herbivores that feed on grapevines. Oecologia 97: 508–511.

Kareiva, P. 1987. Habitat fragmentation and the stability of predator-prey interactions. Nature 326: 388–390.

Kareiva, P. 1990. Population dynamics in spatially complex environments: theory and data. Phil. Trans. Roy. Soc. Lond. B 330: 175–190.

Keddy, P. 1992. Assembly and response rules: two goals for predictive community ecology. Journal of Vegetation Science 3: 157–164.

Keever, C. 1950. Causes of succession on old fields of the Piedmont, North Carolina. Ecological Monographs 20: 230–250.

Kendeigh, S. C. 1948. Bird populations and biotic communities in northern lower Michigan. Ecology 29: 101–114.

King, A. and S. Pimm. 1983. Complexity, diversity, and stability: reconciliation of theoretical and empirical results. Amer. Natur. 122: 229–239

Kingsland, S. E. 1985. *Modeling Nature*. University of Chicago Press, Chicago.

Kitching, R. L. 1987. Spatial and temporal variation in food webs in water-filled tree holes. Oikos 48: 280–288.

Kochmer, J. P. and S. N. Handel. 1986. Constraints and competition in the evolution of flowering phenology. Ecological Monographs 56: 303–325.

Kolmogorov, A. N. 1936. Sulla Teoria di Volterra della Lotta per l'Esisttenza. Giorn. Instituto. Ital. Attuari 7: 74–80

Krebs, J. R. 1973. Social learning and the significance of mixed-species flocks of chickadees (*Parus* spp.). Can. J. Zoology 51: 1275–1288.

Lack, D. 1947. *Darwin's Finches*. Cambridge University Press, Cambridge.

Lack, D. 1966. *Population Studies of Birds*. Clarendon Press, Oxford.

Lack, D. and L. S. V. Venables. 1939. The habitat distribution of British woodland birds. J. Anim. Ecol. 8: 39–70.

Lampert, W. and U. Schober. 1980. The importance of "threshold" food concentrations. Pages 264–267 in W. C. Kerfoot (ed.), *Evolution and Ecology of Zooplankton Communities*. University Press of New England, Hanover.

Larkin, P. A. 1978. Fisheries management—an essay for ecologists. Ann. Rev. Ecol. Syst. 9: 57–73.

Laska, M. S. and J. T. Wootton. 1998. Theoretical concepts and empirical approaches to measuring interaction strength. Ecology 79: 461–476.

Laurance, W. F., S. G. Laurance, V. F. Ferreira, J. M. Rankin-de Merona, C. Gascon, and T. E. Lovejoy. 1997. Biomass collapse in Amazonian forest fragments. Science 278: 1117–1118.

Law, R. and R. D. Morton 1993. Alternative permanent states of ecological communities. Ecology 74: 1347–1361.

Lawler, S. P. 1989. Behavioral responses to predators and predation risk in four species of larval anurans. Animal Behaviour 38: 1039–1047.

Lawler, S. P. 1993a. Direct and indirect effects in microcosm communities of protists. Oecologia 93: 184–190.

Lawler, S. P. 1993b. Species richness, species composition and population dynamics of protists in experimental microcosms. J. Anim. Ecol. 62: 711–719.

Lawler, S. P. and P. J. Morin. 1993a. Temporal overlap, competition, and priority effects in larval anurans. Ecology 74: 174–182.

Lawler, S. P. and P. J. Morin. 1993b. Food web architecture and population dynamics in laboratory microcosms of protists. Amer. Natur. 141: 675–686.

Lawlor, L. R. 1978. A comment on randomly constructed model ecosystems. Amer. Natur. 112: 445–447.

Lawton, J. H. 1989. Food webs. Pages 43–78 in J. M. Cherrett (ed.), *Ecological Concepts*, Blackwell Scientific, Oxford.

Lawton, J. H. and V. K. Brown. 1993. Redundancy in ecosystems. Pages 255–270 in E. D. Schulze and H. A. Mooney (eds.), *Biodiversity and Ecosystem Function*, Springer-Verlag, New York.

Lawton, J. H. and K. J. Gaston. 1989. Temporal patterns in the herbivorous insects of bracken: a test of community predictability. J. Anim. Ecol. 58: 1021–1034.

Lawton, J. H. and M. P. Hassell. 1981. Asymmetrical competition in insects. Nature 289: 793–795.

Lawton, J. H. and P. H. Warren. 1988. Static and dynamic explanations for patterns in food webs. TREE 3: 242–245.

Leibold, M. A. 1989. Resource edibility and the effects of predators and productivity on the outcome of trophic interactions. Amer. Natur. 134: 922–949.

Leibold, M. A. 1996. A graphical model of keystone predators in food webs: trophic regulation of abundance, incidence, and diversity patterns in communities. Amer. Natur. 147: 784–812.

Leibold, M. A., J. M. Chase, J. B. Shurin, and A. L. Downing. 1997. Species turnover and the regulation of trophic structure. Ann. Rev. Ecol. Syst. 28.

Leslie, P. H. and J. C. Gower. 1960. The properties of a stochastic model for the predator-prey type of interaction between two species. Biometrika 47: 219–234.

Levine, S. H. 1976. Competitive effects in ecosystems. Amer. Natur. 110: 903–910.

Levins, R. 1968. *Evolution in Changing Environments*. Princeton University Press, Princeton, NJ. 120 pages.

Lewontin, R. C. 1969. The meanings of stability. Brookhaven Symp. Biology 22: 13–24.

Lindeman, R. L. 1942. The trophic-dynamic aspect of ecology. Ecology 23: 399–418.

Loreau, M. 1998. Ecosystem development explained by competition within and between material cycles. Proc. R. Soc. Lond. B 265: 33–38.

Losos, J. B., T. R. Jackman, A. Larson, K. de Queiroz, and L. Rodriguez-Schettino. 1998. Contingency and determinism in replicated adaptive radiations of island lizards. Science 279: 2115–2118.

Lotka, A. J. 1925. *Elements of Physical Biology*. Williams & Wilkins, Baltimore.

Louda, S. M. 1982. Distribution ecology: variation in plant recruitment over a gradient in relation to insect seed predation. Ecological Monographs 52: 25–41.

Lubchenco, J. 1978. Plant species diversity in a marine intertidal community: importance of herbivore food preferences and algal competitive abilities. Amer. Natur. 112: 23–39.

Luckinbill, L. S. 1973. Coexistence in laboratory populations of *Paramecium aurelia* and its predator *Didinium nasutum*. Ecology 54: 1320–1327.

Luckinbill, L. S. 1974. The effects of space and enrichment on a predator-prey system. Ecology 55: 1142–1147.

Lynch, M. 1979. Predation, competition, and zooplankton community structure: an experimental study. Limnology and Oceanography 24: 253–272.

MacArthur, R. 1955. Fluctuations of animal populations, and a measure of community stability. Ecology 36: 533–536.

MacArthur, R. H. 1958. Population ecology of some warblers of northeastern coniferous forests. Ecology 39: 599–619.

MacArthur, R. H. 1972. *Geographical Ecology*. Princeton University Press, Princeton.

MacArthur, R. H. and R. Levins. 1967. The limiting similarity, convergence, and divergence of coexisting species. Amer. Natur. 101: 377–385.

MacArthur, R. H. and J. W. MacArthur. 1961. On bird species diversity. Ecology 42: 594–598.

MacArthur, R. H. and E. O. Wilson. 1967. *The Theory of Island Biogeography*. Princeton University Press, Princeton, NJ. 203 pages.

Magurran, A. E. 1988. *Ecological Diversity and Its Measurement*. Princeton University Press, Princeton, NJ.

Marquis, R. J. and C. J. Whelan. 1994. Insectivorous birds increase growth of white oak through consumption of leaf-chewing insects. Ecology 75: 2007–2014.

Martinez, N. D. 1991. Artifacts or attributes? effects of resolution on the Little Rock Lake food web. Ecological Monographs 61: 367–392.

Martinez, N. D. 1992. Constant connectance in community food webs. Amer. Natur. 139: 1208–1218.

Mathcad Plus 5.0 Users Guide. 1994. MathSoft, Inc., Cambridge, MA.

Matson, P. A., W. J. Parton, A. G. Power, and M. J. Swift. 1997. Agricultural intensification and ecosystem properties. Science 277: 504–509.

May, R. M. 1972. Will a large complex system be stable? Nature 238: 413–414.

May, R. M. 1973. *Stability and Complexity in Model Ecosystems*. Princeton University Press, Princeton, New Jersey USA.

May, R. M. 1975. Patterns of species abundance and diversity. Pages 81–120 in M. L.

Cody and J. M. Diamond (Eds.), *Ecology and Evolution of Communities.* Belknap Press, Cambridge.

May, R. M. 1976a. Models for single populations. Pages 4–25 in R. M. May (ed.), *Theoretical Ecology: Principles and Applications.* Saunders, Philadelphia.

May, R. M. 1976b. Models for two interacting populations. Pages 47–71 in R. M. May (ed.), *Theoretical Ecology: Principles and Applications.* Saunders, Philadelphia.

May, R. M. 1990. How many species? Phil. Trans. R. Soc. Lond. B. 330: 293–304.

McArdle, B. H., K. J. Gaston, and J. H. Lawton. 1990. Variation in the size of animal populations: patterns, problems and artefacts. J. Animal Ecology 59: 439–454.

McCann, K. and P. Yodzis. 1994. Nonlinear dynamics and population disappearances. Amer. Natur. 144: 873–879.

McGowan, J. A. and P. A. Walker. 1979. Structure in the copepod community of the North Pacific central gyre. Ecological Monographs 49: 195–226.

McGrady-Steed, J. P. M. Harris, P. J. Morin. 1997. Biodiversity regulates ecosystem predictability. Nature 390: 162–165.

McIntosh, R. P. 1985. The background of ecology: concept and theory. Cambridge University Press, Cambridge, UK.

McNaughton, S. J. 1976. Serengeti migratory wildebeest: facilitation of energy flow by grazing. Science 191: 92–94.

McNaughton, S. J. 1977. Diversity and stability of ecological communities: a comment on the role of empiricism in ecology. Amer. Natur. 111: 515–525.

McNaughton, S. J. 1979. Grazing as an optimization process: grass ungulate relationships in the Serengeti. Amer. Natur. 113: 691–703.

McQueen, D. J., M. R. S. Johannes, J. R. Post, T. J. Stewart, and D. R. S. Lean. 1989. Bottom-up and top-down impacts on freshwater pelagic community structure. Ecol. Monogr. 59: 289–309.

Meffe, G. K. and C. R. Carroll. 1994. *Principles of Conservation Biology.* Sinauer, Sunderland, Massachusetts. 600 pages.

Menge, B. and J. P. Sutherland. 1976. Species diversity gradients: synthesis of the roles of predation, competition, and temporal heterogeneity. Amer. Natur. 110: 351–369.

Menge, B. and J. P. Sutherland. 1987. Community regulation: variation in disturbance, competition, and predation in relation to environmental stress and recruitment. Amer. Natur. 130: 730–757.

Menge, B. A., E. L. Berlow, C. A. Blanchette, S. A. Navarrete, and S. B. Yamada. 1994. The keystone species concept: variation in interaction strength in a rocky intertidal habitat. Ecological Monographs 64: 249–286.

Miller, T. E. and W. C. Kerfoot. 1987. Redefining indirect effects. Pages 33–37 in W. C. Kerfoot and A. Sih (eds.), *Predation: Direct and Indirect Impacts on Aquatic Communities.* University Press of New England, Hanover, NH.

Minshall, G. W. 1967. Role of allochthonous detritus in the trophic structure of a woodland springbrook community. Ecology 48: 139–149.

Mitchell, D. S., T. Petr, and A. B. Viner. 1980. The water fern *Salvinia molesta* in the Sepik River, Papua New Guinea. Environ. Conserv. 7: 115–122.

Monod, J. 1950. La technique de la culture continue: theorie et applications. Ann. Inst. Pasteur Lille 79: 390–410.

Montalvo, A. M., S. L. Williams, K. J. Rice, S. L. Buchmann, C. Cory, S. N. Handel, G. P. Nabhan, R. Primack, and R. H. Robichaux. 1997. Restoration biology: a population biology perspective. Restoration Ecology 5: 277–290.

Morin, P. J. 1983. Predation, competition, and the composition of larval anuran guilds. Ecol. Monogr. 53: 119–138.

Morin, P. J. 1984a. The impact of fish exclusion on the abundance and species composition of larval odonates: results of short-term experiments in a North Carolina farm pond. Ecology 65: 53–60.

Morin, P. J. 1984b. Odonate guild composition: experiments with colonization history and fish predation. Ecology 65: 1866–1873.

Morin, P. J. 1986. Interactions between intraspecific competition and predation in an amphibian predator-prey system. Ecology 67: 713–720.

Morin, P. J. 1987. Predation, breeding asynchrony, and the outcome of competition among treefrog tadpoles. Ecology 68: 675–683.

Morin, P. J. 1989 New directions in amphibian community ecology. Herpetologica 45(1): 124–128.

Morin, P. J. and E. A. Johnson. 1988. Experimental studies of asymmetric competition among anurans. Oikos 53: 398–407.

Morin, P. J. and S. P. Lawler. 1995. Food web architecture and population dynamics: theory and empirical evidence. Annu. Rev. Ecol. Syst. 26: 505–529.

Morin, P. J. and S. P. Lawler. 1996. Effects of food chain length and omnivory in experiment food webs. Pages 218–230 in G. A. Polis and K. Winemiller (eds.), *Food Webs: Integration of Patterns & Dynamics*. Chapman & Hall, London.

Morin, P. J., S. P. Lawler, and E. A. Johnson. 1988. Competition between aquatic insects and vertebrates: interaction strength and higher order interactions. Ecology 69: 1401–1409.

Morin, P. J., S. P. Lawler, and E. A. Johnson. 1990. Breeding phenology and the larval ecology of *Hyla andersonii*: the disadvantages of breeding late. Ecology 71: 1590–1598.

Motten, A. F. 1983. Reproduction of *Erythronium umbillicatum* (Liliaceae): pollinator success and pollinator effectiveness. Oecologia 59: 351–359.

Motten, A. F., D. R. Campbell, D. E. Alexander, and H. L. Miller. 1981. Pollination effectiveness of specialist and generalist visitors to a North Carolina population of *Claytonia virginica*. Ecology 62: 1278–1287.

Muller, C. H., W. H. Muller, and B. L. Haines. 1964. Volatile growth inhibitors produced by aromatic shrubs. Science 143: 471–473.

Munger, J. C. and J. H. Brown. 1981. Competition in desert rodents: an experiment with semipermeable enclosures. Science 211: 510–512.

Murdoch, W. W. 1966. Community structure, population control, and competition—a critique. The American Naturalist 100: 219–226.

Murdoch, W. W. and A. Stewart-Oaten. 1989. Aggregation by parasitoids and predators: effects on equilibrium and stability. Amer. Natur. 134: 288–310.

Muscatine, L. and J. W. Porter. 1977. Reef corals: mutualistic symbioses adapted to nutrient-poor environments. Bioscience 27: 454–460.

Naeem, S. and S. Li. 1997. Biodiversity enhances ecosystem reliability. Nature 390: 507–509.

Naeem, S., L. J. Thompson, S. P. Lawler, J. H. Lawton, and R. M. Woodfin. 1994. Declining biodiversity can alter the performance of ecosystems. Nature 368: 734–737.

Naeem, S., L. J. Thompson, S. P, Lawler, J. H. Lawton, and R. M. Woodfin. 1995. Empirical evidence that declining species diversity may alter the performance of terrestrial ecosystems. Phil. Trans. R. Soc. Lond. 347, 249–262.

Naeem, S., K. Hakansson, J. H. Lawton, M. J. Crawley, and L. J. Thompson. 1996. Biodiversity and plant productivity in a model assemblage of plant species. Oikos 76: 259–264.

Neill, W. E. 1974. The community matrix and the interdependence of the competition coefficients. Amer. Natur. 108: 399–408.

Nicholson, A. J. and V. A. Bailey. 1935. The balance of animal populations, Part I. Proc. Zool. Soc. London 3: 551–598.

Nisbet, I. T. C. and M. J. Welton. 1984. Seasonal variations in breeding success of common terns: consequences of predation. Condor 86: 53–60.

Nishikawa, K. C. 1985. Competition and the evolution of aggressive behavior in two species of terrestrial salamanders. Evolution 39: 1282–1294.

Odum, E. P. 1950. Bird populations of the Highlands (North Carolina) Plateau in relation to plant succession and avian invasion. Ecology 31: 587–605.

Odum, E. P. 1969. The strategy of ecosystem development. Science 164: 262–270.

Oksanen, L., S. D. Fretwell, J. Arruda, and P. Niemela. 1981. Exploitation ecosystems in gradients of primary productivity. Amer. Natur. 118: 240–261.

Orians, G. H. and R. T. Paine. 1983. Convergent evolution at the community level. Pages 431–453 in D. J. Futuyma and M. Slatkin (eds.), *Coevolution*, Sinauer Associates, Sunderland, MA.

Orloci, L. 1966. Geometric models in ecology. I. The theory and application of some ordination methods. J. Ecology 54: 193–215.

Pacala, S. 1986a. Neighborhood models of plant population dynamics. 2. Multispecies models of annuals. Theor. Pop. Biol. 29: 262–292.

Pacala, S. 1986b. Neighborhood models of plant population dynamics. IV. Single and multi-species models of annuals with dormant seed. Amer. Natur. 128: 859–878.

Pacala, S. 1987. Neighborhood models of plant population dynamics. III. Models with spatial heterogeneity in the physical environment. Theor. Pop. Biol. 31: 359–392.

Pacala, S. W., C. D. Canham, J. Saponara, J. A. Silander, R. K. Kobe, and E. Ribbens. 1996. Forest models defined by field measurements: estimation, error analysis, and dynamics. Ecological Monographs 66: 1–44.

Pacala, S. and J. Roughgarden. 1982. Resource partitioning and interspecific competition in two-species insular *Anolis* lizard communities. Science 217: 444–446.

Pacala, S. and J. Roughgarden. 1985. Population experiments with the Anolis lizards of St. Maarten and St. Eustatius. Ecology 66: 129–141.

Pacala, S. W. and J. A. Silander Jr. 1985. Neighborhood models of plant population dynamics. I. Single-species models of annuals. Amer. Natur. 125: 385–411.

Pacala, S. W. and J. A. Silander Jr. 1990. Field tests of neighborhood population dynamic models of two annual weed species. Ecol. Monogr. 60: 113–134.

Paine, R. T. 1966. Food web complexity and species diversity. Amer. Natur. 100: 65–75.

Paine, R. T. 1969a. A note on trophic complexity and community stability. Amer. Natur. 103: 91–93.

Paine, R. T. 1969b. The *Pisaster-Tegula* interaction: prey patches, predator food preference, and intertidal community structure. Ecology 50: 950–961.

Paine, R. T. 1971. A short-term experimental investigation of resource partitioning in a New Zealand rocky intertidal habitat. Ecology 52: 1096–1106.

Paine, R. T. 1974. Intertidal community structure: experimental studies on the relationship between a dominant competitor and its principal predator. Oecologia 15: 93–120.

Paine, R. T. 1980. Food webs: linkage, interaction strength and community infrastructure. J. Anim. Ecol. 49: 667–685.

Paine, R. T. 1984. Ecological determinism in the competition for space. Ecology 65: 1339–1348.

Paine, R. T. 1988. Food webs: road maps of interactions or grist for theoretical development. Ecology 69: 1648–1654.

Paine, R. T. 1992. Food web analysis through field measurement of per capita interaction strength. Nature 355: 73–75.

Pajunen, V. I. 1982. Replacement analysis of non-equilibrium competition between rock pool corixids (Hemiptera, Corixidae). Oecologia 52: 153–155.

Palmer, M. A., R. F. Ambrose, and N. L. Poff. 1997. Ecological theory and community restoration ecology. Restoration Ecology 5: 291–300.

Park, T. 1962. Beetles, competition, and populations. Science 138: 1369–1375.

Park, T., D. B. Mertz, W. Grodzinski, and T. Prus. 1965. Cannibalistic predation in populations of flour beetles. Physiological Zoology 38: 289–321.

Parker, V. T. 1997. The scale of successional models and restoration objectives. Restoration Ecology 5: 301–306.

Parrish, J. D. and S. B. Saila. 1970. Interspecific competition, predation, and species diversity. J. Theor. Biol. 27: 207–220.

Pearl, R. L. and L. J. Reed. 1920. On the rate of growth of the population of the United States since 1790 and its mathematical representation. Proc. Nat. Acad. Sci. USA 6: 275–288.

Peterson, C. H. 1984. Does a rigorous criterion for environmental identity preclude the existence of multiple stable points? Amer. Natur. 124: 127–133.

Peterson, C. H. and S. V. Andre. 1980. An experimental analysis of interspecific competition among marine filter feeders in a soft-sediment environment. Ecology 61: 129–139.

Pianka, E. R. 1966. Latitudinal gradients in species diversity: a review of the concepts. Amer. Natur. 100: 33–46.

Pianka, E. R. 1986. *Ecology and Natural History of Desert Lizards*. Princeton University Press, Princeton, NJ. 208 pages.

Pianka, E. R. 1988. *Evolutionary Ecology*, 4th Ed. Harper & Row, New York. 468 pages.

Pickett, S. T. A. 1989. Space-for-time substitution as an alternative to long-term studies. Pages 110–135 in G. E. Likens (ed.), *Long-term Studies in Ecology: Approaches and Alternatives*. Springer-Verlag, New York.

Pickett, S. T. A. and M. J. McDonnell. 1989. Changing perspectives in community dynamics: a theory of successional forces. TREE 4: 241–245.

Pickett, S. T. A. and J. N. Thompson. 1978. Patch dynamics and the design of nature reserves. Biol. Conserv. 13: 27–37.

Pielou, E. C. 1977. *Mathematical Ecology*. John Wiley & Sons, New York. 385 pages.

Pielou, E. C. 1984. *Analysis of Ecological Data*. J. Wiley & Sons, New York.

Pimm, S. L. 1982. *Food Webs*. Chapman and Hall, London.

Pimm, S. L. and R. L. Kitching. 1987. The determinants of food chain lengths. Oikos 50: 302–307.

Pimm, S. L. and J. H. Lawton. 1977. Number of trophic levels in ecological communities. Nature 268: 329–331.

Pimm, S. L. and J. H. Lawton. 1978. On feeding on more than one trophic level. Nature 275: 542–544.

Pimm, S. L. and J. H. Lawton. 1980. Are food webs divided into compartments? Journal of Animal Ecology 49: 879–898.

Pimm, S. L., J. H. Lawton, and J. E. Cohen. 1991. Food web patterns and their consequences. Nature 350: 669–674.

Polis, G. A. 1991. Complex desert food webs: an empirical critique of food web theory. Amer. Natur. 138: 123–155.

Polis, G. A. and D. R. Strong Jr. 1996. Food web complexity and community dynamics. Amer. Natur. 147: 813–846.

Polis, G. A., C. A. Myers, and R. D. Holt. 1989. The ecology and evolution of intraguild predation: potential competitors that eat each other. Annu. Rev. Ecol. System. 20: 297–330.

Pomerantz, M. J. 1981. Do "higher order interactions" in competition systems really exist? Amer. Natur. 117: 583–591.

Poole, R. W. and B. J. Rathcke. 1979. Regularity, randomness, and aggregation in flowering phenologies. Science 203: 470–471.

Porter, J. W. 1972. Predation by *Acanthaster* and its effect on coral species diversity. Amer. Natur. 106: 487–492.

Power, M. E., W. J. Matthews, and A. J. Stewart. 1985. Grazing minnows, piscivorous bass, and stream algae: dynamics of a strong interaction. Ecology 66: 1448–1456.

Price, P. W. 1984. *Insect Ecology*, 2nd Edition. Wiley Interscience, New York.

Price, P. W., C. E. Bouton, P. Gross, B. A. McPheron, J. N. Thompson, and A. E. Weis. 1980. Interactions among three trophic levels: influence of plants on interactions between insect herbivores and natural enemies. Annu. Rev. Ecol. System. 11: 41–65.

Ratcliffe, F. N. 1959. The rabbit in Australia. Pages 545–564 in A. Keast, R. L. Crocker, and C. S. Christian (eds.), *Biogeography and Ecology in Australia*, Monographiae Biologicae VIII, Dr. W. Junk, The Hague.

Ratcliffe, F. N., K. Myers, B. V. Fennessy, and J. H. Calaby. 1952. Myxomatosis in Australia. A step towards the biological control of the rabbit. Nature 170: 7–11.

Redfearn, A. and S. L. Pimm. 1988. Population variability and polyphagy in herbivorous insect communities. Ecological Monographs 58: 39–55.

Resetarits, W. J., Jr. and H. M. Wilbur. 1989. Oviposition site choice in *Hyla chrysoscelis:* role of predators and competitors. Ecology 70: 220–228.

Resetarits, W. J., Jr. and H. M. Wilbur. 1991. Calling site choice by *Hyla chrysoscelis*: effect of predators, competitors, and oviposition sites. Ecology 72: 778–786.

Richards, O. W. 1926. Studies on the ecology of British heaths. III. Animal communities of the felling and burn successions at Oxshott Heath, Surrey. J. Ecology 14: 244–281.

Ricker, W. E. 1975. Computation and interpretation of biological statistics of fish populations. Department of Environment Fisheries and Marine Service, Ottawa, Canada. 382 pages.

Ricklefs, R. E. 1989. Speciation and diversity: the integration of local and regional processes. Pages 599–622 in D. Otte and J. A. Endler (eds.), *Speciation and Its Consequences*, Sinauer Associates, Sunderland, MA.

Ricklefs, R. E. 1990. *Ecology*, 3rd Edition. W. H. Freeman, New York.

Ricklefs, R. E. and G. W. Cox. 1972. Taxon cycles in the West Indian avifauna. Amer. Natur. 106: 195–219.

Ricklefs, R. E. and D. Schluter. 1993. Species diversity: regional and historical influences. Pages 350–363 in R. E. Ricklefs and D. Schluter (eds.*)*, *Species Diversity in Ecological Communities*, University of Chicago Press.

Ricklefs, R. E. and J. Travis. 1980. A morphological approach to the study of avian community organization. The Auk 97: 321–338.

Robichaud, B. and M. F. Buell. 1973. *Vegetation of New Jersey*. Rutgers University Press, New Brunswick.

Room, P. M., K. L. S. Harley, I. W. Forno, and D. P. A. Sands. 1981. Successful biological control of the floating weed *Salvinia*. Nature 294: 78–80.

Root, R. B. 1967. The niche exploitation pattern of the blue-gray gnatcatcher. Ecol. Monogr. 37: 317–350.

Rosenthal, G. A. and M. R. Berenbaum (eds.). 1992. Herbivores: Their interactions with secondary plant metabolites. 2nd Edition. Academic Press, San Diego.

Rosenzweig, M. L. 1971. The paradox of enrichment: destabilization of exploitation ecosystems in ecological time. Science 171: 385–387.

Rosenzweig, M. L. 1973. Exploitation in three trophic levels. Amer. Natur. 107: 275–294.

Rosenzweig, M. L. 1995. *Species Diversity in Space and Time*. Cambridge University Press, Cambridge. 436 pages.

Roughgarden, J. and M. Feldman. 1975. Species packing and predation pressure. Ecology 56: 489–492.

Safina, C. 1990. Bluefish mediation of foraging competition between roseate and common terns. Ecology 71: 1804–1809.

Sale, P. F. 1977. Maintenance of high diversity in coral reef fish communities. Amer. Natur. 111: 337–359.

Sale, P. F. 1980. The ecology of fishes on coral reefs. Ann. Rev. Oceanogr. Mar. Biol. 18: 367–421.

Salt, G. W. 1967. Predation in an experimental protozoan population (*Woodruffia—Paramecium*). Ecol. Monogr. 37: 113–144.

Sanders, H. L. 1968. Marine benthic diversity: a comparative study. Amer. Natur. 102: 243–282.

Schmitt, R. J. 1987. Indirect interactions between prey: apparent competition, predator aggregation, and habitat segregation. Ecology 68: 1887–1897.

Schoener, T. W. 1968. The *Anolis* lizards of Bimini: resource partitioning in a complex fauna. Ecology 49: 704–726.

Schoener, T. W. 1974. Resource partitioning in ecological communities. Science 185: 27–39.

Schoener, T. W. 1983. Field experiments on interspecific competition. Amer. Natur. 122: 240–285.

Schoener, T. W. 1984. Size differences among sympatric, bird-eating hawks: a worldwide survey. Pages 254–281 in D. R. Strong, Jr., D. Simberloff, L. G. Abele, and A. B. Thistle (eds.), *Ecological Communities: Conceptual Issues and the Evidence*, Princeton University Press, Princeton, NJ.

Schoener, T. W. 1986. Overview: Kinds of ecological communities—ecology becomes pluralistic. Pages 467–479 in J. Diamond and T. J. Case (eds.), *Community Ecology*, Harper & Row, New York.

Schoener, T. W. 1988. Leaf damage in island buttonwood, *Conocarpus erectus*: correlations with pubescence, island area, isolation and the distribution of major carnivores. Oikos 53: 253–266.

Schoener, T. W. 1989. The ecological niche. Pages 79–113 in J. M. Cherrett (ed.), *Ecological Concepts*, Blackwell, Oxford.

Schoenly, K. and J. E. Cohen. 1991. Temporal variation in food web structure: 16 empirical cases. Ecological Monographs 61: 267–298.

Schwartz, M. W. and J. D. Hoeksema. 1998. Specialization and resource trade: biological markets as a model of mutualisms. Ecology 79: 1029–1038.

Sebens, K. P. 1981. Recruitment in a sea anemone population: juvenile substrate becomes adult prey. Science 213: 785–787.

Shaw, R. G. 1987. Density-dependence in *Salvia lyrata*: experimental alteration of densities of established plants. J. Ecology 75: 1049–1063.

Shaw, R. G. and J. Antonovics. 1986. Density-dependence in *Salvia lyrata*, an herbaceous perennial: experimental alteration of seed densities. J. Ecology 74: 797–813.

Shelford, V. E. 1929. *Laboratory and Field Ecology. The Responses of Animals as Indicators of Correct Working Methods*. The Williams & Wilkins Company, Baltimore. 608 pages.

Shorrocks, B. and M. Bingley. Priority effects and species coexistence: experiments with fungal-breeding *Drosophila*. J. Anim. Ecol. 63: 799–806.

Shugart, H. H. and D. C. West. 1977. Development of an Appalachian deciduous forest succession model and its application to assessment of the impact of the chestnut blight. J. Environm. Management 5: 161–179.

Shugart, H. H. and D. C. West. 1980. Forest succession models. Bioscience 30: 308–313.

Shulman, M. J., J. C. Ogden, J. P. Ebersole, W. N. McFarland, S. L. Miller, and N. G. Wolf. 1983. Priority effects in the recruitment of juvenile coral reef fishes. Ecology 64: 1508–1513.

Sih, A. 1982. Foraging strategies and the avoidance of predation by an aquatic insect, *Notonecta hoffmanni*. Ecology 63: 786–796.

Sih, A., P. Crowley, M. McPeek, J. Petranka, and K. Strohmeier. 1985. Predation, competition, and prey communities: a review of field experiments. Ann. Rev. Ecol. Syst. 16: 269–311.

Silvertown, J. S. 1987. Ecological stability: a test case. Amer. Natur. 130: 807–810.

Simberloff, D. S. 1976. Experimental zoogeography of islands: effects of island size. Ecology 57: 629–648.

Simberloff, D. S. 1988. The contribution of population and community biology to conservation science. Annu. Rev. Ecol. System. 19: 473–511.

Simberloff, D. S. and W. Boecklen. 1981. Santa Rosalia reconsidered: size ratios and competition. Evolution 35: 1206–1228.

Simberloff, D. S. and E. O. Wilson. 1969. Experimental zoogeography of islands: the colonization of empty islands. Ecology 50: 278–296.

Slobodkin, L. B. 1960. Ecological energy relationships at the population level. Amer. Natur. 94: 213–236.

Slobodkin, L. B., F. E. Smith, and N. G. Hairston. 1967. Regulation in terrestrial ecosystems, and the implied balance of nature. Amer. Natur. 101: 109–124.

Smith, D. C. 1981. Competitive interactions of the striped plateau lizard (*Sceloporus virgatus*) and the tree lizard (*Urosarus ornatus*). Ecology 62: 679–687.

Solomon, M. E. 1949. The natural control of animal populations. J. Anim. Ecol. 18: 1–35.

Sousa, W. P. 1979a. Experimental investigations of disturbance and ecological succession in a rocky intertidal algal community. Ecol. Monogr. 49: 227–254.

Sousa, W. P. 1979b. Disturbance in marine intertidal boulder fields: the nonequilibrium maintenance of species diversity. Ecology 60: 1225–1239.

Spiller, D. A. and T. W. Schoener. 1989. An experimental study of the effect of lizards on web-spider communities. Ecological Monographs 58: 57–77.

Spiller, D. A. and T. W. Schoener. 1990. A terrestrial field experiment showing the impact of eliminating top predators on foliage damage. Nature 347: 469–472.

Sprules, W. G. 1972. Effects of size-selective predation and food competition on high altitude zooplankton communities. Ecology 53: 375–386.

Sprules, W. G. 1977. Crustacean zooplankton communities as indicators of limnological conditions: an approach using principal component analysis. J. Fish. Res. Board Can. 34: 962–975.

Sprules, W. G. and J. E. Bowerman. 1988. Omnivory and food chain length in zooplankton food webs. Ecology 69: 418–426.

Steneck, R. S., S. D. Hacker, and M. N. Dethier. 1991. Mechanisms of competitive dominance between crustose coralline algae: an herbivore-mediated competitive reversal. Ecology 72: 938–950.

Stenhouse, S. L., N. G. Hairston, and A. E. Cobey. 1983. Predation and competition in *Ambystoma* larvae: field and laboratory experiments. Journal of Herpetology 17: 210–220.

Sterner, R. W., A. Bajpai, and T. Adams. 1997. The enigma of food chain length: absence of theoretical evidence for dynamic constraints. Ecology 78: 2258–2262.

Stiles, F. G. 1975. Ecology, flowering phenology, and hummingbird pollination of some Costa Rican *Heliconia* species. Ecology 56: 285–301.

Stiles, F. G. 1977. Coadapted competitors: the flowering seasons of hummingbird pollinated plants in a tropical forest. Science 198: 1177–1178.

Stouffer, P. C. and R. O. Bierregaard Jr. 1995. Use of Amazonian forest fragments by understory insectivorous birds. Ecology 76: 2429–2445.

Strauss, S. Y. 1991. Indirect effects in community ecology: their definition study and importance. TREE 6: 206–210.

Strong, D. R., Jr. 1992. Are trophic cascades all wet? Differentiation and donor-control in speciose ecosystems. Ecology 73: 747–754.

Strong, D. R., L. A. Szyska, and D. Simberloff. 1979. Tests of community-wide character displacement against null hypotheses. Evolution 33: 897–913.

Strong, D. R., Jr., D. Simberloff, L. G. Abele, and A. B. Thistle (eds.) 1984. *Ecological Communities: Conceptual Issues and the Evidence*, Princeton University Press, Princeton, NJ.

Strobeck, C. 1973. N species competition. Ecology 54: 650–654.

Sugihara, G. 1980. Minimal community structure: an explanation of species abundance patterns. Amer. Natur. 116: 770–787.

Sugihara, G., K. Schoenly, and A. Trombla. 1989. Scale invariance in food web properties. Science 245: 48–52.

Summerhayes, V. S. and C. S. Elton. 1923. Contributions to the ecology of Spitzbergen and Bear Island. J. Ecology 11: 214–286.

Sutherland, J. P. 1974. Multiple stable points in natural communities. Am. Nat. 108: 859–873.

Sutherland, J. P. 1981. The fouling community at Beaufort, North Carolina: a study in stability. Am. Nat. 118: 499–519.

Tanner, J. T. 1975. The stability and the intrinsic growth rates of prey and predator populations. Ecology 56: 855–867.

Tansley, A. G. and R. S. Adamson. 1925. Studies of the vegetation of the English chalk. III. The chalk grasslands of the Hampshire-Sussex border. J. Ecol. 13: 177–223.

Temple, S. A. 1977. Plant-animal mutualism: coevolution with Dodo leads to near extinction of plant. Science 197: 885–886.

Thomas, K. J. 1981. The role of aquatic weeds in changing the pattern of ecosystems in Kerala. Environmental Conserv. 8: 63–66.

Thompson, D. J. 1975. Towards a predator-prey model incorporating age structure: the effects of predator and prey size on the predation of *Daphnia magna* by *Ischnura elegans*. J. Anim. Ecol. 44: 907–916.

Thornton, I. 1996. *Kraktau: The Destruction and Reassembly of an Island Ecosystem.* Harvard University Press, Cambridge, MA, USA.

Tilman, D. 1977. Resource competition between planktonic algae: an experimental and theoretical approach. Ecology 58: 338–348.

Tilman, D. 1978. Cherries, ants and tent caterpillars: timing of nectar production in relation to susceptibility of caterpillars to ant predation. Ecology 59: 686–692.

Tilman, D. 1982. *Resource Competition and Community Structure.* Princeton University Press, Princeton, NJ. 296 pp.

Tilman, D. 1983. Plant succession and gopher disturbance along an experimental gradient. Oecologia 60: 285–292.

Tilman, D. 1984. Plant dominance along an experimental nutrient gradient. Ecology 65: 1445–1453.

Tilman, D. 1985. The resource ratio hypothesis of succession. Amer. Natur. 125: 827–852.

Tilman, D. 1987. Secondary succession and the pattern of plant dominance along experimental nitrogen gradients. Ecol. Monogr. 57: 189–214.

Tilman, D. 1996. Biodiversity: population versus ecosystem stability. *Ecology* 77, 350–363.

Tilman, D. 1997. Community invasibility: recruitment limitation, and grassland biodiversity. Ecology 78: 81–92.

Tilman, D. and J. A. Downing. 1994. Biodiversity and stability in grasslands. Nature 367: 363–365.

Tilman, D. and S. Pacala. 1993. The maintenance of species richness in plant communities. Pages 13–25 in R. E. Ricklefs and D. Schluter (eds.), *Species Diversity in Ecological Communities*, University of Chicago Press.

Tilman, D., D. Wedin, and J. Knops 1996. Productivity and sustainability influenced by biodiversity in grassland ecosystems. Nature 379: 718–720.

Tilman, D., C. L. Lehman, and K. T. Thompson. 1997. Plant diversity and ecosystem productivity: theoretical considerations. Proc. Natl Acad. Sci. USA 94:1857–1861.

Uhl, C. 1987. Factors controlling succession following slash and burn agriculture in Amazonia. J. Ecol. 75: 377–407.

Underwood, A. J. 1986. The analysis of competition by field experiments. Pages 240–268 in J. Kikkawa and D. J. Anderson, *Community Ecology: Pattern and Process.* Blackwell, London.

Underwood, A. J., E. J. Denley, and M. J. Moran. 1983. Experimental analysis of the structure and dynamics of mid-shore rocky-intertidal communities in New South Wales. Oecologia 56: 202–219.

Vandermeer, J. H. 1969. The competitive structure of communities: an experimental approach with Protozoa. Ecology 50: 362–371.

Vandermeer, J. H. 1980. Indirect mutualism: variations on a theme by Stephen Levine. Amer. Natur. 116: 441–448.

Vandermeer, J. H. 1981. *Elementary Mathematical Ecology.* John Wiley & Sons, New York.

Vermeij, G. J. 1987. *Evolution and Escalation: An Ecological History of Life.* Princeton University Press, Princeton, NJ.

Verhulst, P. F. 1838. Notice sur la loique la population suit dans son accroissement. Correspondances Mathematiques et Physiques 10: 13–121.

Vitousek, P. M. 1990. Biological invasions and ecosystem processes: towards an integration of population and ecosystem studies. Oikos 57: 7–13.

Vitousek, P. M., H. A. Mooney, J. Lubchenco, and J. M. Melillo. 1997. Human domination of the Earth's ecosystems. Science 277: 494–499.

Volterra, V. 1926. Variations and fluctuations in the numbers of individuals in animal species living together. (Reprinted in 1931. In Chapman, R. N., Animal Ecology, McGraw-Hill, New York.)

Wallace, A. R. 1878. *Tropical Nature and Other Essays.* London and New York, MacMillan.

Walker, L. R. and F. S. Chapin III. 1987. Interactions among processes controlling successional change. Oikos 50: 131–135.

Wangersky, P. J. and W. J. Cunningham. 1957. Time lag in prey-predator population models. Ecology 38: 136–139.

Warner, R. R. and P. L. Chesson. 1985. Coexistence mediated by recruitment fluctuations: a field guide to the storage effect. Amer. Natur. 125: 769–787.

Warren, P. H. 1989. Spatial and temporal variation in the structure of a freshwater food web. Oikos 55: 299–311.

Watt, A. S. 1940. Studies in the ecology of Breckland. IV. The grass heath. J. Ecology 42–70.

Watt, A. S. 1947. Pattern and process in the plant community. Journal of Ecology 35: 1–22.

Watt, A. S. 1955. Bracken versus heather, a study in plant sociology. Journal of Ecology 43: 490–506.

Watt, A. S. 1957. The effect of excluding rabbits from grassland B (Mesobrometum) in Breckland. J. Ecology 45: 861–878.

Watt, A. S. 1981. A comparison of grazed and ungrazed grassland A in East Anglian Breckland. J. Ecology 69: 499–508.

Werner, E. E. 1986. Amphibian metamorphosis: growth rate, predation risk, and the optimal size at transformation. Amer. Natur. 128: 319–341.

Werner, E. E. and B. R. Anholt. 1993. Ecological consequences of the trade-off between growth and mortality rates mediated by foraging activity. Amer. Natur. 142: 242–272.

Werner, E. E. and J. F. Gilliam. 1984. The ontogenetic niche and species interactions in size-structured populations. Annu. Rev. Ecol. System. 15: 393–425.

Werner, E. E., J. F. Gilliam, D. J. Hall, and G. G. Mittelbach. 1983a. An experimental test of the effects of predation risk on habitat use in fish. Ecology 64: 1540–1548.

Werner, E. E. and D. J. Hall. 1976. Niche shifts in sunfishes: experimental evidence and significance. Science 191: 404–406.

Werner, E. E., G. G. Mittelbach, D. J. Hall, and J. F. Gilliam. 1983b. Experimental tests of optimal habitat use in fish: the role of relative habitat profitability. Ecology 64: 1525–1539.

Wheeler, W. M. 1910. *Ants: Their Structure, Development, and Behavior.* Columbia University Press, New York. 663 pages.

Whittaker, R. H. 1952. A study of summer foliage insect communities in the Great Smoky Mountains. Ecological Monographs 22: 1–44.

Whittaker, R. H. 1956. Vegetation of the Great Smoky Mountains. Ecological Monographs 26: 1–80.

Whittaker, R. H. 1967. Gradient analysis of vegetation. Biological Reviews 42: 207–264.

Whittaker, R. H. 1975. *Communities and Ecosystems*, 2nd Edition. MacMillan, New York.

Whittaker, R. H. and W. A. Niering. 1965. Vegetation of the Santa Catalina Mountains, Arizona: a gradient analysis of the south slope. Ecology 46: 429–452.

Wickler, W. 1968. *Mimicry in Plants and Animals*. World University Library, London.

Wiens, J. A. 1977. On competition and variable environments. Amer. Sci. 65: 590–597.

Wiens, J. A. 1983. On understanding a nonequilibrium world: myth and reality in community patterns and processes. Pages 439–457 in D. R. Strong et al. (eds.), Ecological Communities: Conceptual Issues and the Evidence. Princeton Univ. Press, Princeton, NJ.

Wiens, J. A. and J. Rotenberry. 1981. Habitat associations and community structure of birds in shrub-steppe environments. Ecol. Monogr. 51: 21–41.

Wilbur, H. M. 1972. Competition, predation, and the structure of the *Ambystoma—Rana sylvatica* community. Ecology 53: 3–21.

Wilbur, H. M. 1987. Regulation of structure in complex systems: experimental temporary pond communities. Ecology 68: 1437–1452.

Wilson, E. O. 1992. *The Diversity of Life*. Belknap/Harvard Press, Cambridge.

Winemiller, K. O. 1990. Spatial and temporal variation in tropical fish trophic networks. Ecological Monographs 60: 331–367.

Winemiller, K. O. and E. R. Pianka. 1990. Organization in natural assemblages of desert lizards and tropical fishes. Ecological Monographs 60: 27–55.

Wit, C. T. de. 1960. On competition. Versl. Landbouwk. Onderz. 66: 1–82.

Wolda, H. 1987. Seasonality and the community. Pages 69–95 in J. H. R. Gee and P. S. Giller (eds.), *Organization of Communities Past and Present*, Blackwell, Boston.

Wolin, C. L. 1985. The population dynamics of mutualistic systems. Pages 248–269, in D. H. Boucher (ed.), *The Biology of Mutualism*. Oxford University Press.

Wootton, J. T. 1992. Indirect effects, prey susceptibility, and habitat selection: impacts of birds on limpets and algae. Ecology 73: 981–991.

Wootton, J. T. 1993. Indirect effects and habitat use in an intertidal community: interaction chains and interaction modifications. Amer. Natur. 141: 71–89.

Wootton, J. T. 1994a. Putting the pieces together: testing the independence of interactions among organisms. Ecology 75: 1544–1551.

Wootton, J. T. 1994b. Predicting direct and indirect effects: an integrated approach using experiments and path analysis. Ecology 75: 151–165.

Wootton, J. T. 1994c. The nature and consequences of indirect effects in natural communities. Annu. Rev. Ecol. Syst. 25: 443–466.

Wootton, J. T. and M. E. Power. 1993. Productivity, consumers, and the structure of a river food chain. Proc. Natl. Acad. Sci. USA 90: 1384–1387.

Worthen, W. B. 1989. Predator-mediated coexistence in laboratory communities of mycophagous *Drosophila* (Diptera: Drosophilidae). Ecological Entomology 14: 117–126.

Worthen, W. B. and T. R. McGuire. 1988. A criticism of the aggregation model of coexistence: non-independent distribution of dipteran species on ephmeral resources. Amer. Natur. 131: 453–458.

Worthen, W. B. and J. L. Moore. 1991. Higher-order interactions and indirect effects: a resolution using laboratory *Drosophila* communities. Amer. Natur. 138: 1092–1104.

Wright, S. J. 1981. Intra-archipelago vertebrate distributions: the slope of the species-area relation. Amer. Natur. 118: 726–748.

Yodzis, P. 1988. The indeterminacy of ecological interactions as perceived through perturbation experiments. Ecology 69: 508–515.

Yodzis, P. 1989. *Introduction to theoretical ecology.* Harper & Row, New York.

Yodzis, P. 1994. Predator-prey theory and management of multispecies fisheries. Ecological Applications 4: 51–58.

Zaret, T. M. 1969. Predation-balanced polymorphism of *Ceriodaphnia cornuta* Sars. Limnology and Oceanography 14: 301–303.

INDEX